Lau Siu-kai

THE PRACTICE

of

"ONE COUNTRY, TWO SYSTEMS" POLICY

in

HONG KONG

The Commercial Press

The Practice of "One Country, Two Systems" Policy in Hong Kong

Author:	Lau Siu-kai
Translators:	Guixia Xie, Xiaoqi Shang, Andrew Herrick
Editor:	Betty Wong
Cover design:	Yi Zhang
Published by:	The Commercial Press (H.K.) Ltd. 8/F, Eastern Central Plaza, 3 Yiu Hing Road, Shau Kei Wan, Hong Kong http://www.commercialpress.com.hk
Distributed by:	SUP Publishing Logistics (H.K.) Ltd. 3/F, C & C Building, 36 Ting Lai Road, Tai Po, N. T., Hong Kong
Printed by:	Elegance Printing & Book Binding Co., Ltd. Block A, 4/F, Hoi Bun Industrial Building, 6 Wing Yip Street, Kwun Tong, H.K.
Edition:	First edition, May, 2017 ©2017 The Commercial Press (H.K.) Ltd.
	ISBN 978 962 07 6593 3
	Printed in Hong Kong
	All rights reserved.

Dedicated to my dear father,

Lau Kang-por（劉鏡波）

PREFACE

This year (2017) marks the 27th anniversary of the enactment of the Basic Law of the Hong Kong Special Administrative Region of the People's Republic of China (hereafter referred to as the Basic Law) and the 20th year following Hong Kong's return to China, still 30 years from 2047, the last year of the 50-year period, within which the "one country, two systems" policy towards Hong Kong is to remain unchanged. Since Hong Kong returned to China in 1997, the "one country, two systems" policy (hereafter referred to as "one country, two systems") has been implemented in Hong Kong, with the Basic Law being the legal means of carrying out the policy. Throughout the subsequent 20 years, despite arguments in favour of "one country, two systems" along with the Basic Law, a portion of Hong Kong people, especially opposition activists, resisted the idea of Hong Kong's return. While Article 23 of the Basic Law regarding the local obligation to legislate on national security has also not yet been fulfilled, Hong Kong yet enjoys prosperity, stability, and continued development, though not without administrative difficulties. Although Western voices are occasionally raised in criticism of alterations in Hong Kong following the return, the international community, by and large, thinks highly of the progress Hong Kong has made. Generally speaking, "one country, two systems" has proven to be a wise and appropriate policy and has been successful on the whole.

Notwithstanding, we must be aware that "one country, two systems" – as a policy submitted during a specific historical moment and aimed at solving a problem left by history – has to take into consideration the interests, demands, and points of view of all parties concerned, as the policy also reflects the process and consequences of the rivalry among them. Therefore, it unavoidably includes content that is contradictory or even "unreasonable". As time goes by and the international order, domestic situation and Hong Kong society continue to change, "one country, two systems" and the Basic Law, aimed at keeping Hong Kong's original system and lifestyle unchanged for 50 years, will inevitably encounter various difficulties and new challenges, some of which might have been anticipated before the return and others of which could not have been. Nevertheless, all of them have to be addressed within the dual frameworks of "one country, two systems" and the Basic Law, frameworks that may not be even slightly adjusted unless absolutely necessary. In this sense, the challenge of how to appropriately apply "one country, two systems" and the Basic Law, in order to resolve problems and to deal with challenges, presents a new, yet persistent task, testing the wisdom and courage of all the parties concerned, including the central authorities, the Government of the Hong Kong Special Administrative Region (hereafter referred as the HKSAR Government) and the Hong Kong people.

In my opinion, an analysis of the experience and lessons available 27 years after the enactment of the Basic Law, and that available 20 years after the establishment of the HKSAR, provides us with an opportunity to make a preliminary summary of the implementation of "one country, two systems" and helps us to put forward some questions that we must consider in order to implement "one country, two systems" more effectively in the years remaining. We can also take this opportunity to explore how to better practise the Basic Law, promote the prosperity, stability, development, and efficient governance of Hong Kong, and improve the relationship between the central authorities and the HKSAR Government. I also hope that this book will provide some reference materials pertaining to the question of whether to keep and how to carry on with the "one country, two systems" policy after the 50 years.

In June of 2014, the State Council Information Office of the People's Republic of China issued a *White Paper* entitled *The Practice of the "One country, Two Systems" Policy in the Hong Kong Special*

Administrative Region[1] (hereafter referred to as the *White Paper*). Based on the report of the 18th Communist Party of China (CPC) National Congress, the *White Paper* comprehensively states and summarises the experience and lessons from the practice of the "one country, two systems" in Hong Kong since Hong Kong's return. In addition, it reiterates and expounds the central authority's policy towards Hong Kong in a solemn and explicit manner by highlighting the power and responsibilities of the central authority under "one country, two systems". As it aims to reiterate and further explain the consistent basic principles, strategic objectives, and core content of the central authority's "one country, two systems" policy, the *White Paper* can be regarded as the most authoritative exposition on the practice of "one country, two systems". On the one hand, the *White Paper* reiterates that the practice of "one country, two systems "has proven the importance of the status, power and responsibilities of the country and the central authorities"; on the other hand, it criticises those views that only focus on the "two systems" aspect while overlooking the "one country" aspect. Undoubtedly, the *White Paper* serves as the most appropriate supplement to the Basic Law, with the two expounding the "one country, two systems" policy from the perspectives of law and policy respectively.

The purpose of writing this book is to explore and analyse how "one country, two systems" has operated in practice and to identify the difficulties it has encountered and the problems deriving from the policy. Whether viewed as a strategy or a policy, the principles, ideas and objectives of "one country, two systems" are bound to conflict with reality, especially in considering that the current, real situation is one of constant flux and considering that "one country, two systems" needs to mitigate the discrepancies among all parties concerned. Even though

---

1   State Council Information Office of PRC (2014). *The Practice of the "One Country, Two Systems" Policy in the Hong Kong Special Administrative Region* (《"一國兩制"在香港特別行政區的實踐》). Beijing: Foreign Languages Press. See also *Guide Readings for the Report of the 18th CCP Congress* (《十八大報告：輔導讀本》) (Beijing: People's Press, 2013), which provides references on Beijing's Hong Kong policy. On pages 339-347 of the guide readings, the article "Enrich the Practice of 'One Country, Two Systems'"("豐富'一國兩制'實踐"by Zhang Xiaoming (then the vice-director of the Hong Kong and Macao Affairs Office of the State Council) is of great help.

central-authority leaders had considered all kinds of possible scenarios when formulating the "one country, two systems" policy, it was difficult to anticipate all the possible future changes. Therefore, when encountering difficulties and problems during the practice of "one country, two systems", we should address them with rationality, tolerance, and a realistic mindset rather than accuse our predecessors of short-sightedness or ill-judgment, or even to jump to the premature conclusion that "one country, two systems" is inappropriate.

Recognising the wisdom and contributions of the planners of "one country, two systems", this book also points out the policy's internal contradictions as well as the problems encountered during its practice. The contradictions and problems result not only from the conservatism, static thinking and compromising tendency of "one country, two systems" , but also from the rapidly changing situation in Hong Kong, mainland China, and the rest of the world. This book further points out that the contradictions and problems not only prevent the strategic objectives of "one country, two systems" from being realised, but also pose obstacles to the prosperity and stability of Hong Kong, the relationship between the central authorities and the HKSAR Government, and the effective governance of Hong Kong.

When contradictions and problems became increasingly apparent and unavoidable, both the central authorities and different sectors of Hong Kong attempted to put forward suggestions and enact measures to address them. However, opinions and actions from different sectors consistently contradicted each other, triggering frequent political conflicts and pronounced social divisions, which were reflected in the perennial political struggles centring on political-system reform. Political struggles would in turn damage Hong Kong's social and economic development. This book also attempts a description of the above situation.

Twenty years have passed since the return, and there are still a fair number of Hong Kong people, especially the political opposition, who hold an understanding of "one country, two systems" that is at odds with that of the central authority. One reason is that these people tend to understand "one country, two systems" from the perspective of "Hong Kong-centrism". Most importantly, the opposition in Hong Kong has always deliberately interpreted "one country, two systems" from the "unique" perspective of "Hong Kong as an independent political entity", and unscrupulously distorted the central authority's exposition of "one country, two systems", claiming that their own "version" was the most authoritative one. For quite a long time after the return, based on the "non-interference" principle, the central authorities did not criticise or correct the opposition's misrepresentative interpretation, which generated widespread misconceptions of "one country, two systems" in Hong Kong, severely distorting the practice of "one country, two systems" and damaging the relationship between the central authority and the Hong Kong people. As a result, when the central authority later took actions toward "rehabilitation", they were criticised by many Hong Kong people for violating "one country, two systems".

In this book, the reason why I so often cite remarks from Deng Xiaoping (鄧小平) and other leaders and officials before Hong Kong's return is that I want to present the original and true intention and understanding of "one country, two systems", hoping to tackle it at the source, clarify doubts, resolve conflicts, restore the truth and and clear up the misconceptions of some Hong Kong people. I am convinced that only after the central authority and Hong Kong people reach a consensus on the understanding of the "one country, two systems" policy will the practice of the policy go smoothly and its strategic objectives be realised. Here I would like to remind readers that in this book I usually refer to the Chinese Government as "the central authority" and the totality of national politicel institutions as "the central authorities". However, in order to indicate the fact that Hong Kong was a colony of Britain before the return, I still use "the Chinese Government" or "the Chinese side" when I describe the situation before the return.

As a preliminary review of and outlook on the Basic Law and the "one country, two systems" policy, which were enacted 27 years ago and implemented 20 years ago respectively, this book aims to analyse experiences and draw conclusions regarding lessons that have been learned in order to pose questions worthy of in-depth discussion. I am sure that there are a large number of people who hold unlike views of and arrive at different conclusions about the topics discussed. This book only represents my personal observations, and I welcome any suggestions and comments.

Lau Siu-kai
Hong Kong, 2017

INTRODUCTION

The "one country, two systems" policy is an important national policy and strategy that has helped maintain the stability and prosperity of Hong Kong, enhance the relationship between the central authorities and the Hong Kong Special Administrative Region (HKSAR), and make full use of Hong Kong's economic value to China after the Chinese Government regained sovereignty over Hong Kong, taking over from Britain in a peaceful manner. At first, "one country, two systems" was put forward for the prospective peaceful reunification of both sides of the Taiwan Straits. But, even without this background, the Chinese Government would have been very likely to come up with a similar policy when addressing the Hong Kong issue left over from history. From the perspective of its essence and objective, "one country, two systems" is a continuation of the "long-term planning and full utilisation" policy which was set forth towards Hong Kong after the establishment of the People's Republic of China. Both policies were intended to support and coordinate the strategy for national development and international relations formulated by a Chinese government led by the CPC [Communist Party of China]. In essence, the two policies are an important part of the strategy. "One country, two systems" has enabled Hong Kong to continue to make contributions to the nation after the central authority took it back in 1997.

However, the specific content of "one country, two systems" was influenced by the international landscape, Sino-British relations, the political challenges facing the CPC, as well as the situation in mainland China and Hong Kong at a time when the issue of Hong Kong's future was emerging. In other words, as a solution for addressing the issue of Hong Kong's return, "one country, two systems" was a special product crafted in a special historic period, its objective being to ensure a peaceful and smooth transition in reclaiming sovereignty over Hong Kong from Britain and to maintain Hong Kong's stability and prosperity in the long term. Its core idea was to preserve the present state of affairs in Hong Kong in the late 1980s for an extended period, to disperse Hong Kong citizens' political doubts and fears, and to boost their confidence in Hong Kong's future. Its core content such as "Hong Kong people governing Hong Kong with a high degree of autonomy and original system and lifestyle remaining unchanged for 50 years" expresses the very solemn commitment of the Chinese Government to accommodate the interests and concerns of all the parties both at home and abroad.

"One country, two systems" undoubtedly demonstrates that the CPC-led Chinese Government is rational, pragmatic, flexible, and innovative in administering the nation. Meanwhile, it is essentially an expression of conservatism and static thinking, for it proposes that the "current state" of Hong Kong is to remain "unchanged" for 50 years. Undoubtedly, since changes take place all the time in Hong Kong, it is hardly possible to "remain unchanged for 50 years". Still, the Chinese Government's promise of "Hong Kong remaining unchanged for 50 years" plays a significant part in comforting Hong Kong people amidst their worries about the future of Hong Kong and their fear of change.

The issue of Hong Kong's future arose in a historic period when China walked out of the shadow of the Cultural Revolution, abandoning the route centring on political struggles. Meanwhile, the CPC set economic development, centring on "reform and opening up" as China's national strategy and strived to rebuild its political prestige through economic growth as well as the improvement of people's livelihood. Hong Kong played a remarkable role in carrying out the national development strategy at that time, so the Chinese Government decided to offer Hong Kong the preferential policy of "one country, two systems", which fully alleviated the worries of Hong Kong people and satisfied their demands. From a historical perspective, the period signified the most favourable time for Hong Kong people to "bargain" with the

Chinese Government whenever the issue of Hong Kong's future arose. From another perspective, it was also a time when the interests of the Chinese Government and Hong Kong people were more closely intertwined. Taking into account the existing conditions of the time, the "one country, two systems" policy served as a wise arrangement in line with the common interests of both sides.

The policy of "one country, two systems" was put forward in the 1980s when China experienced amicable relations with Western countries, especially the U.S.. At that point, China's reform and opening up policy needed to obtain recognition and support of the Western world. It was a time when Britain and the U.S. were trapped in diplomatic, economic and political predicaments. China and the U.S. intended to team up to cope with the threat from the Soviet Union; both Britain and the U.S. were disinclined to see Hong Kong's return to China, but neither did they harbour a strong intention to prevent China from taking it back. Instead, they hoped that Hong Kong, after returning to China, would push China onto the path of "peaceful evolution". Meanwhile, China was not willing to fall foul with Western countries or to weaken their mutual strategic partnerships over the Hong Kong issue. The country understood that support from Western countries was indispensable to the continuous stability and prosperity of Hong Kong. The relatively positive political atmosphere and relationships between China and Western countries provided a favourable environment for all stakeholders to accept "one country, two systems".

Since "one country, two systems" is attentive to the interests, points of view, concerns and hopes of China, Britain, Western allies, as well as mainland and Hong Kong people from various classes and backgrounds, there are unavoidable contradictions and a tendency to compromise on principles. In other words, some arrangements, promises or policies in "one country, two systems" conflict with others or cannot be achieved simultaneously with others. From time to time, the contradictions have provoked political conflicts among the parties concerned, which hampered the practice of "one country, two systems", weakening all parties' confidence in "one country, two systems" and diminishing the effects of its practice.

What is worse, the British, the political opposition in Hong Kong and cross section of Hong Kong people refused to accept the CPC and "one country, two systems"; furthermore, their understanding of "one country, two systems" was at odds with that of the central authority. These parties deliberately distorted the central authority's "one country, two systems" by interpreting it in a "special" way and successfully implanted the "special" version in the minds of Hong Kong people. The essential gist of this "special" "one country, two systems" lies in that it regards Hong Kong as an independent political entity, denying the central authority's jurisdiction over Hong Kong as well as the power and responsibilities that it was obliged to shoulder in the practice of "one country, two systems".

The UK's pursuit of "exit with honour", in tandem with its need to maintain effective governance in Hong Kong before the return, impelled them to carry out a series of political reforms, some of which reinforced the political opposition with its "special" understanding of "one country, two systems", and meanwhile promoted a continuation of their political power following the return. Actually, in the "long" transition period, the British had enough time to change Hong Kong's current conditions by installing various political arrangements in Hong Kong according to their own understanding of "one country, two systems", and they managed to force the Chinese Government to accept the changes to a certain extent. As a result, immediately following the return, Hong Kong was faced with a number of conditions incompatible with the central authorities' vision of "one country, two systems". Residual elements collided with some content of "one country, two systems" and intensified its internal contradictions while exacerbating the friction between some Hong Kong people and the central authority as well as between different factions in Hong Kong.

Even without the deliberate political reforms, the international landscape, national development and Hong Kong's situation would still have undergone tremendous and continuous changes following the middle 1980s. The changes that did occur further intensified the internal contradictions of "one country, two systems", rendering the possibility of "remaining unchanged for 50 years" increasingly incompatible with reality. After the return, the swift rise of the nation and the relatively slow development of Hong Kong triggered a series of conflicts between the "two systems", most of which were reflected in the discordant relationship between a segment of Hong Kong people and the central authority, difficulties in the administration of the HKSAR Government, and the lack of harmony between Hong Kong people and their main-land compatriots. Due to the intensified contradictions, the demands for democracy grew ever stronger among Hong Kong people, and the oppo-sition's desire to seize power increased accordingly, yet these demands and desires cannot be satisfied owing to the anxieties and doubts of the central authority towards Hong Kong's democratic development.

In the "colonial" period, to a certain extent, the colonial govern-ment served as a third party to "separate" Hong Kong from the Chinese Government and mainland China, and it effectively alleviated disagree-ments between the two sides. Under "one country, two systems", there is no replacement for the role played by the colonial government, since the Hong Kong SAR Government is not competent enough to play that role and the central authority is also hesitant to step forward, which makes it impossible to create a new third party to harmonise the contra-dictions between the "Two Systems".

China's growing national strength and international influence has brought it numerous opportunities as well as challenges all over the world and also significantly impacted Hong Kong's international position and global role. The international landscape under which China proposed the "one country, two systems" policy changed quickly after the issue of Hong Kong's future was resolved, largely because Western countries facing domestic and foreign challenges sensed a threat from the rise of China and increasingly took measures to counter it. Thus, overt and potential contradictions and conflicts between China and the Western countries continued to escalate. With increasing potential and overt disagreements and conflicts between China and the Western coun-tries, resulting from misunderstandings of each other's intentions and actions, the rivalry between China and the Western countries, especially between China and the U.S., sent forth ripples around the world. For the U.S. and some other Western countries, Hong Kong's role in their strategies towards China inevitably changed. The possibility of Hong Kong becoming a bargaining chip for Western countries to counter-balance China has drawn the attention of the central authorities and some Hong Kong people. Besides, it is also uncertain whether the lasting stability and prosperity of Hong Kong would be in the strategic interests of Western countries. Under the new circumstances, Hong Kong's function as a "bridge" connecting China and the world will also change accordingly. Therefore, how to play the role of "bridge" between China and the world efficiently without causing hidden trouble for national security has become a crucial issue closely linked to the existence and development of the "one country, two systems" policy, which deserves due attention from both the central authorities and Hong Kong.

Changes in various aspects gave rise to new circumstances, challenges, and contradictions that the conservative "one country, two systems" policy has had to resolve. These new contradictions and challenges arose from two aspects. One is the inconsistency between the conservative and static nature of "one country, two systems" and the ever-changing nature of Hong Kong itself. The specific content of "one country, two systems" is complicated, and a large part of it is stipulated by legal provisions; thereby, effectively coping with the new challenges and tasks deriving from various changes of Hong Kong became problematic. The other aspect is that, because of the unceasing changes in both the international landscape and the domestic situation, some came to doubt the correctness, function, effect, and sustainability of "one country, two systems". Such apprehension diminished their confidence in "one country, two systems" and caused them to grow uncertain about Hong Kong's future development.

Initially, respecting the principle of "Hong Kong people governing Hong Kong" with a high degree of autonomy, the central authorities tried not to intervene in Hong Kong's affairs. Nevertheless, as time went on, certain contradictions and conflicts arose and escalated. When the national interests and security were under threat, the power of the central authority and the HKSAR Government was often challenged, and the problems in administering Hong Kong became ever more prominent. When internal disagreements slowed Hong Kong's development and affected its security and stability to a great degree, the central authority began to deal with Hong Kong by exerting its power as stipulated by "one country, two systems" and trying to push forward the development of the policy according to its original "blueprint".

Objectively speaking, to effectively handle the internal contradictions of "one country, two systems" along with those resulting from the rapid changes in mainland China, Hong Kong and the international community, the central authorities and Hong Kong people needed to join hands in sincerity, making concerted efforts to cope with various problems in a down-to-earth and rational manner within the framework of "one country, two systems" and the Basic Law (being the driving force promoting the thorough and correct practice of "one country, two systems") so as to benefit both Hong Kong and the whole country. However, the opposition in Hong Kong and their supporters who considered the arrangements of "one country, two systems" as their "family issues" or the "internal affairs" of Hong Kong rejected any interference by the central authority. The central authority, on the other hand, was obliged to assume a relatively active role in dealing with Hong Kong's affairs for the reasons of safeguarding the national security and interests; performing their due responsibilities in carrying out "one country, two systems"; ensuring the thorough and correct practice of "one country, two systems" in Hong Kong; and maintaining Hong Kong's prosperity, stability and development. Thus, over a period of time, the central authority together with "people who love China and Hong Kong" clashed with the opposition and external forces, which inevitably triggered further contradictions and conflicts of various kinds.

Such political turbulence could hardly be avoided and indeed might benefit the practice of "one country, two systems" in the long run. If the central authorities' policy towards Hong Kong proved efficient, and Hong Kong people responded as pragmatically as is their habit, then based on the clear-cut division of labour, mutual respect and close cooperation, the central authorities and Hong Kong people could flexibly apply "one country, two systems" and the Basic Law to maintain Hong Kong's prosperity, stability and development after the turbulence. The opposition force would then be marginalised, or would change its course of action, thus significantly reducing Hong Kong's conflicts with mainland China and the central authority and preventing Hong Kong from becoming a tool used by external powers to threaten China. Under such conditions, the practice of "one country, two systems" in Hong Kong would proceed on a new and suitable ground and would realise its strategic objective, benefiting both Hong Kong and the nation as a whole.

"One country, two systems" arose at a time when Hong Kong people, who were still nostalgic for the colonial administration of Britain, lacked the confidence in the central authority and were worried about Hong Kong's future and when the central authority was not, in turn, completely at ease about Hong Kong. Although it stood as the best solution for tackling the issue of Hong Kong's future under the circumstances at that time, the problems stemming from lack of confidence and distrust among all parties could not be resolved overnight. Moreover, as an unprecedented political invention, "one country, two systems" is bound to contain quite a few poorly-defined items, contradictions and uncertainties, which is why it is no easy task to successfully implement "one country, two systems" over a long period of time. To ensure the successful practice of this significant national strategy, the central authority along with people in mainland China as well as Hong Kong people must have goodwill, broad perspective and vision, mutual respect and understanding, dynamic thinking as well as a problem-solving attitude in order to combat the problems deriving from "one country, two systems".

CHAPTER ONE

Domestic and International Situations of "One Country, Two Systems"

The issue of Hong Kong's future arose in a historical period when domestic and international circumstances favoured an arrangement acceptable to all "stakeholders". During this period, China, Britain and the U.S. were seeking strengthened strategic cooperation in order to cope with military and diplomatic threats from the Soviet Union. At the moment, China urgently needed to actively participate in and take advantage of the western-dominated economic globalisation and marketisation in order to achieve its economic modernisation and "Reform and Opening up"; the U.S. and Britain were mired in grave international situations, domestic political instability, economic difficulties and popular discontent and needed to boost their economies by revitalising the market. As a result, China and Western countries both cherished warmer ties and hoped to discover a perfect solution that would resolve the issue of Hong Kong's future on the premise that it would not do harm to their cooperative relationship, and take into account interests of all parties. The perfect solution meant policies that could help to maintain and enhance Hong Kong's prosperity and stability, consolidate residents' confidence in Hong Kong's future, secure the interests of Western countries in Hong Kong, and maintain Hong Kong's role as an economic bridge connecting China and the world following the return of Hong Kong to China.

International Situations

The issue of Hong Kong's future appeared between the late 1970s and the early 1980s. For over a decade, the New Cold War had been ongoing, with the Soviet Union and the Western Bloc engaged in fierce contests around the world. At that time, the national power, especially the armed forces, of the Soviet Union was on an upward trajectory, and the pitfalls of its model of economic development had yet to emerge. In contrast, the capitalist systems of the U.S. and Britain were undergoing a period of poor economic growth and stagflation. The Soviet Union was ambitious in diplomatic and military affairs, while the U.S. and U.K. were both plagued by self-doubt and pessimism. Overall, the early 1970s and 1980s was a time of dramatic global transformations that led to a favourable international environment for the peaceful and smooth resolution of the issue of Hong Kong's future.[1]

To give a historical perspective, not long before the close of the Second World War, the Soviet Union found itself at odds with the U.S. and its allies, particularly in Poland, Eastern Europe, Turkey, and Iran. The widening rift led to the breakdown of the Yalta Agreement and the subsequent Potsdam Agreement (co-designed by the U.S., Britain and the Soviet Union and aimed at rebuilding the postwar world milieu) which eventually led to the outbreak of the Cold War.[2] Later on, with the founding of new China and the outbreak of the Korean War, the Cold War intensified. However, eventually, the East and the West began to acknowledge each other's sphere of influence and basic interests in order to reduce conflicts. With the emergence of "détente" between the U.S. and the Soviet Union, the international "order" formed by the Cold

---

1 Thomas, Borstelmann (2012). *The 1970s: A New Global History from Civil Rights to Economic Inequality.* Princeton: Princeton University Press; and Daniel J, Sargent (2015). *A Superpower Transformed: The Remaking of American Foreign Relations in the 1970s.* New York: Oxford University Press.

2 Fraser J. Harbutt (2010). *Yalta 1945: Europe and America at the Crossroads.* Cambridge: Cambridge University Press; John Lewis Gaddis (2005). *The Cold War: A New History.* New York: Penguin; and Michael Neiberg (2015). *Potsdam: The End of World War II and the Remaking of Europe.* New York: Basic Books.

War entered a stable period that lasted for as long as 50 years.

In form, the Helsinki Accords between the Soviet Union and the West in 1975 marks the peak of the "détente". The Accords officially confirmed the Soviet Union's sphere of influence in Europe in return for its promise to promote human rights within its territory and in the Eastern European countries. Unfortunately, due to the widening of the Soviet Union's sphere of influence around the world, the Accords soon amounted to hollow promises. Following the breakup of the Helsinki Accords came the Second Cold War and the second round of the arms race between East and West.

In the heyday of the "Second Cold War", the U.S.'s disastrous defeat in the Vietnam War not only damaged its national reputation and gave rise to economic difficulties, but it also weakened Americans' confidence and desire to intervene in international affairs. Therefore, the U.S. entered its "retrenchment" stage in diplomatic and military affairs. Sestanovich, a former senior American official responsible for foreign affairs, pointed out that "The story of American foreign policy... is not one of dogged continuity but of regular, repeated, and successful efforts to change course."[3] Such a tendency can be seen from strategies of 'maximalism' and 'retrenchment' that bear an obvious, cyclical relation to each other."[4]

During President Nixon's administration, the U.S. entered a new round of "retrenchment". In the 1970s, "It was less and less about whether America was too strong, and more and more about whether it had become too weak. Those concerned about American weakness were not just imagining things. During the Nixon, Ford, and Carter administrations, the U.S. economy was in recession almost one-fourth of the time. It was battered by 'energy crises,'...and by chronic inflation.... With the dollar's decline, American troops in Europe began to have trouble making ends meet....The 1970s turned into one of the sourest,

---

3   Stephen Sestanovich (2014). *Maximalist: America in the World from Truman to Obama.* New York: Alfred A. Knopf. P7.

4   Ibid, P9.

most frustrating, least successful periods of U.S. foreign policy."[5] From the international perspective, "Checking Soviet influence in the Middle East was not the only challenge for American policy. The war produced an embargo by Arab oil producers on exports to the United States and other Western countries.... Suddenly Americans were aware of a new kind of vulnerability–a threat to their economic confidence and well-being that was, for many, far more worrying than Soviet military might."[6]

Iran had been the most important and the most reliable ally of the U.S. in the Middle East, safeguarding the oil interests of the U.S. and protecting its national interests in the region. In 1979, the Iranian Revolution broke out, unseating the then extremely pro-American Shah and establishing a new theocratic regime rooted in Islamic Fundamentalism, the first of its kind after the Second World War. This new regime not only held a completely negative attitude towards modern Western civilization, but was utterly anti-American. The Iran Hostage Crisis that occurred shortly after the revolution greatly humiliated the U.S.. The Iranian Revolution represented a defeat of the West in its contest with the East. Later, it spread throughout the Middle East and other Islamic areas, thoroughly changing the political situation and the balance of powers in the Middle East and the world as a whole. The revolution also marked a serious defeat in the diplomacy and military of the U.S., delivering a heavy blow to its global influence.[7]

Although the deployment of American military forces still focused on Europe and Northeast Asia, Southeast Asia was displaced by the Middle East in terms of strategic importance because of the tensions in the Middle East. Subsequently, the influence of the U.S. declined in Southeast Asia, creating conditions for China to strengthen its influence there in the years ahead. America's diplomatic and military "retrenchment" and the expansion of the Soviet Union enabled many countries

5   Ibid, P193.
6   Ibid, P195.
7   Christian Caryl (2013). *Strange Rebels: 1979 and the Birth of the 21st Century*. New York: Basic Books.

in the Third World to take "aggressive" actions for their respective interests by taking advantage of the contradiction between the U.S. and the Soviet Union. More often than not, the U.S. and the Soviet Union were dragged into the conflicts or internal struggles of other countries, frequently causing the U.S. to end up in the dock or to suffer embarrassment.

In terms of economics, the U.S., and its "special" ally, Britain, were both trapped in financial predicaments. The strong performance of the U.S. dollar came to an end due to the enormous financial costs of the Vietnam War. In 1971, the U.S. dollar had little choice but to unpeg from gold, leading to the collapse of the Bretton Woods system established by the U.S. after the Second World War to protect its core interests. Thereafter, American global economic and financial dominance was greatly weakened.

Meanwhile, the U.S. and Britain were facing enormous challenges from the Federal Republic of Germany, Japan and other European countries. Western countries had various disputes concerning the monetary, fiscal and economic policies, with individual countries seeking to maximise their own interests while at the same time preserving an open international trade system. In the early 1970s, the U.S. began to be challenged by a series of economic problems, including economic stagnation, inflation, high unemployment, and particularly, the surge in oil prices caused by war in the Middle East. The British economic situation at that time was as frustrating as that of the U.S. after the Second World War; Britain had been carrying out British socialism, which led to the excessive power of labour unions, incontrollable welfare expenses, economic slowdown, and a withering entrepreneurial spirit, together making the U.K. become "the sick man of Europe". The presence of various problems in the U.K. engendered intense discontent with government among the British people. In order to find its way out of the economic turmoil, Britain had to carry out thorough structural adjustment in its economic system.

To overcome economic difficulties and boost confidence, the U.S. and the U.K. moved to cut economic intervention and to shake off the constraints of Keynesianism, stimulating economic growth and relying on the private sector and market mechanisms. Britain made even bigger strides than the U.S. in this regard. Its government not only reduced its intervention in economic activities, but also loosened its grip on enterprises and individuals while streamlining the laws and rules regulating economic activities. As such, there was adequate room for private capitalists and entrepreneurs to play their roles.

In such a context, liberal economics rapidly spread from the U.S. and Britain to European countries and a number of other countries. As countries consecutively opened their economies, capital began to circulate across the globe, combining with advances in information technology and a lowered cost of transportation to create an increasingly closer international economic cooperation. At the same time, transnational corporations flourished, acting as a new driving force for global economic growth. The 1970s witnessed the onset of economic globalisation which afterwards spread swiftly and aggressively throughout the globe, reshaping the global economic landscape.

In contrast, after suffering a setback in the Cuban Missile crisis and then losing its close relationship with China, particularly after the Soviet leaders claimed that it was unlikely for the socialist revolution to take place in the Third World, the Soviet Union restrained itself to some extent in its diplomatic and military activities throughout the 1960s.

In the 1970s, however, the U.S. retreated from Vietnam in dismay and lost its important ally–Iran–in the Middle East due to the outbreak of the Iranian Revolution in 1979. As a result, its foreign policy towards the Third World entered into the "retrenchment" stage. In contrast, the Soviet Union was entering into a period of diplomatic and military expansion. The collapse of the Portuguese empire in Africa offered a prime opportunity for the Soviet Union to send troops to Angola and help the pro-Soviet forces within Angola to win the civil war and seize power. The country also actively supported the pro-Soviet forces in Yemen and Mozambique and intensified its influence in Iraq and Syria. After the revolution in Ethiopia, the new government came to the side of the Soviet Union, which helped to extend the country's influence to strategically important places such as the Red Sea and the Indian Ocean, thus posing a grave threat to the supply line of oil to the West. The Soviet influence could also be seen throughout Rhodesia (later renamed Zimbabwe), Southeast Africa (Namibia), South Africa and other African areas.

In spite of condemnation from the international community, the Soviet Union invaded Afghanistan in 1979 and occupied the country in order to protect the "Communist" regime in Afghanistan. This directly threatened the interests of the West in the Middle East.[8] Despite opposition from the West, the Soviet Union took frequent actions in Europe and Latin America. For instance, the Soviet Union deployed medium-range missiles in its western territory pointing at Western Europe. What posed a grave threat to the U.S. regime was the emergence of a new left-wing in Nicaragua, located in the American "backyard", with considerable support from the Soviet Union.

In order to reduce threats from the Soviet Bloc and curb its diplomatic and military expansion of the Soviet Union, the U.S. needed to draw China onto its side. With the improvement of Sino-American relations, the U.S. could also successfully withdraw from Vietnam. From a strategic perspective, President Nixon's visit to China in 1972 and the reestablishment of Sino-American relations was a political masterstroke. Such strategic action completely changed the international order and greatly benefited the interests of both countries. According to an American analyst, "China also got protection from America against the Soviet Union, as well as the economically resurgent Japan. This would provide China with the security it needed to liberalise its economy to the great benefit of the entire region a few short years later."[9]

Australian diplomat and scholar White holds a similar view: "At the heart of the [Sino-American 1972] deal were assurances given by Washington to both Beijing and Tokyo. In return for acceptance of U.S.

---

8 Odd Arne Westad (2007). *The Global Cold War: Third World Interventions and the Making of Our Times.* Cambridge: Cambridge University Press.

9 Robert D. Kaplan (2014). *Asia's Cauldron: the South China Sea and the End of a Stable Pacific.* New York: Random House. P28-29.

primacy, America assured China concerning its security from Japan and the Soviets, and Japan about its security from the Soviets and China. One might call it a double-double assurance deal." "It left Asia set to follow a Western-oriented economic and political path, which made it, under America's leadership, the world's most vibrant region. After 1975, the Soviets never achieved–or even attempted–any substantial strategic gain in Asia east of Afghanistan." In the eyes of westerners, "In 1972, China tacitly relinquished its claim to great power status in Asia."[10]

Besides the threat from the Soviet Union, the U.S.-led Western Bloc also faced other challenges. As some Western allies of the U.S. increased their national strength, they began to raise questions about America's determination and capacity to protect Europe and contain the Soviet Union. For the sake of their own national interests, they would therefore not always follow in America's footsteps on diplomatic issues. Two typical examples were the Ostpolitik policy of the Federal Republic of Germany aimed at improving its relationship with the Soviet Union and the German Democratic Republic, as well as the independent foreign policy of France which contradicted American foreign policy.[11]

Likewise, the internal affairs of Western countries were also undergoing new tests and changes. Despite their overall strength, Western political systems and development patterns also brought a great deal of unfairness and injustice. Young people's dissatisfaction with the state of affairs was on the rise, best demonstrated in the workers' and students' movements that swept the entire Western world in 1968.[12] The

---

10 Hugh White (2012). *The China Choice: Why We Should Share Power*. Oxford: Oxford University Press. P20, 22 and 61.

11 Hans Kundnani (2015). *The Paradox of German Power*. New York: Oxford University Press; Daniel J. Sargent (2015). *A Superpower Transformed: The Remaking of American Foreign Relations in the 1970s*. New York: Oxford University Press.

12 Ronald Fraser (1988). *1968: A Student Generation in Revolt*. London: Pantheon; Mark Kurlansky (2004). *1968: The Year that Rocked the World*. New York: Random House; Kenneth J., Heineman (1993). *Campus Wars: The Peace Movement at American State Universities in the Vietnam Era*. New York: New York University Press; Robert Gildea et al. (eds.) (2013). *Europe's 1968: Voices of Revolt*. New York: Oxford University Press; David Wyatt (2014). *When America Turned: Reckoning with 1968*. Amherst: University of

emergence of Eurocommunism in France, Italy, and Spain in the 1970s also reflected people's dissatisfaction with the state of the world. In fact, throughout the 1970s, protests from all sectors of society emerged one after another, which called into question the universality of the system and values advocated by the West.[13]

In the 1970s, the issue of human rights played an increasingly significant role in international politics. The Helsinki Accords marked the rise of the politics of human rights. In 1979, John Paul II took up his post as the new pope of the Roman Catholic Church. As the first pope from Poland, he devoted himself to promoting human rights and democracy all over the world, especially in Eastern Europe. The emergence of human rights further challenged the international reputation and image of the U.S. and Britain. Considering their traditions, national interests, and the need for seeing through the Cold War, the two countries lacked initiative in coping with affairs such as decolonisation and pushing for the independence of Third World countries. It was hard for Britain to separate itself from its imperial history, clearly shown by its support for the whites in the conflicts in South Africa. The U.S. initiated the Vietnam War and supported as well as protected some dictatorial and military regimes in the Third World. As the issue of human rights began to draw increasing attention, the two countries became the object of fierce criticism by the international community.[14]

Generally speaking, prior to the emergence of the issue of Hong Kong's future, the U.S. and Britain, both closely connected to the issue, were facing both internal and external challenges. Americans and Britons were concerned about the reality and prospects of their countries and were dissatisfied with their leaders. A large number of people from the two countries and indeed the West as a whole felt desperate for reforms to ameliorate economic hardships. Such needs led to market

Massachusetts Press.

13 Martin Klimbe and Joachim Scharloth (eds.) (2008). *1968 in Europe: A History of Protest and Activism, 1957-1977*. New York: Palgrave Macmillan.

14 James E. Cronin (2014). *Global Rules: America, Britain and a Disordered World*. New Haven and London: Yale University Press.

liberalisation and market orientation which later catalysed economic and financial globalisation.

The U.S.-led Western bloc became obligated to seek strategic cooperation with China in order to contend against threats from the Soviet Union and its allies. Though it was more or less out of expediency that the U.S. was willing to cooperate with China, the West in general did in fact need to maintain a positive relationship with China. Under such circumstances, the issue of Hong Kong could be peacefully resolved in a generally favourable international environment. While Britain was able to defeat Argentina in the Falklands War, thus improving its national prestige, the victory could neither alter Britain's fundamental weakness nor promote its political advantages vis-à-vis China in dealing with the issue of Hong Kong.

The Domestic and International Situation of China

Like the Western bloc, just prior to the emergence of the "1997" issue, China also found herself in a dire domestic and international situation. Shortly after the end of the traumatic Cultural Revolution, which lasted a decade, China needed to start more or less from scratch. After defeating "the Gang of Four", the new leaders of the CPC had no other choice but to consider reorganising the CPC, restoring the architecture and operation of the state apparatus, promoting the government's authority, and rebuilding the legitimate foundation of the CPC's administration as the most urgent strategic tasks.

Throughout the 1970s, China's economy was basically in recession: agricultural products could not meet people's demands, and food supply crises could arise at any time; the industrial system, with heavy industry as its pillar, was burdened with low efficiency and severe waste; the salaries of workers in cities had remained unchanged for an extended period; farmers' incomes were still below subsistence level; people were enduring substandard livelihoods, especially in rural areas;

infrastructure in cities was old and inadequate, let alone those in rural areas. Such poverty was all the more depressing when compared with the socialist countries in Eastern Europe and the "four Asian tigers" (South Korea, Singapore, Taiwan, and Hong Kong), highlighting China's inefficient economic development pattern and the economic devastation caused by various political movements.[15] Frustrated by such a grim state of affairs, Chinese people felt resentment towards the CPC-led government, which posed a serious threat to the status and legitimacy of the CPC.

At the end of 1978, the Third Plenary Session of the Eleventh Central Committee of the Communist Party of China convened, in which a series of landmark decisions were made. The meeting began on December 18 and came to an end on December 23, and Deng Xiaoping was selected as the central figure of the second generation of the CPC leadership. In addition, the doctrine of the "Two Whatevers" ("We will resolutely uphold whatever policy decisions Chairman Mao made, and unswervingly follow whatever instructions Chairman Mao gave") was abandoned, and the principle of "Seeking truth from the facts" was promoted in this meeting. Later, the gist of this meeting was simplified as "one central task and two basic points". "One central task" meant "the central task of developing productivity" while the "two basic points" meant "adherence to the Four Cardinal Principles and implementation of the Reform and Opening up policy" respectively. At this meeting, the shift from the "Two Whatevers" to "Seeking truth from the facts" and the change in work priorities marked an essential ideological and political turning point for the CPC. The paramount decision of this meeting, clearly written into the first paragraph of The Announcement by the Third Plenary Session, was to put an end to class struggle and transfer the CPC's priority in work to economic construction and carrying out the "Reform and Opening up" policy.

The "Reform" was aimed at stimulating the domestic economy,

---

15 Andrew G. Walder (2015). *China under Mao: A Revolution Derailed*. Cambridge, MA: Harvard University Press. P315-344.

and the "Opening up" meant to open up economically to the outside world. The "reform" policy was made to replace the current national development strategies that were "centring on class struggle in the country, ignoring economic productivity, and producing policies incompatible with the early stage of socialism". The "opening up" was put forward to change the country's "isolation from the outside world".

The Third Plenary Session of the Eleventh Central Committee of the CPC required the economy to grow in a rapid and steady manner. Its announcement recommended that the country "take a series of important economic measures, carry out prudent reforms in the economic management system and management methods, act according to laws of economics, and pay careful attention to the law of value".[16]

The historical significance of the Third Plenary Session of the Eleventh Central Committee of the CPC lies in its transformation from centring on political and ideological struggles to modernisation and national rejuvenation. To facilitate the economic growth and improve people's livelihood, the CPC discarded its previous strategy of planned economy and self-reliance and launched a series of economic policies to implement the "Reform and Opening up" policy, including ushering in the market competition mechanism, optimising the operational patterns of the agriculture and industry, attracting foreign capital, promoting exports, establishing special economic zones, and joining in the international trade system.

Since the founding of new China, Hong Kong had been playing an important role in terms of national development, especially when China was in economic difficulties. In the new historical chapter, Hong Kong played a pivotal role in China's strategy with regard to the "Reform and Opening up" policy. Before the issue of Hong Kong's future emerged, Hong Kong had developed into a highly modernised and mature capitalist economy as well as a significant international hub for finance, trade, information, telecommunications, and transportation. In

---

16 Ye Yonglie (2008). *Deng Xiaoping Changed China–1978, the Turning Point of China's Destiny* (《鄧小平改變中國—1978：中國命運大轉折》). Nanchang: Jiangxi Publishing Group. P419.

addition, it possessed a large number of internationally advanced talents, enterprises, laws and regulations, as well as institutional infrastructure that complied with international conventions. Therefore, it could assist the nation in joining the international economic system. Actually, in the above-mentioned aspects, no cities in mainland China were comparable to Hong Kong, which is why maintaining Hong Kong's various advantages was of extreme importance to China during its modernisation.

Just prior to the emergence of the issue of Hong Kong's future, China was also facing a grave international situation. In the late 1950s, the relationship between China and the Soviet Union soured because of ideological disagreements and competition for the leadership of the Socialist bloc. Not only did the Soviet Union withdraw its experts and economic aid from China, but it also considered initiating a large-scale nuclear attack against China in order to destroy its nuclear weapons once and for all. Both had disputes along their mutual borders from time to time, and both found themselves engaged in extended contests for influence in the Third World.

In reality, the Soviet Union was not the only threat at China's periphery. According to an observation of Kissinger, the former U.S. Secretary of State, China's borders were far from safe. In 1979, "Viewed from Beijing, a strategic nightmare was evolving along China's borders. In the north, the Soviet built-up continued unabated: Moscow still maintained nearly fifty divisions along the border. To China's west, Afghanistan had undergone a Marxist coup and was subjected to increasingly overt Soviet influence. Beijing also saw Moscow's hand in the Iranian revolution, which culminated with the flight of the Shah on January 16, 1979. Moscow continued to push an Asian collective security system with no other plausible purpose than to contain China"[17]. "Preventing an Indochina bloc linked to the Soviet Union became the dominant preoccupation of Chinese foreign policy under Deng and a link to increased cooperation with the United States",[18] "but the U.S.-China rapproche-

---

17 Henry Kissinger (2011). *On China*. New York: The Penguin Press. P340.
18 Ibid, P346-347.

ment had set up a significant bulwark against further expansion".[19] In 1972, Nixon urged the European allies of the U.S. to push forward China's economic development together with the U.S. and Japan. "He [Nixon] had a vision of an entirely new international order emerging based essentially on using China's influence to build the Third World into an anti-Soviet coalition".[20]

In 1979, China launched the Sino-Vietnamese War. "The invasion [of Vietnam] served its fundamental objective: when the Soviet Union failed to respond it demonstrated the limitations of its strategic reach. From that point of view, it can be considered a turning point of the Cold War, though it was not fully understood as such at the time. The Third Vietnam War was also the high point of Sino-American strategic cooperation during the Cold War."[21]

As the external environment changed for the better, the reunification of China was once again on the agenda of Chinese leaders. In 1978 and 1979, Japan and the U.S. established comprehensive diplomatic relations with China respectively. Over the following decade, China's cooperation with Japan and the U.S. went smoothly: Japan's technological and financial supports to China benefited China's economic construction to a large extent[22]; the Sino-American cooperation in intelligence and military improved the relationship between the two countries.

Nixon's visit in 1972 showed his willingness to see China's peaceful reunification.[23] As a link in rebuilding the political authority of the CPC, the issue of Taiwan's return to China had to be addressed soberly. On behalf of the Chinese Government, Ye Jianying (葉劍英), a

---

19 Ibid, P387.
20 Ibid, P393.
21 Ibid, P340.
22 Sheila A. Smith (2015). *Intimate Rivals: Japanese Domestic Politics and a Rising China.* New York: Columbia University Press. P33-39. After the Second World War, both the KMT (Kuomintang)-led Government and the CPC-led Government gave up claiming war reparations from Japan. Some Japanese regard Japan's financial support to China as "another form" of war reparations, but few Chinese agree with such view.
23 Margaret MacMillan (2007). *Nixon and Mao: The Week that Changed the World.* New York: Random House.

former leader of the central authority, came up with "Nine Points for the Peaceful Reunification of Mainland China and Taiwan", which marked a great step forward on China's path to reunification.

When the issue of Hong Kong's future arose, the Chinese people began to eagerly and passionately discuss China's ideological development. The development model and experience of the Soviet Union were considered deficient, as were the economic-development policies of Mao Zedong's administration. In contrast, capitalism, market economy, free trade, and Western-style globalization gained popularity. The market competition mechanism was no longer seen as a privilege of capitalist societies–socialist countries could also wisely make use of it. For a moment, a cross section of Chinese people, particularly some prominent intellectuals, tended to favour the theories and ideas from the West, creating a favourable environment for East-West cooperation in terms of ideology and psychology.[24]

Generally speaking, China's internal and external environment generated favourable conditions for the CPC's innovations in the way China was governed. While Britain, the U.S., Japan and some other Western countries were reluctant to see China take over Hong Kong, a former useful member of the Western bloc, and were worried about Hong Kong's future after its return, on the other hand, in order to maintain the East-West Partnership against the Soviet Union, Britain and other Western countries did not want to damage their relationship with China over the Hong Kong issue.

At the same time, the U.S. and Western countries preferred to continue their support of and care for Hong Kong's development after its return to China. In particular, they hoped Hong Kong would maintain its capitalist system, the legal system deriving from the West, as well as its extensive links with the West and the international community. In order to show its continuous support for and ensure its various interests in Hong Kong, the U.S. enacted the U.S. Hong Kong Policy Act in

24 Rana Mitter (2004). *A Bitter Revolution: China's Struggle with the Modern World*. Oxford: Oxford University Press. P246-272.

1992, granting preferential treatment to Hong Kong, an advantage only enjoyed by some independent and friendly countries. The preferential treatment included an enlarged immigrant quota and the introduction of high and new technologies from the West, provided that human rights and freedom would not deteriorate in Hong Kong after the return. The intentions of Western countries were straightforward; they did not even try to conceal potentially divisive aspects. They hoped that Hong Kong would serve as a catalyst of China's "peaceful evolution" by helping end the CPC's one-party regime and encouraging China to adopt the political and economic development patterns from the West, so that "pro-West" political forces would come to lead China, and the CPC's threat to the West would be thoroughly eliminated in the long run.

Hong Kong's Situation

Before the issue of Hong Kong's future arose, Hong Kong had already been an industrial society and was transforming rapidly into a service-oriented economic system by taking advantage of the "Reform and Opening up" policy of China. Hong Kong people had basically shaken off their refugee mentality by the end of the 1970s, meaning that they already regarded Hong Kong as a place where they could settle down and pursue a living rather than viewing Hong Kong as a temporary residence. Also, mainland China was no longer their ultimate home. To the Hong Kong people who were born and grew up after the Second World War, Hong Kong was their only "home". Since a large number of people moved to Hong Kong to escape from the war and the authority of the CPC, these people and their descendants were anti-Communist or Communist-phobic to some degree. After 1949, Hong Kong and mainland China headed along different development paths, and Hong Kong's economic achievements and the livelihood of its people were far superior to those of mainland China in the early 1980's, giving quite a number of Hong Kong people the impression that they were far different from the people in mainland China in a great many aspects. As a result, they not only harboured a sense of superiority, but also were deeply prejudiced against their compatriots in mainland China.

As a matter of fact, the rising self-identification of Hong Kong people was based on their anti-Communist awareness and a deliberate comparison with their mainland compatriots. Their mentality and pessimism towards China's future were trapped in a constant state of anxiety and uncertainty and contributed to their inadequate psychological preparation for the "1997" question.

For more than a decade prior to the emergence of the "1997" question, the colonial government was devoted to improving its governance model in a considered manner. It explored new channels for communication between officials and citizens to demonstrate a more open-minded political style, which enabled the Hong Kong people, especially the growing number of middle-class elites, to have more space to exert political influence, thereby shaping the image of government as acting in the service of Hong Kong people. The government also increased investment in public welfare and services step by step, deliberately fostering Hong Kong people's sense of belonging to Hong Kong along with their feeling of being the "masters of Hong Kong".

Hong Kong people took pride in their economic achievements after the Second World War, attributing Hong Kong's "economic miracle" to the British administration and the "colonial" system, the rule of law, and to a policy of "non-intervention in economy". Though the income gap between the rich and the poor began to widen in the 1970s, it was not serious enough to provoke social attention or disputes. All levels of society reached a consensus on Hong Kong's institutional structure and public policies. The voices calling for changing the status quo on a large scale could be heard from time to time but were not loud enough to capture wide attention. Although protests for various social and political reforms took place quite frequently, they lacked a solid popular base as well as political energy. Generally speaking, when the "1997" question emerged, Hong Kong was enjoying a high level of stability and its people were optimistic about its future.[25]

To most Hong Kong people, the emergence of the "1997" question was a big surprise, even something out of the blue. Although Britain had been aware of the Chinese Government's intention to take back Hong Kong a few years before, it could not confirm such an intention and still hoped to rule Hong Kong in some aspects after 1997. Britain was also reluctant to let Hong Kong people know that the colonial administration would possibly end in 1997 because they feared it would give rise to panic and political turmoil in Hong Kong. Besides, it was politically unwise to put an uncertain future in front of Hong Kong people without knowing what the Chinese Government would plan to do after taking Hong Kong back. On the other hand, the Chinese Government would not prematurely trigger the issue of Hong Kong's future in Hong Kong because it would not want to affect Hong Kong's prosperity and stability.

Therefore, when the issue of Hong Kong's future emerged without any advance notice, it inevitably stirred up fear, concern, insecurity and anxiety among the Hong Kong people. Emigration rates rose swiftly, especially among elites. Some enterprises, including Britain-funded enterprises, began to withdraw their capital and staff from Hong Kong. In the shadow of the "1997" question, Hong Kong's economic development and social stability were challenged by various factors. From either the perspective of safeguarding the effective governance of the colonial government or that of maintaining Hong Kong's economic value to China, resolving the issue of Hong Kong's future as soon as possible became a top priority. Under such circumstances, Britain as well as the Chinese Government, the protector of Hong Kong's long-term prosperity and stability in the future, faced enormous pressure.

To resolve the issue of Hong Kong's future as soon as possible, China and Britain launched a two-year negotiation in 1982 and offi-

---

25 Lau Siu-kai (1982). *Society and Politics in Hong Kong*. Hong Kong: The Chinese University Press; Lui Tai-lok (2012). *Those were the days: Hong Kong in the 70s* (《那似曾相識的七十年代》). Hong Kong: Chung Hwa Book Company.

cially signed the Joint Declaration of the Government of the United Kingdom of Great Britain and Northern Ireland and the Government of the People's Republic of China on the Question of Hong Kong with Annexes in 1984 (the Joint Declaration). Both sides agreed to put an end to the issue by implementing "one country, two systems".[26]

---

26 There are not many serious studies on the Sino-British negotiation. The following books can serve as references: Li Hou (1997). *The Return of Hong Kong* (《回歸的歷程》). Hong Kong: Joint Publishing; Qian Qichen (2004). *Ten Episodes in China's Diplomacy* (《外交十記》). Hong Kong: Joint Publishing; Lu Ping (2007). *Lu Ping's Verbal Recollections of the Reunification of Hong Kong* (《魯平口述香港回歸》). Hong Kong: Joint Publishing; Zhang Chunsheng & Xu Li (eds.) (2012). *Zhou Nan Declassifies the Returns of Hong Kong and Macao: the Background of China-Britain Negotiation and China-Portugal* (《周南解密港澳回歸---中英及中葡談判台前幕後》). Hong Kong: Chinese Publishing House; Gao Wanglai (2012). *The Negotiation Strategies of Great Powers: the Scenes Behind the Sino-British Negotiation on the Hong Kong Issue* (《大國談判謀略：中英香港談判內幕》). Beijing: Current Affairs Press; Chen Dunde (2013). *Abolishing the Treaties: the China-UK Negotiation of Hong Kong Issue* (《廢約：中英香港問題談判始末》). Beijing: China Youth Publishing Group.

CHAPTER TWO

The Basic Connotations and Objectives of "One Country, Two Systems"

The policies of the CPC and the CPC-led Chinese Government towards Hong Kong are based on a rational, objective, and thorough analysis of history and facts. They have taken into full consideration the domestic and international situation, the relationship between China and Western countries, Hong Kong's potential contributions to China, as well as the overall national interests. They were seldom influenced by revolutionary enthusiasm, ideologies, and motives related to temporary changes of external environment. From the establishment of the People's Republic of China until the present, the Chinese Government's policies towards Hong Kong have been stable, consistent, and highly predictable, which helps maintain Hong Kong's value to China, Hong Kong people's confidence in Hong Kong, and the support from the international community for Hong Kong.

The Chinese Government had temporarily decided not to take back Hong Kong in the period directly before the establishment of the PRC, with the aim of avoiding unnecessary conflict with Britain and the U.S. over the Hong Kong issue. At that time, China could also restrain the U.S. by taking advantage of the conflict of interest between the U.K. and the U.S. and maintain Hong Kong's role as a bridge to connect China with Western countries. Considering the new China's isolation and containment by the West shortly after its establishment and the unfriendly treatment it received from the Soviet Union and the Socialist bloc, the CPC leaders' policies towards Hong Kong are wise and forward-looking indeed.[1]

In the early 1950s, the Chinese Government's policy towards Hong Kong became explicit, systematised and normalised, featuring "long-term planning and full utilisation". Under this guideline, the Chinese Government recognised neither the three unequal treaties signed by the Qing Government and Britain nor Britain's proclaimed sovereignty over Hong Kong. However, in the early years after new China was established in 1949, Beijing was not in a hurry to take back Hong Kong. Instead, it allowed Britain to administer Hong Kong on the condition that Hong Kong would not become independent or semi-independent, and that Hong Kong would not become a threat to China's national security. In particular, the Chinese Government forbade the Kuomintang (also referred to as the KMT) to undertake destructive activities against mainland China via Hong Kong. On the other hand, it was hard to stop the U.S. and some Western countries from exploiting Hong Kong by launching espionage and intelligence activities against China.

---

1   Li Hou (李後), the former deputy director of Hong Kong and Macao Affairs Office of the State Council made a brief interpretation of the CPC's policies towards Hong Kong in his book. He said, "The national leaders decided not to take back Hong Kong for a pretty long time based on the following considerations: first, with the fierce competition between the two [capitalist and socialist] camps, it is impossible to resolve the Hong Kong issue in a peaceful way. It had to resort to force to take Hong Kong back. Britain knew that it could not win a war with China on its own. Inevitably, it would join hands with the U.S. to safeguard its control of Hong Kong. That scenario was not what China wanted to see. In the eyes of the Chinese Government and its leaders, it was better to keep Hong Kong in the hands of Britons than to fight against Britain and the U.S. for Hong Kong; second, it had not been a long time since the establishment of new China and there were not many countries with which China had established diplomatic relations. Under the leadership of the U.S., western countries imposed economic blockade on China. Against such background, if the status quo was maintained, Hong Kong would serve as a bridge to connect China and the outside world, and thereby China would obtain things that could not be obtained from other channels. See Li Hou (1997). *The Return of Hong Kong* (《回歸的歷程》). Hong Kong: Joint Publishing. P47. Although Britain and the U.S. had negotiations and disputes on whether they should make "joint efforts to protect Hong Kong", the U.S. had never had the determination to provide military aid to Britain to "protect" Hong Kong. See Chi-kwan Mark (2004). *Hong Kong and the Cold War: Anglo-American Relations 1949-1957*. Oxford: Clarendon Press. P12-82. Also, see the article "Hong Kong's Role in China's International Strategy" ("香港在中國國際戰略中的角色") in Lau Siu-kai (2013). *Hong Kong Politics since the Handover* (《回歸後的香港政治》). Hong Kong: The Commercial Press (H. K.) Ltd. P77-109., and Chen Dunde (2013). *Abolishing the Treaties: the China-UK Negotiation of Hong Kong Issue* (《廢約：中英香港問題談判始末》). Beijing: China Youth Publishing Group. P2-14.

In order to alleviate China's worries, Britain imposed restrictions on the activities of the Soviet Union in Hong Kong. The Soviet Union was not allowed to establish a consulate in Hong Kong. Moreover, it promised that Hong Kong would not become independent like other British colonies, and promised not to "allow the people of Hong Kong to govern themselves", or take other actions which would turn Hong Kong into "an independent political entity". In order to maintain good relations between Hong Kong and China, the colonial government also made great efforts to forestall incidences that would anger the Chinese Government. Anti-communist demonstrations of the Hong Kong people were restrained to a considerable extent, but anti-communist opinions were usually allowed or even encouraged by the colonial government in order to weaken the CPC's political and ideological influence over Hong Kong people.

The colonial government was especially afraid of the leftist patriotic force which supported the CPC in Hong Kong. In order to prevent the leftist force from growing and developing into a palpable threat, it made every effort and took various actions to crack down on the pro-Beijing patriots in Hong Kong. Riots which broke out between 1967 and 1968, organised by the leftists against the colonial government as an extension of the Cultural Revolution in Hong Kong, were violently suppressed. The unfortunate leftist force was since then politically marginalised as it was excluded and rejected by Hong Kong people who were unequivocally against the riots. Hong Kong people's attitude towards the CPC also worsened, and anti-Communist sentiments and general Communist-phobia greatly intensified in Hong Kong.

After new China was founded, a large number of people from mainland China moved to Hong Kong through legal or illegal means to escape from the continuous political struggles and movements on the mainland. Some of these were social activists and dissidents targeted by the Chinese Government. These immigrants and their offspring saw Hong Kong as a political sanctuary. Though they resented the CPC, they had no intention to use Hong Kong as a base to launch activities against the CPC regime. In fact, under the administration of the colonial government, they had neither the capacity nor the opportunity to do so. Therefore, under the colonial administration, Hong Kong could remain at peace with the central authority and mainland China, even though a large number of Hong Kong people were dissatisfied with the CPC and often expressed their dissatisfaction through the media or other channels.

As for the Chinese Government, it had neither the intention nor the capability to intervene in Hong Kong's affairs under Britain's administration. Britain, for its part, worried that the Chinese Government would contest the colonial administration and extended every means in order to prevent the mainland's intervention in Hong Kong's affairs. In this sense, the colonial government played the role of a "wall" in "separating" Hong Kong people and the Chinese Government.

In effect, from 1949 up to the end of the 1970s, the relationship between Hong Kong and mainland China was not close. The two places were following enormously different development paths, and with infrequent contact, their people were increasingly estranged from one another. The Chinese Government had difficulties influencing Hong Kong affairs, and Hong Kong people also rarely involved themselves in the affairs of mainland China, nor did they pose any political challenge to the central authority.

Under the policy of "long-term planning and full utilisation", Hong Kong played some unique and irreplaceable roles for new China. Britain also allowed it to play these roles to some extent, because they thought that by doing so China's anxieties about the continuing colonial administration of Hong Kong would be relieved. For example, after the Korean War broke out in the early 1950s, with the absence of the Soviet Union, the U.S. obtained the consent of the United Nations to impose an economic blockade on China. However, China could still acquire some war materials from Hong Kong with Britain's implicit acquiescence. Later, contained by Western countries and isolated by the Soviet Union, China had perforce to "seclude itself from the outside world" and take the path of "self-reliance". Due to its isolation and the failures of the rash economic policies of the leftists, China struggled with economic difficulties for several decades. During the hard times in China, Hong Kong served as the most important export market and transit port, the main source of foreign exchanges, and as a window connecting China with the outside world, as well as a channel for obtaining external information and technologies. China could also make use of Hong Kong to conduct some necessary and sensitive exchanges with the West.

At the end of 1978, the Third Plenary Session of the Eleventh Central Committee of the CPC made a decision to abolish the developmental model "centring on the class struggle", and instead lay a priority upon "Reform and Opening up" along with general economic development. As a modern international city, Hong Kong occupied a significant position in the modernisation of China under the new national policy. China could absorb various elements indispensable to its modernisation from or via Hong Kong, such as capital, talents, laws and regulations, institutional and industrial systems, science and technology, commercial information, all kinds of productive services, and modern management methods. In addition, Hong Kong's excellent port facilities and highly-developed shipping industry also provided indispensable services benefiting the export-oriented economic policy of China. At the same time, the declining labour-intensive manufacturing industry in Hong Kong thrived once more, as it was relocated to mainland China, which in turn promoted the industrialisation and modernisation of the Pearl River Delta. The reform and opening up of China helped Hong Kong realise its economic transformation swiftly, and allowed it to develop into an advanced economic system centring on finance in tandem with a modern service industry. Likewise, Hong Kong could inject a lasting driving force into the Chinese national economy. It can be said that China's policy of "long-term planning and full utilisation" emphatically proved its worth and effectiveness during the late 1970s.

The issue of Hong Kong's future arose during this "peak" moment, greatly impacting the Chinese Government's policy towards Hong Kong. Having fairly good relations with the West, the Chinese Government obtained favourable conditions and opportunities to put forward a new Hong Kong policy, based on the ongoing "long-term planning and full utilisation", and accommodating the interests of all parties concerned. The Chinese Government would use this new policy to strive for the peaceful return of Hong Kong from Britain, to enhance Hong Kong people's confidence in Hong Kong's future, and to maintain the support from the international community for Hong Kong. This policy was entitled "one country, two systems".

In essence, "one country, two systems" was the extension and further development of "long-term planning and full utilisation" employed after Hong Kong returned to China. The policy aimed to prevent the changes in Hong Kong's political and legal status from affecting Hong Kong's value to the nation. However, if Hong Kong was transformed from a British colony into a "special administrative region" of China, a number of elements were bound to change. Therefore, it was to a certain extent uncertain and unpredictable as to whether "one country, two systems" would achieve its goal of maintaining Hong Kong's value to China.

Due to these uncertain and unpredictable factors, the Chinese Government was faced with a difficult decision as to whether to take over responsibility for the sovereignty of Hong Kong or not. Some thought it would be easy to take back Hong Kong but difficult to maintain Hong Kong's prosperity and stability. Huang Wenfang (黃文放), a former official of the Hong Kong Branch of Xinhua News Agency, noted that China's deciding to take back Hong Kong was a painful process. In 1979, when Britain raised the issue, China, having just walked out of the crisis of the Cultural Revolution, was facing an enormous challenge in terms of cultural and economic restoration. The Chinese Government, without neither a persuasive solution nor the psychological space to come to terms with this question of Hong Kong's future, had repeatedly cautioned Britain not to raise the issue for the time being. Nonetheless, Britain applied continuous pressure on the Chinese Government.[2]

---

2   Huang Wenfang (1997). *China's Resumption of Sovereignty over Hong Kong* (《中國恢復對香港行使主權的決策歷程與執行》). Hong Kong: David C. Lam Institute for East-West Studies, Hong Kong Baptist University. P2-3.

According to Huang, the Chinese Government hoped to resolve the issue of Hong Kong's future by copying the Macao model. He stated, "When China established diplomatic relations with Portugal in 1979, both arrived at four consensuses in their negotiations: First, Macao was part of the territory of China and the Chinese Government held sovereignty over Macao; second, the Macao issue was an issue left over from history; third, the Macao issue would be resolved through negotiation by the governments of both countries at an appropriate time; fourth, Macao would maintain its status quo before the Macao issue was resolved.... In this way, it was natural for China to consider the feasibility of solving the '1997' question with the Macao model after Britain forced China to negotiate on the question of sovereignty over Hong Kong.... We had lots of unofficial exchanges with Britain on the Hong Kong issue, hoping that we would negotiate with each other at a proper time and come to an agreement that Hong Kong had always been a part of the territory of China and its sovereignty belonged to the Chinese Government, and Hong Kong would maintain its status quo until the proper conditions arose.

However, Britain disagreed with the Chinese Government's assessment. From Britain's view, the legal basis of their administration in Hong Kong was laid out in the three treaties, and furthermore, the legal basis of colonial rule would not exist without the three treaties. They hoped that China would allow them to administer Hong Kong for another 30 to 50 years through formal negotiation. The British perspective saw China as having just put an end to the Cultural Revolution and needing to make full use of Hong Kong to develop its economy quickly. Therefore, Britain would have the leverage to force China to make a decision in Britain's interests. However, they underestimated China's strong nationalism, and the CPC was as persistent as usual when it came to the issue of territory. In this sense, Britain left no leeway for China, so China had no other choice but to reconsider its plan."[3]

"When the conference for unifying ideas was convened in Beijing

---

3   Ibid, P8-9.

in May 1982, Hong Kong officials still held the view that it would be easy to take back Hong Kong but difficult to maintain its prosperity and stability. Normally, it was expected that everybody in the CPC should align their views with the position of the centre. Nevertheless, in that conference several heads of the leftist force in Hong Kong expressed views opposing the transfer of sovereignty of Hong Kong."[4] "Actually, from 1979, when Britain raised the issue of Hong Kong's sovereignty, to 1981, the whole process of making a decision on the issue was a painful experience for China. Within these three years, even Deng Xiaoping did not dare to make the decision. The core problem affecting the decision was that it would be easy to take back Hong Kong but hard to maintain its prosperity and stability; and Hong Kong people lacked confidence in China at that time as China herself was facing a severe crisis in its economy and over the consolidation of political power."[5]

The recollections of Huang Wenfang were confirmed by Lu Ping (魯平), former director of Hong Kong and Macao Affairs Office of the State Council. He said, "We did not reach consensus on whether to take back Hong Kong or not. In particular, some of our comrades working in Hong Kong preferred to maintain the status quo of Hong Kong and adopt the Macao model in which China maintained sovereignty over Macao while Portugal continued to administer Macao. Nevertheless, Liao Chengzhi (廖承志) thought that, from the perspective of the whole country, we should consider how to take back Hong Kong instead of whether to take it back or not. "[6]

The standpoint of the Chinese Government was as follows: Since the lease of the New Territories would expire in 1997, Britain would have to return the New Territories, which accounted for a major part of Hong Kong, that year. If China would not take the opportunity to take back Hong Kong as a whole in 1997, it would be extremely difficult to take it

---

4   Huang Wenfang (1997). *China's Resumption of Sovereignty over Hong Kong* (《中國對香港恢復行使主權的決策歷程與執行》). Hong Kong: David C. Lam Institute for East-West Studies, Hong Kong Baptist University. P12.

5   Ibid, P22.

6   Lu Ping (2007). *Lu Ping's Verbal Recollections of the Reunification of Hong Kong* (《魯平口述香港回歸》). Hong Kong: Joint Publishing. P14.

back afterwards, and the Chinese Government would thereby disappoint the Chinese people. The national leaders might even be condemned as was the second Li Hongzhang, the official in the late Qing Dynasty who signed the three unequal treaties with Britain.[7] After making the decision to take back Hong Kong in 1997, what needed to be done next was to formulate a set of policies aimed at maintaining Hong Kong's prosperity and stability after its return. Such policies would be used to forge the Hong Kong people's and investors' confidence in Hong Kong's future, and force Britain to give back Hong Kong in a "decent", "willing" and peaceful manner and to agree to administer to Hong Kong in the 15 years prior to 1997 in line with China's Hong Kong policy.

In the Chinese Government's earlier plan for national reunification, cross-straits reunification was a priority. On September 30, 1981, Ye Jianying, a member of the Standing Committee of the National People's Congress, announced the 9 points for the peaceful reunification of mainland China and Taiwan, which declared the Chinese Government's stance towards cross-straits reunification. The contents of the 9 points were as follows:

(1) In order to end the split state of the Chinese nation as soon as possible, we suggest that the CPC and KMD negotiate with each other and cooperate for the third time to realise the great reunification of China. Both sides can dispatch representatives at first for exchanging views;

(2) People across the straits are eager to contact each other, reunite with families, trade with each other, and promote mutual under-

---

7   When Deng Xiaoping met Thatcher, the Prime Minister of Britain, on September 24, 1982, he said, "If China could not take back Hong Kong in 1997, 48 years after the establishment of the People's Republic of China, no Chinese leader and no Chinese Government could explain that satisfactorily to the Chinese people or to the world. If China failed to take back Hong Kong, it would mean that the Chinese Government was just another late Qing Government and the Chinese leader was another Li Hongzhang. We have been waiting for 33 years. Plus 15 years, it will be 48 years. It is only because we are trusted by the Chinese people that we can wait for such a long time. If China failed to take back Hong Kong in 1997, people would have no reason to trust us. Any Chinese Government should step down and voluntarily leave the political arena. There would be no other choice." See Deng Xiaoping (1993). *How Deng Xiaoping Views Hong Kong Issue* (《鄧小平論香港問題》). Hong Kong: Joint Publishing. P1-2.

standing. We suggest both sides draw agreements to provide convenience for the direct links of post, trade, and air, visiting relatives, tourism, and for conducting academic, cultural and sports exchanges;

(3) After reunification is realised, Taiwan can serve as a special administrative region with a high degree of autonomy and will be entitled to preserve its military forces. The central authority will not intervene in Taiwan's affairs;

(4) The current social and economic systems in Taiwan will remain unchanged, and so will the lifestyle in Taiwan and its economic and cultural relationships with foreign countries; the personal property, houses, land, enterprise ownership, legitimate right of inheritance, and foreign investment in Taiwan will also be under protection;

(5) Taiwan authorities and representatives from all sectors in Taiwan will be qualified to hold leadership positions in national political bodies and take part in the administration of the country;

(6) When the local finance of Taiwan runs into trouble, the central authorities will provide financial subsidies to Taiwan accordingly depending on needs;

(7) If people from all walks of life and ethnicities in Taiwan want to come back to mainland China and settle down, they will receive good care and live at liberty, without suffering from discrimination;

(8) People from the industrial and commercial circles in Taiwan will be welcome to invest in mainland China and undertake all kinds of business activities. Their legal rights, interests and profits will be protected;

(9) Everyone is obligated to contribute to the reunification of our country. All nationalities, individuals and groups in Taiwan are welcome to contribute suggestions on national affairs through various channels.

Despite the difference in name, Ye Jianying's 9-point policy on Taiwan shared much common ground with "one country, two systems". Though it did not obtain the anticipated official reply from Taiwan, it laid the foundation for "one country, two systems" as the central authority's solution to the issue of Hong Kong's future. It can be said that "one country, two systems" derives from these 9 points. Shortly after the announcement of the 9 points, the Chinese Government formulated a 12-point policy for Hong Kong in the same year. This policy was designed to maintain Hong Kong's prosperity and stability after Hong Kong's return to China and was used as a bargaining chip when China negotiated with Britain over the issue of Hong Kong's future.

The 12 points include:

(1) Hong Kong will return to China on July 1, 1997;

(2) After returning to China, Hong Kong will become a special administrative region with a high degree of autonomy and will be under the direct administration of the central authority;

(3) Hong Kong's status as a free port and financial centre will be maintained;

(4) The Chief Executive of Hong Kong will be appointed by the central authority and can be a Hong Kong citizen;

(5) The current lifestyle, social, economic and welfare systems in Hong Kong will remain unchanged;

(6) The personal property, houses, land, enterprise ownership, legitimate right of inheritance in Hong Kong will be under protection;

(7) Foreign enterprises and businesses will be free from infringement;

(8) Hong Kong will establish mutually beneficial economic relations with Britain;

(9) Hong Kong's dollar will remain unchanged;

(10) The existing laws, decrees, and regulations in Hong Kong will be retained on the whole;

(11) Hong Kong Special Administrative Region will be responsible for its own security;

(12) The Chinese and foreign staff in the governmental agencies of Hong Kong will be paid equally if they want to continue their work. If necessary, the Government of the Hong Kong Administrative Region will hire foreigners as consultants.

It can be observed from historical facts that "one country, two systems" was not the product of negotiation between China and Britain, but a unilateral proposal from the Chinese Government. As an important policy that complemented the "Reform and Opening up" policy, "one country, two systems" was already solemnly reflected in the Constitution of the People's Republic of China in 1982. In the Basic Law, a text reads as follows: "To facilitate national reunification, safeguard territorial integrity, and maintain prosperity and stability in Hong Kong, based on the historical and current conditions in Hong Kong and according to the Article 31 of the Constitution of the People's Republic of China, after regaining the sovereignty over Hong Kong, China will establish the Hong Kong Special Administrative Region but will not implement socialist systems and policies there under 'one country, two systems'." As a fundamental policy of China towards Hong Kong, "one country, two systems" was further expounded on in the Joint Declaration signed by China and Britain in 1984.

Although the Chinese Government had resorted to different channels to express its intention to take back Hong Kong in 1997, such moves failed to draw careful attention from Britain. The British continued to believe that the Chinese Government was increasingly dependent on Hong Kong economically and that it would never kill this "goose that lays the golden eggs". Thus, they hoped that China would not rush to take back Hong Kong. They took the upcoming "1997" question as a pretext for placing continuous pressure on China through different means, hoping to continue their administration in Hong Kong in some fashion after 1979. Only when they knew that they would have to end colonial governance in Hong Kong by 1997 and that the "one country, two systems" policy was favoured by many Hong Kong people did they reluctantly agree to give Hong Kong back to China. Still, they tried to influence the Chinese Government's policies towards Hong Kong in order to maximise their own interests. They did their best to maintain their effective administration in Hong Kong in the transitional period and groom those "Hong Kong people" who in their view would play a great part in "administering" Hong Kong after its return to China.

Despite all the twists and turns in the Sino-British negotiation over the issue of Hong Kong's future, both sides managed to resolve the issue left from history in a peaceful way and to maintain good relations between China and Britain and even between China and the Western world as a whole. Evidently, the Sino-American strategic partnership to contain the Soviet Union, the support of the West for China's "Reform and Opening up" policy and its "one country, two systems" policy for Hong Kong were conducive to the smooth resolution of the issue. During the Sino-British negotiation and the consultation between the Chinese Government and the Hong Kong people, the Chinese Government revised the original 12 points in response to demands from various parties. For instance, in the revised version, Hong Kong was allowed to develop its democratic politics in a gradual and orderly way. Still, the "one country, two systems" policy remained basically intact.

It can be said that the "one country, two systems" policy was successfully installed with the "good wishes" of China, the West and the Hong Kong people. To consolidate the confidence of all parties concerned in this policy and in Hong Kong's future, the Chinese Government wrote its policy towards Hong Kong into an international agreement, the Joint Declaration between China and Britain. In addition, it legalised the policy through the constitutional document of the Basic Law.

Rao Geping (饒戈平), a legal scholar of mainland China, observed, "Obviously, the national policy on Hong Kong and the related constitutional provisions are prior to China's international promise about Hong Kong's legal status after its return. The Basic Law is based on the Constitution of the People's Republic of China and China's policy towards Hong Kong, not only on its international obligations in the Joint Declaration. This is the true nature of the relations between them."[8] Therefore, "The Joint Declaration is an international agreement between China and Britain designed to resolve the issue of Hong Kong

---

8   Rao Geping (2014). The Legal Property and Status of Hong Kong's Power in Dealing with Foreign Affairs (香港地區對外事務權的法律性質和地位). In Rao Geping (eds.) *The External affairs of Hong Kong SAR from the Perspectives of the International Law, and the Political System Development and Legal Reform of the Macao SAR* (《燕園論道看港澳 — 香港特區對外事務權的國際法視角澳門特區的政治發展與法律改革》). Beijing: Peking University Press. P3-22, P13.

while the Basic Law is China's constitutional law on administering Hong Kong. The Joint Declaration is to the Basic Law what international law is to domestic law to some extent. They share a great deal of content and are complementary to each other. The Joint Declaration embodies China's international promise about its policy on Hong Kong while the Basic Law legalises and institutionalises such policy, completely fulfilling that promise. Although both facilitate the implementation of 'one country, two systems', they differ from each other, perform different functions and belong to different legal systems."[9]

"As a sovereign country, China laid down a domestic law, namely the Basic Law, in accordance with its own constitution and legislative procedures. Certainly, China had to fully consider and perform its international obligations when formulating a law. China kept its promise under an international agreement to formulate the Basic Law in order to carry out its policy towards Hong Kong. From this point of view, China did fulfil its international obligations. The Joint Declaration constitutes a necessary legislative condition for the Basic Law and makes international law the legal guarantee that Hong Kong administer itself with a high degree of autonomy and that its powers to conduct foreign affairs remain intact. According to the formulation process and content of the Basic Law, China completely respected international law as specified in the Joint Declaration and abided by the announcements thereof, which shows that China was fulfilling its international obligations."[10]

In fact, the preamble of the Basic Law clearly announces that the Basic Law is the legal foundation of "one country, two systems". It says, "The basic policies of the People's Republic of China regarding Hong Kong have been elaborated on by the Chinese Government in the Sino-British Joint Declaration. In accordance with the Constitution of the People's Republic of China, the National People's Congress hereby enacts the Basic Law of the Hong Kong Special Administrative Region of the People's Republic of China, prescribing the systems to be

9   Ibid, P14.
10 Ibid, P15.

practised in the Hong Kong Special Administrative Region, in order to ensure the implementation of the basic policies of the People's Republic of China regarding Hong Kong."

The Strategic Significance of "One country, Two Systems"

The "one country, two systems" policy was formulated to facilitate national reunification, safeguard national territorial integrity, and maintain Hong Kong's prosperity and stability. In other words, "one country, two systems" serves national interests and the development of socialism with Chinese characteristics, and it assumes that maintaining the capitalist system in Hong Kong is most conducive to national development.

Deng Xiaoping once said, "'One country, two systems' includes capitalism and socialism. The major part of China with more than a billion people will take the path of socialism unswervingly. It is a necessary precondition for carrying out 'one country, two systems'. Under this precondition, the practice of capitalism in a small area is allowed. We are convinced that the practice of capitalism in a small area will be more favourable to developing our socialistic undertaking."[11]

Under "one country, two systems", the nation, the central authority and socialism take precedence over Hong Kong, the Government of HKSAR and capitalism. Without a socialist China led by the CPC, without the "Reform and Opening up" policy or the grand plan for modernising China, "one country, two systems" would not have been formulated. The intent of one country, two systems" is that it calls upon Hong Kong to continue to serve the national interests and needs after its return to China. Deng Xiaoping clearly said, "Our policies on Hong Kong, Macao, and Taiwan are based on the premise that the major part of China would abide by the Four Cardinal Principles (sticking to the

---

11 Deng Xiaoping (1993). *How Deng Xiaoping Views Hong Kong Issue* (《鄧小平論香港問題》). Hong Kong: Joint Publishing. P29.

socialist path, the people's democratic dictatorship, Marxism-Leninism and Mao Zedong Thought and the leadership of the Communist Party of China). Without the CPC or the practice of socialism, nobody else could have formulated these policies. No individual and no party had the courage and resourcefulness to do so. Do you think I am right? Courage and resourcefulness are necessary, and they need a foundation which is the socialist system and the socialist China led by the CPC. It is the socialism with Chinese characteristics that enables us to put forward 'one country, two systems' and allow the co-existence of socialism and capitalism. Our courage also derives from people's support for the national socialist system and the leadership of the CPC. In this sense, our adherence to the Four Cardinal Principles is related to our policies. If we overlook the Four Cardinal Principles, our policies would change accordingly, and there would be no continuous prosperity and stability in Hong Kong."[12]

From the perspective of national development strategy, the "one country, two systems" policy is an integral component of the CPC's strategy in eradicating the detrimental influence of the Cultural Revolution and launching the Reform and Opening up policy. This policy has carefully taken the following factors into consideration: the judgment of the international situation that "peace and development" would be the main pursuit at the present time and for a long time in future; the opportunities arising for China to build cooperation with the West; the urgent need to "learn from the West"; the consideration of China's long-term national security (against the Soviet Union), interests and development; and the consolidation and reinforcement of the CPC's ruling status and abilities.

From another perspective, the "one country, two systems" policy indicates that the CPC and the Hong Kong people shared common interests at the beginning of the 1980s, a special historical period in history when the CPC needed Hong Kong to promote China's modernisation,

---

12 Deng Xiaoping (1993). *How Deng Xiaoping Views Hong Kong Issue* (《鄧小平論香港問題》). Hong Kong: Joint Publishing. P32-33.

and Hong Kong was able to obtain generous and favourable treatment under "one country, two systems". It can also be said that if the issue of Hong Kong's future appeared today, whether Hong Kong people might still achieve such an advantageous arrangement so highly regarded by mainland compatriots is a difficult question.

The starting point of "one country, two systems" is to maintain Hong Kong's stability and prosperity, keep Hong Kong's status as a priority within the scheme of China's development and prevent Hong Kong from becoming a burden to China. The basic principle of 'one country, two systems' is that Hong Kong retains its capitalist system after its return to China. The core reason and consideration is that China needs it."[13] However, "one country, two systems" has some provisos, and the most important one is that it should be beneficial to the fundamental interests of the nation and to Hong Kong, with the former being particularly important. "One country, two systems" is not an arrangement utterly based on principle, ideal, faith or emotion, or out of the central authority's special affection for Hong Kong people. Actually, it greatly shows CPC's pragmatism and understanding of history. Whether to practise "one country, two systems" or not and how long it will be practised depends on whether "one country, two systems" really "works" and whether it will be beneficial to China and Hong Kong. Hong Kong must rely on the advantages of "its system", and continually play its role in promoting China's modernisation, especially in the economic dimension, in order for "one country, two systems" to become an important long-term national policy. With the continuous development of the country, Hong Kong's importance is bound to decline. As long as Hong Kong maintains its prosperity and stability, without being a burden for China, "one country, two systems" can continue to exist. On the other hand, if Hong Kong suffers from long-term economic depression and social instability and constantly needs "blood transfusion" and "external relief", or even hinders China's development, "one

---

13 Huang Wenfang (1997). *China's Resumption of Sovereignty over Hong Kong* (《中國對香港恢復行使主權的決策歷程與執行》). Hong Kong: David C. Lam Institute for East-West Studies, Hong Kong Baptist University. P30.

country, two systems" will be called into question. Under those circumstances, even Hong Kong people will be sceptical about the feasibility of "one country, two systems", and it will be even harder for mainland compatriots to accept the fact that Hong Kong people can still enjoy the favourable treatment which is denied to them.

The Core of "One country, Two systems"

In the past two decades and more, discussions in mainland China, Hong Kong and overseas on "one country, two systems" are limited and most are explanations and interpretations of Beijing's policies on Hong Kong or provide theoretical foundation or rationale for Beijing's Hong Kong policies. Yet, despite the inadequate discussions about "one country, two systems" and the Basic Law, their contents are quite consistent[14]. Therefore, the following part will make a brief introduction of the core contents of "one country, two systems". My personal observation and experience will also be included in the discussion.

Specifically, the core contents of "one country, two systems" are: maintaining Hong Kong's capitalist system, original institutions, policies and lifestyle, "Hong Kong people governing Hong Kong", "a high degree of autonomy," "Hong Kong's capitalist system and way of life remaining unchanged for fifty years" and preventing Hong Kong from being "a base for subversion" threatening the security of the

---

14 See Xiao Weiyun (1990). *One Country, Two systems and Hong Kong's Basic Law System* (《一國兩制與香港基本法制度》). Beijing: Peking University Press; Wang Shuwen (1990). *Introduction to the Basic Law of the Hong Kong Special Administrative Region* (《香港特別行政區基本法導論》). Beijing: Party School of the Central Committee of C.P.C Press; Hong Kong and Macao Research Centre of Development Research Centre of the State Council of the People's Republic of China (2009). *Handbook of the Basic Law of Hong Kong* (《香港基本法讀本》). Beijing: The Commercial Press; Wang Zhenmin (2014). *The Relationship between the Central Government and Hong Kong Special Administrative Region* (《中央與特別行政區關係》). Beijing: Tsinghua University Press; Song Xiaozhuang (2003). *On "One Country, Two systems" the Relationship between the Central Government and Hong Kong* (《論"一國兩制"中央和香港特區的關係》); Dong Likun (2014). *Relationship between the Governance Power of the Central Government and the High Degree of Autonomy of Hong Kong Special Administrative Region* (《中央管治權和香港高度自治權》) Beijing: Law Press.

nation and the CPC regime. The above are considered to be beneficial to Hong Kong's prosperity and social stability, maintaining Hong Kong's economic value to China, and preventing Hong Kong from becoming a threat to the nation and the CPC regime. I try to quote narrations and conversations of Deng Xiaoping and other leaders as well as central government officials, in order that readers will have a better understanding of the Chinese Government's intention when designing "one country, two systems", while simultaneously refuting the currently popular fallacy that the central authority is "treacherous". In fact, Hong Kong people are quite familiar with the principles of "one country, two systems", "Hong Kong people governing Hong Kong", "a high degree of autonomy" and "Hong Kong's capitalist system and lifestyle remaining unchanged for fifty years". But the principle that Hong Kong cannot become "a base for subversion" has not been paid particular attention to by Hong Kong people, so it is often neglected and the opposition even disdains it.

1. Maintain Hong Kong's original capitalist system

At the end of the 1970s, the West and the East were still embroiled in the "New Cold War" and China's "Reform and Opening up" had just started and was still in the exploratory stage. It was popular to roughly divide the world into two blocs or systems–"capitalism" and "socialism", and excessively highlight their differences and incompatibility, while neglecting their similarities. With the mainland's deepening understanding of capitalism, the viewpoint that market mechanism and fair competition did not equate capitalistic monopoly was widely accepted, and people had a better understanding that the state could better maintain capitalism only when it enjoyed a high degree of autonomy from the capitalists. Nevertheless, these "new" viewpoints appeared only after "the issue of Hong Kong's future" was solved.

Indeed, the Chinese Government's determination to maintain Hong Kong's original capitalist system not only conformed to Hong

Kong people's opinions and was instrumental in keeping capital and talents to remain in Hong Kong, but also sincerely demonstrated its appreciation of and reliance on Hong Kong's capitalism. Maintaining Hong Kong's capitalism was regarded as the cornerstone of Hong Kong's prosperity and stability. With capitalism, Hong Kong could still be an important member of the global economy and financial order dominated by the West, so that the mainland could continue drawing capital, talents, information, technology, managerial experience, institutions and laws and regulations, as well as other elements critical to China's "reform and opening up" from Hong Kong.

In 1990, Zhou Nan (周南), Director of the Hong Kong Branch of Xinhua News Agency, made a statement about Hong Kong's role in China's development, "After the return of Hong Kong to China, Hong Kong will enjoy a significant position in China's new opening-up framework. Within the framework of 'one country, two systems', the motherland, which has built up a solid socialist economic base, and Hong Kong, which preserves a capitalist economic system, will promote mutual benefits and draw on each other's strengths, leading to mutual prosperity. When China pursues its economic development, Hong Kong will continue to be of importance to China. The current international economic structure is clearly changing, and Asian economic strengths are rising. China plays a significant role in the economic development in the Asia-Pacific region and will become the pillar of peace, stability and prosperity there. Hong Kong in the future will not only be a vital part of China's open economy, but will also play an important role in economic exchange within the Asia-Pacific region. However, Hong Kong also has its shortcomings, for example, the lack of natural and human resources, the weakness of its technological base, especially in terms of high-technology, and a fragile economy because of overdependence on the world market. The above aspects are precisely the strengths of the mainland. Support of the mainland is indispensable for Hong Kong's continued prosperity. The mainland provides Hong Kong with manpower, material resources as well as a broad market, and vital scientific and technological cooperation. More importantly, the mainland has a stable political situation, and in turn creates a favourable environment for Hong Kong's economic development."[15]

China maintains the time-honoured tradition of "grand unification", which stresses that the whole country should be consistent or "homogeneous" in system, thought and policy. Besides, CPC believes that socialism is superior to capitalism and represents the future of human history. On the whole, to allow the existence of a totally different system within China is not only difficult in practice, but can also provoke criticism both at home and abroad. Putting forward the policy of "one country, two systems" therefore shows CPC's rationalism, pragmatism, breadth of mind and strategic thinking. It also reflects the CPC's belief in the ultimate success of China's modernisation and confidence in the superiority of socialism with Chinese characteristics.

Yet, the Chinese Government's understanding of Hong Kong's capitalist system is at the same time influenced by the historical environment. The leaders and officials had an insufficient understanding of the uniqueness of Hong Kong's capitalist system. They tended to consider Hong Kong's capitalist system as capitalism in general, and as a result, their summation of Hong Kong's system was delineated according to the three following understandings: first, in politics, a capitalist system needed the guidance of capitalists, who should therefore enjoy more political power than other members of society. The hypothesis underneath this observation was that even if the bourgeoisie had internal contradictions, what was most important was that, they as members of the same class still shared common interests and a common fate, so the bourgeoisie would tend to be strongly united. Second, even though there was divergence of interest between the bourgeoisie and other social classes, generally speaking, what was more important and salient were their common interests, so protecting the interests of the bourgeoisie did not contradict the overall interests of Hong Kong. Third, the govern-

---

15　Zhao Rui & Zhang Mingyu (1997). *Chinese Leaders' Comments on Hong Kong* (《中國領導人談香港》). Hong Kong: Ming Pao Publications Ltd. P283-284.

ment and the bourgeoisie were close partners in politics, a great proof of which was the close relationship between the colonial government and major British corporations. These understandings of Hong Kong's capitalism in fact very much underestimated the colonial government's political autonomy vis-à-vis the bourgeoisie, its vital role in reconciling the interests of various capitalists and those of all social classes, and its key role in managing the bourgeoisie and maintaining social stability.

Based on the above understandings of Hong Kong's capitalism, the Chinese Government thought that it was necessary to authorise Chinese capitalists to take over the status and roles of the British capitalists and grant them the necessary power to protect their interests. The objectives were to prevent the Hong Kong Special Administrative Region from having a "big government", and to preserve "capitalism with Hong Kong characteristics", including its free and intense market competition, low and simple tax regime, "limited government", as well as mechanisms tending toward capital accumulation, all for the sake of preventing a populism threatening to Hong Kong's current capitalist system. Under "one country, two systems", that Hong Kong's capitalist system should "remain unchanged" actually means, ironically, maintaining a very rare and unique economic system, and the precondition for maintaining this economic system is to ensure Hong Kong capitalists' superior political status. Now that Hong Kong's capitalism is very uncommon in the world, its appearance and existence must be related to a constellation of unique historical and structural factors and special conditions. Unfortunately, there was no nuanced and deep thinking about whether changes caused by the reunification of Hong Kong would hamper the effective operation and viability of such kind of capitalism.

2. "Hong Kong people governing Hong Kong"

"Hong Kong people governing Hong Kong" is a simple and clear statement. The Chinese Government promised that after the return of Hong Kong, Hong Kong people would replace the British in taking over Hong Kong's governance. Hong Kong people endowed with the right of "governing Hong Kong" refers primarily to permanent residents of Hong Kong. That is to say, the Chinese Government will not send mainland Chinese to run Hong Kong after its return to China. With the purpose of relieving the anxieties of Hong Kong people, those in charge of governing the HKSAR must have lived in Hong Kong for an extended period and must possess no right of abode elsewhere. It is therefore impossible for the Chinese Government to arrange for mainland people to settle down in Hong Kong before 1997 so as to become the Hong Kong people who would govern Hong Kong after its return to China. For example, according to Article 44 of the Basic Law, the Chief Executive of the HKSAR shall be a permanent resident of the Region with no right of abode in any foreign country who has resided in Hong Kong for a continuous period of not less than 20 years. Article 61 of the Basic Law stipulates that the principal officials of the Hong Kong Special Administrative Region shall be Chinese citizens who are permanent residents of the Region with no right of abode in any foreign country and who have resided in Hong Kong for a continuous period of not less than 15 years.

Yet, these requirements or restrictions are not the most important conditions. Those "Hong Kong people governing Hong Kong" must be from among those fully trusted by the Chinese Government so as to be able to implement entirely and accurately "one country, two systems" in accordance with the Chinese Government's blueprint, and in addition, build a good relationship between the central authorities and the HKSAR as well as between Hong Kong and the mainland, and finally, to prevent Hong Kong from becoming "a base for subversion". From the perspective of the central authority, now that it had given Hong Kong an extremely high degree of autonomy, the central authorities didn't have much legal authority to directly control Hong Kong affairs. It maintained limited power to correct any infringements by Hong Kong during the practice of "one country, two systems". Of course, under extreme conditions, the central authorities could resort to a range of more direct means to rectify a situation, but those would inevitably weaken domestic and international confidence in "one country, two systems" and damage China's international image. Under normal circumstances, the Chinese Government hoped that Hong Kong people could govern Hong Kong, leaving no need for Beijing to "interfere" in Hong Kong affairs.

To ensure that those people chosen to run the HKSAR could fulfil all expectations, Chinese leaders initially came up with the idea that the Hong Kong people who would govern Hong Kong after 1997 must be "patriots". Subsequently "patriots" were changed into "people who love China and Hong Kong". In 1984, Deng Xiaoping put forward the proposition of "patriots governing Hong Kong", and elaborated on it in detail. He stated that "Some requirements or qualifications should be established with regard to the administration of Hong Kong affairs by Hong Kong people. It must be required that patriots form the main body of administrators of the future government of the Hong Kong Special Administrative Region. [...] What is a patriot? A patriot is one who respects the Chinese nation, sincerely supports the motherland's resumption of sovereignty over Hong Kong and will not impair Hong Kong's prosperity and stability. Those who meet these requirements are patriots, whether they believe in capitalism or feudalism or even slavery. We don't demand that they be in favour of China's socialist system; we only ask them to love China and Hong Kong."[16] On another occasion in the same year, Deng Xiaoping further explicated the meaning of "patriot": "Hong Kong people must take part in [Hong Kong affairs] [during the period of transition]; if not, they would not be able to understand the manner in which Hong Kong is governed. There are opportunities to identify and choose talents through their participation [in Hong Kong affairs], and Hong Kong after 1997 can thus be better governed [by people who have experience in governance prior to Hong Kong's return to China]. Participants [in Hong Kong affairs] are only asked to meet one requirement; that is to say, they must be patriots who

---

16 Deng Xiaoping (1993). *How Deng Xiaoping Views Hong Kong Issue* (《鄧小平論香港問題》). Hong Kong: Joint Publishing. P8.

love China and Hong Kong. After 1997, Hong Kong people governing Hong Kong will continue with their capitalist system as usual, as long as they do not impair the interests of the motherland and Hong Kong compatriots. Both left and right wingers should be included of course, but not too many. It's preferable to choose more people who are politically moderate."[17]

Almost at the same time, Hu Yaobang (胡耀邦), former General Secretary of the Communist Party of China, stated that Hong Kong people governing Hong Kong "should meet two requirements: love China and love Hong Kong. Love of China is to support national unity and not division; love of Hong Kong is to serve Hong Kong people and maintain Hong Kong's prosperity".[18] Some leaders even specifically pointed out those who were the "people who love China and Hong Kong". Jiang Zemin (江澤民), former General Secretary of the Communist Party of China, described measures implemented in 1994 as follows, "We established the preparatory committee and appointed advisors on Hong Kong affairs charged with providing opportunities and channels for Hong Kong people to take part in the peaceful and stable transition of Hong Kong's reunification with China, laying the foundation of 'Hong Kong people governing Hong Kong' after 1997."[19] Qian Qichen (錢其琛), former Chairman of the Preparatory Committee for the HKSAR, affirmed in 1985 that Hong Kong's civil servants were an important force for "Hong Kong people governing Hong Kong". He said, "I once said that Hong Kong's civil servants are a reliable force for realising 'Hong Kong people governing Hong Kong' after 1997. That is to say, the post-1997 Hong Kong administration will still rely very much on current civil servants."[20]

There is no doubt that among all "Hong Kong people governing Hong Kong", the position of the Chief Executive of the HKSAR is

---

17 Ibid, P13.

18 Zhao Rui & Zhang Mingyu (1997). *Chinese Leaders' Comments on Hong Kong* (《中國領導人談香港》). Hong Kong: Ming Pao Publications Ltd. P48.

19 Zhao Rui & Zhang Mingyu (1997). *Chinese Leaders' Comments on Hong Kong* (《中國領導人談香港》). Hong Kong: Ming Pao Publications Ltd. P50.

20 Ibid, P53.

particularly important. The Chief Executive is the lynchpin of the "two systems" and, as far as the Chinese Government is concerned, the Chief Executive is the key to the implementation of the Basic Law and "one country, two systems" in Hong Kong. The central authority affords a lot of power to the Chief Executive, so that he or she can effectively govern Hong Kong and safeguard national and Hong Kong's security and interests. The Chinese Government has set up stringent standards and conditions for the candidates for the Chief Executive. As Qian Qichen, former Chairman of the Preparatory Committee for the HKSAR once said, "In order to realise 'Hong Kong people governing Hong Kong' and 'a high degree of autonomy', the most important require-ment for candidates for the Chief Executive obviously is that he or she must comprehensively and correctly understand and firmly implement 'one country, two systems', and loyally enforce the Basic Law [in Hong Kong]." The future Chief Executive should be accepted by various parties, and should be qualified to guide Hong Kong people to safeguard Hong Kong's long-term prosperity and stability."[21]

From the above quotations we can see that CPC leaders' defini-tion of "patriots" or "people loving the motherland and Hong Kong" tends to be abstract and general, so it is understandable that different people have different interpretations. For example, opposition figures in Hong Kong consistently claim that they themselves are qualified "patriots", but the country they "love" is China in the abstract sense, and definitely not the People's Republic of China founded by the CPC. What they propose is to make China a duplicate of a Western country, the core contents of which are a Western electoral system with a rota-tion of ruling parties. Nonetheless, if we take the strategic goal of "one country, two systems" as the starting point, identifying the "patriots" and the "people who love China and Hong Kong", is hardly a complex matter. Such individuals must be those who will: recognise and defend the People's Republic of China; acknowledge and accept the fact that the CPC is the party in power in China; recognise the Chinese Govern-

---

21 Ibid, P235-236.

ment's interpretation and understanding of "one country, two systems" and the Basic Law; be willing to cooperate with the Chinese Government in an all-round way and faithfully uphold "one country, two systems" and the Basic Law; proactively and consciously safeguard the national security and interests and that of the CPC security and interests; refrain from confronting the central authority; refrain from conspiring with foreign countries and external powers against the PRC and the CPC; prevent Hong Kong from becoming "a base for subversion". Still, it is certainly difficult to provide an exact legal definition for "patriot", a delineation that indeed also escapes Western countries.

3. A high degree of autonomy

Through the ages, a number of empires and countries have lent a high degree of autonomy to their local governments. For instance, the dominions of the British Empire (Canada, Australia, and New Zealand) had a great deal of autonomous power. Different territories of the Austro-Hungarian Empire had their own parliaments, ministers, bureaucracies, courts and military forces.[22] In countries with federal systems, states or provinces tend to retain partial sovereignty, and local governments enjoy a high degree of autonomy. Some unitary states may allow certain areas to enjoy an especially high degree of autonomy, such as Puerto Rico in the U.S., and Azores in Portugal.

Clark put forward two criteria for measuring the level of autonomy enjoyed by the localities: "the power of initiation" and "the

---

22 Each part of the Austro-Hungarian Empire ran its own affairs with its own parliament, ministers, bureaucracy, law courts, and armed forces. The sole remaining shared activities were foreign affairs and defence as well as the finances to pay for them, each with its own minister who met together as the three common ministers, and the only other remaining link was the Emperor himself, or as he was known in Hungary, the King. Of all the major European powers, Austria-Hungary had the poorest mechanisms for sharing information among ministries and coordinating policies. See Margaret MacMillan (2013). *The War that Ended Peace: The Road to 1914*. New York: Random House. P224-225.

power of immunity".[23] "The power of initiation" is the power of the local government to carry out its legal functions. The power of initiation of a local government can be broad or narrow. "The power of immunity is essentially the power of localities to act without fear of oversight by the state's higher authorities. In this sense immunity allows local governments to act however they wish within the limits imposed by their initiative powers."[24] Based on these two criteria, under "one country, two systems", Hong Kong's "power of initiation" and "power of immunity" are actually quite extensive so long as Hong Kong doesn't break the Basic Law or intrude upon affairs under the jurisdiction of the central authority, notwithstanding that the central authority retains the right not to appoint the Chief Executive elected by Hong Kong people.

Under "one country, two systems", Hong Kong's highly autonomous powers are quite extensive. Ji Pengfei (姬鵬飛), Chairman of the Basic Law Drafting Committee, introduced the draft of the Basic Law to the National People's Congress in 1990, and elaborated on the autonomous powers of the HKSAR: "Regarding the executive power, the draft law, while stipulating that the Special Administrative Region shall, on its own, conduct the administrative affairs of Hong Kong in accordance with the Basic Law, specifically defines the Special Administrative Region's autonomy in areas such as finance, economy, industry, commerce, trade, transport and communications, development and management of land and natural resources, education, science and technology, culture, sports, public order and control of entry and exit activities. For instance, the draft law stipulates that the Special Administrative Region shall have independent finances, its revenues shall not be handed over to the Central Government, and the Central Government shall not levy taxes in the Region; and the Special Administrative Region may, on its own, formulate monetary and financial policies, the Hong Kong dollar shall be the legal tender in the Region, and the authority to issue Hong Kong currency shall be vested in the Special

---

23 Gordon L. Clark (1984) "A Theory of Local Autonomy". *Annals of the Association of American Geographers*. Vol. 74, No.2 (Jun. 1984), P195-208.

24 Ibid, P198.

Administrative Region Government. Also, the draft stipulates that representatives of the Special Administrative Region Government may act as members of delegations of the Chinese Government to participate in negotiations at the diplomatic level affecting Hong Kong; the Special Administrative Region may on its own, using the name 'Hong Kong, China,' maintain and develop relations and conclude and implement agreements with foreign states and regions and relevant international organisations in economic, trade, financial and monetary, shipping, communications, tourism, cultural, sports and other appropriate fields." What is equally remarkable is that Hong Kong has independent legislative and judicial power, including that of final adjudication.

To conduct "one country, two systems" and assure Hong Kong and the international community, the Chinese Government legislated the Basic Law, a constitutional document, to provide a legal basis for Hong Kong's high degree of autonomy. Ji Pengfei also said, "[the draft Basic Law] contains stipulations on the economic system and policies of the Hong Kong Special Administrative Region divided into eight fields of endeavour: public finance, monetary affairs, trade, industry, commerce, land leases, shipping and civil aviation. These stipulations are indispensable to ensuring the normal operation of Hong Kong's capitalist economic mechanism and maintaining its status as an international financial centre and free port."... [The draft Basic Law] carries stipulations on the maintenance and development of Hong Kong's current systems and policies concerning education, science, culture, sports, religion, labour and social services. These stipulations involve the interests of Hong Kong residents in many aspects of public life and are important for social stability and development. There are quite a number of articles concerning policies [in the draft Basic Law]. The Chinese Government has undertaken, in the Sino-British Joint Declaration, to write its basic principles and policies on Hong Kong and their detailed explanations as given in Annex I of the Joint Declaration into the Basic Law, and Hong Kong residents from all walks of life have a strong desire for the Basic Law to reflect and protect their interests. Therefore, it was decided in the end that these articles concerning policies should remain in the draft Basic Law, despite the differing opinions expressed over the brevity of articles during the drafting of the law." "The rights, freedoms and duties of Hong Kong residents are prescribed in the draft [Basic Law] in accordance with the principle of 'one country, two systems' and in the light of Hong Kong's actual situation. They include such specific provisions as protection of private ownership of property, the freedom of movement and freedom to enter or leave the Region, the right to raise a family freely and protection of private persons' and legal entities' property."

Hong Kong has no sovereignty and its autonomous power is granted by the central authority and established by the Basic Law– a constitutional document. Under the policy of "a high degree of autonomy", Hong Kong has greater autonomous power than any other autonomous region in any country in the world. Except for national defence, diplomacy and some other affairs (for example, Hong Kong's political system and method of selecting the Chief Executive and the Legislative Council), Hong Kong people can manage the affairs of the HKSAR all by themselves. The uniqueness of Hong Kong's high degree of autonomy as compared with other regions in the world can also be seen in other areas, for example, Hong Kong has its own Court of Final Appeal; Hong Kong doesn't need to pay taxes to the central government; Hong Kong has its own currency; it can legislate on national security; and Hong Kong people won't be enlisted for military services. Most of these powers and privileges are not enjoyed by sub-national units elsewhere across the globe. Of course, the central authorities can cut down Hong Kong's high degree of autonomy by law under extreme or dangerous situations, but such situations are highly unlikely to arise.

In Hong Kong, though people appreciate and understand that under "one country, two systems" Hong Kong enjoys a high degree of autonomy, they don't however know much about and are not interested in the powers of the central authorities. Indisputably, Hong Kong's high degree of autonomy is not "total" autonomy, nor "absolute" autonomy, or "the highest degree of" autonomy, so the central authority also has its own powers. The powers of the central authorities and HKSAR's high degree of autonomy coexist, which shows that Hong Kong is not a sovereign state and its high degree of autonomy is granted by the central authority. The central authority plays an irreplaceable role in safeguarding national sovereignty, security as well as development interests and ensuring the successful implementation of "one country, two systems" and the Basic Law. Ji Pengfei made a clear statement on this in his introduction of the draft Basic Law to the National People's Congress: "The Hong Kong Special Administrative Region, as an inalienable part of the People's Republic of China, will be a local administrative region directly under the Central People's Government, and at the same time, it will be a special administrative region enjoying a high degree of autonomy and practising a system and executing policies different from those of the mainland. Therefore, the draft Basic Law contains both provisions embodying the unity and sovereignty of the country and provisions empowering the Special Administrative Region with a high degree of autonomy in the light of Hong Kong's special circumstances. The power to be exercised by, or the affairs which are the responsibility of the Standing Committee of the National People's Congress or the Central People's Government, as prescribed in the draft law, is indispensable to maintaining the state sovereignty. For example, the Central People's Government will be responsible for the Special Administrative Region's defence and foreign affairs; the Chief Executive and other principal officials of the Special Administrative Region will be appointed by the Central People's Government; a small number of national laws relating to defence and foreign affairs as well as other matters beyond the limits of the autonomy of the Special Administrative Region will be applied locally by way of promulgation or legislation by the Special Administrative Region; and in the event that the National People's Congress Standing Committee decides to declare a state of war or, by reason of turmoil within the Hong Kong Special Administrative Region Government, decides that the Region is in a state of emergency, the Central People's Government may issue an order applying the relevant national laws in Hong Kong. The draft law also stipulates that the Hong Kong Special Region shall enact laws on its own to prohibit any act of treason, secession, sedition, or subversion against the Central People's Government, or theft of state secret; to prohibit foreign political organisations or bodies from conducting political activities in the Region; and to prohibit political organisations or bodies of the Region from establishing ties with foreign political organisations or bodies. These stipulations are entirely necessary for maintaining the state sovereignty, unity and territorial integrity as well as for preserving Hong Kong's long-term stability and prosperity."

Many Hong Kong people are largely uneducated with regard to the central authorities' powers, assuming that Hong Kong has complete autonomous power with the exception of national defence and foreign affairs. Some even think that the powers of the HKSAR stand equal to that of the central authorities and that "residual power" also belongs to the HKSAR. Dong Likun (董立坤), a mainland legal scholar points out that, in accordance with the Constitution of the People's Republic of China and the Basic Law, major powers relating to sovereignty that are exercised by the central authority includes: power to (1) enact, amend and interpret the Basic Law; (2) the initiative to establish the SAR; (3) review the law; (4) determine constitutional development; (5) appoint and remove principal officials; (6) defend the nation and handle foreign affairs; (7) decide on vital matters (such as declaration of state of emergency, application of national laws in Hong Kong, unified leadership and coordination of the relationship between Hong Kong and other regions of China).[25] In addition, the central authority enjoys the right of supervision over the HKSAR's high degree of autonomy, specified as follows: (1) to supervise Hong Kong's handling of those powers which originally belong to the central authorities but are delegated to Hong Kong, including supervision of the manner in which the Basic Law is interpreted by the courts of the HKSAR, and (2) to supervise the manner in which autonomous powers delegated to the HKSAR are exercised.[26]

---

25 Dong Likun (2014). *Relationship between the Governance Power of the Central Government and the High Degree of Autonomy of Hong Kong Special Administrative Region* (《中央管治權和香港高度自治權》). Beijing: Law Press. P41-56.

26 Ibid, P60-66.

Regarding the significance of keeping the central authorities' powers under "one country, two systems", Deng Xiaoping had commented on the matter on several occasions: (1) "One more thing to be clarified: do not think that everything will be great if all Hong Kong affairs are managed by Hong Kong people, and that the central authorities needn't ever do anything. This is impractical. In general, the central authorities will not intervene in Hong Kong's specific affairs, and it is also unnecessary. But is it impossible that something might happen in the HKSAR that could endanger fundamental national interests? It might happen one day. In that case, should Beijing intervene or not? Isn't it possible that something might occur in Hong Kong that would be detrimental to its own fundamental interests some day? Can we assume that there won't be forces interfering with Hong Kong's operation? Is it beyond possibility that there be intervention and destructive elements within Hong Kong? I don't think such self-comfort is well-founded. If the central authority gave up all powers, turmoil might arise in Hong Kong and impair Hong Kong's interests. Thus, keeping some powers in the hands of the central authority will only benefit Hong Kong. Let's think it over calmly. Will problems appear in Hong Kong someday which can only be resolved with Beijing's involvement? In the past, it was Britain that Hong Kong relied on when it had troubles."[27]

(2) "Some people are worried about intervention [by the central authorities]. We shouldn't be afraid of all interventions, because some interventions are necessary. It all depends on whether the intervention is beneficial to the Hong Kong people, Hong Kong's prosperity and stability or whether it impairs the interests of the Hong Kong people and Hong Kong's prosperity and stability. [...] Never believe that there is no possibility of destructive influence in Hong Kong. It may come from any sector. If some disturbance happens, the central government will intervene and put an end to it. Should such intervention be welcomed or rejected? Of course it should be welcomed."[28]

---

27 Deng Xiaoping (1993). *How Deng Xiaoping Views Hong Kong Issue* (《鄧小平論香港問題》). Hong Kong: Joint Publishing. P36.

28 Ibid, P12-13.

(3) "Apart from change and intervention, what else are they [the Hong Kong people] afraid of? Some people refer to turmoil. If there is turmoil, intervention would be necessary. Not only the central authorities but also Hong Kong people should intervene. There must be some people who will make trouble, but we must stop them at the very beginning."[29]

(4) "The People's Liberation Army [PLA] Hong Kong Garrison can also prevent turmoil. Those who want to create turmoil will consider twice if they know there is the PLA Hong Kong Garrison. Even when turmoil happens, it can be solved without delay." [30]

In a word, the central authorities' power is essential for the successful practice of "one country, two systems".

4. "Executive-Led Political System"

In the eyes of national leaders and central government officials, the colonial political system of Hong Kong is an "executive-led" political system where power is concentrated in the hands of the governor of Hong Kong. The advantage of this system is that through the formation of a strong and efficient government, Hong Kong's prosperity and stability can be achieved; that it constitutes an aggressive government with a high level of administrative efficiency. The executive-led political system bears the advantages of ensuring Hong Kong's prosperity and stability, safeguarding its laissez faire capitalism, and keeping populism and welfarism in check. More importantly, the governor of Hong Kong retains enough power to maintain and advance the interests of Britain as the "suzerain". Therefore, the HKSAR's political system retains the features of the pre-1997 colonial political system to a great extent.

The Basic Law grants the Chief Executive a high political and

---

29   Ibid, P13.
30   Ibid, P14.

administrative status as well as a great deal of power. In accordance with Article 43, "the Chief Executive of the Hong Kong Special Administrative Region shall be the head of the Hong Kong Special Administrative Region." The Chief Executive is also the head of the Government of the HKSAR. In the political system of the Region, the Chief Executive is powerful, and most importantly, he or she has authority of policy-making, legislative initiative as well as appointing or removing administrative officials. But the Chief Executive only has partial authority in terms of the selection of principal officials. He or she has the power to nominate principal officials, but the power to appoint and remove them is in the hands of the central authority.[31]

Indeed, there is no explicit mention of the term "executive-led" in any of the articles of the Basic Law. But from the perspective of comparative politics, the political system of the HKSAR is definitely "executive-led". Xiao Weiyun (蕭蔚雲), a mainland legal scholar and member of the Basic Law Drafting Committee, put forward the concept of "executive-led" long before the reunification. Scholar Wang Yu (王禹), well-informed concerning Xiao's theory, said once that there have been controversies over "the executive-led" system after the Basic Law was implemented. One view is that an "executive-led" Hong Kong is not the intention of the Basic Law. As the term "executive-led" does not appear in the Basic Law, so we can't conclude that the HKSAR's political system is an "executive-led" system. They even remark that it is the mainland scholars who have changed their tone after the reunification and that "executive-led" was added later. But this perspective is untenable. As a matter of fact, on April 9th, 1992, Professor Xiao Weiyun published an article entitled "Political system and Transition of the Basic Law" in *People's Daily* (Overseas Edition), wherein the concept of "executive-led" was mentioned. In Xiao's words, "The Government of the HKSAR in future must be highly efficient, so it needs to maintain the current executive-led mode." In Chapter 24 of his book entitled A

---

31 Lau Siu-kai (2015). "Central Government's Power of Appointment and Removal of Hong Kong SAR's Principal Officials May Become the New Normal," ("中央對特區主要官員的實質任免權和監督權將成為新常態"). *Hong Kong & Macao Studies* (《港澳研究》). 7(2). P15-16.

Guide to Hong Kong Basic Law, published in 1996, Xiao spoke of "the checks and balances between executive and legislature in an 'executive-led' system."[32]

Based on the various stipulations of the Basic Law, the political system of the HKSAR is definitely an "executive-led" system. "In accordance with the Basic Law, the Chief Executive shall be the head of the HKSAR and shall represent the Region. The Chief Executive shall be accountable to the Central People's Government and the HKSAR. Thus, the Chief Executive plays a very important role in the relationship between the HKSAR and the Central People's Government. The Chief Executive is the link between the Central People's Government and the HKSAR."[33]

Wang Shuwen (王叔文), a mainland legal scholar and member of the Drafting Committee of the Basic Law, pointed out the similarities and differences between the "executive-led" system in the HKSAR and that of the "colony" era. He said, "The HKSAR's political system has the characteristics of being 'executive-led', but it is different from the governor system, in which the governor's power is above the Executive Council and the Legislative Council. The relationship between the executive and the legislature of the HKSAR are a relationship of checks and balances as well as cooperation. They are independent from each other and no one is superior to the other. They have equal legal status and perform different functions."[34]

The Chief Executive has an elevated position and possesses authority, but his or her constitutional status and power is lower than that of the colonial governor of Hong Kong. Xiao Weiyun pointed out that "the Chief Executive is not the representative of the Central People's Government in the HKSAR. He or she is totally different from

---

32 Wang Yu, "Editor's Words". See Wang Yu (2015). *Xiao Weiyun's View on the Political Systems of Hong Kong and Macao* (《蕭蔚雲論港澳政治體制》). Macao: Joint Publishing. P5-10, P1-2.

33 Wang Shuwen (ed.) (1990). *Introduction to the Basic Law of the Hong Kong Special Administrative Region* (《香港特別行政區基本法導論》). Beijing: Central Party School Press. P207.

34 Ibid, P209.

of the original political system and gradually develop a democratic system suitable for Hong Kong."[38]

Qian Qichen, former Chairman of the Preparatory Committee for the HKSAR, placed Hong Kong's reunification and its democratic development on a par. He said, "Democracy will be realised only after the colonial rule ends; and after Hong Kong returns to China as a special administrative region; and after the setting up of the HKSAR; and after the implementation of 'Hong Kong people governing Hong Kong' and the establishment of 'a high degree of autonomy'. Only then will Hong Kong people become the masters of Hong Kong. […] Of course, democracy should be realised through certain forms and institutions. Considering Hong Kong's particular situation, in order to ensure smooth transition and long-term prosperity, HKSAR's democracy and democratic development must be based on Hong Kong's realities. The prerequisites and goals of realising democracy in Hong Kong are the setting up of the HKSAR, Hong Kong people governing Hong Kong and maintaining a high degree of autonomy. It would be not only insufficient but defective to take the electoral process, whether direct or indirect, or some other procedure, as the symbol of democracy. Mechanically applying Western democracy to Hong Kong doesn't suit Hong Kong's reality and won't meet the interests of all ranks of society, and thus it won't promote Hong Kong's stability and prosperity."[39]

Li Hou, Deputy Director of the Hong Kong and Macao Affairs Office of the State Council shares the views of the national leaders with regard to the development of the political system of the HKSAR, noting that "HKSAR's political system should achieve the following goals: first, stay in line with the spirit of Sino-British Joint Declaration and the principle of 'one country, two systems'; second, maintain national unity while achieving a 'high degree of autonomy'; third, promote Hong Kong's stability and prosperity as well as the development of its capitalist economy; fourth, consider the interests of all ranks of society;

38 Zhao Rui & Zhang Mingyu (1997). *Chinese Leaders' Comments on Hong Kong* (《中國領導人談香港》). Hong Kong: Ming Pao Publications Ltd., P227-228.
39 Ibid, P29-30.

fifth, maintain, to a great degree, characteristics of Hong Kong's original political system, for example its highly efficient administration,
broad consultation system as well as its civil servant system; and finally,
gradually develop democratic participation suitable for Hong Kong."[40]
"Popular opinion during the drafting of the Basic Law is that the legislature will be elected through a hybrid electoral system, which will
maintain the high efficiency and effectiveness of the government. [...]
If the executive is not effective, it will affect Hong Kong's stability and
prosperity. While the relationship between the executive and legislature
of the HKSAR in the future should be that of checks and balances as
well as cooperation, and cooperation must take ultimate priority. [...] In
the last analysis, the economy is Hong Kong's fundamental issue. If its
economy can't remain prosperous, all Hong Kong that has now will be
lost. Generally speaking, all political systems should ensure continuous
economic prosperity. Politics serves economy. The above opinions are
in line with that of Marx: economy is the base, and politics is the super-
structure. In other words, the economic base determines the superstruc-
ture."[41]

5. "Remain unchanged for fifty years"

In order to reassure the Hong Kong people, the Chinese Government promised that Hong Kong's original system and lifestyle should
remain unchanged for 50 years after its return. Fifty years is quite a
long period. But the intention behind the principle is to provide the next
two or three generations of Hong Kong people with a clear indication
of what Hong Kong will be like in the years ahead setting them at ease
about Hong Kong's future. This commitment to allow Hong Kong to
"remain unchanged for fifty years" can at least put "the issue of Hong
Kong's future" on the shelf for quite some time.

---

40   Ibid, P226.
41   Ibid, P227.

Allowing Hong Kong to "remain unchanged for fifty years" as put forward by the Chinese Government actually has a firm theoretical foundation based for the most part on a comprehensive analysis of the changing international situation and trends in China's development. Deng Xiaoping explained that "We make a solemn promise that our policies towards Hong Kong will remain unchanged for fifty years. Why fifty years? There is a solid rationale behind this policy. The fifty years is not only to set Hong Kong people's minds at ease, but also takes into consideration the close relationship of Hong Kong's prosperity and stability as it relates to China's development strategy. China's development strategy requires twelve years in this century and fifty years in the next to achieve its goals. So how can we allow Hong Kong to change within fifty years? Right now we have one Hong Kong, but we will create several 'Hong Kongs' on the mainland. That is to say, to realise our development strategy, China needs to open up more. As such, why should we change our policy towards Hong Kong? Fifty years is really just a description, and our policy should not require an overhaul after fifty years. In the first fifty years, it can't change, and then there should be no need to change it after fifty years. Thus, what we are saying has merit."[42] Deng Xiaoping recalled that when he met with Japanese friends, "A Japanese friend asked me why I set Hong Kong's period of stasis at 'fifty years', that is, why Hong Kong's capitalist system is to remain unchanged for fifty years? He wondered about the rationale behind it and whether I had some other underlying considerations. I told him: yes. The rationale is based on China's realities. China has set up a grand goal that her GP will quadruple and reach the 'well-off' (小康) level within two decades, that is, the end of this century. Even if China has achieved this goal, the Chinese people will not be rich enough. China will still not be a developed country. That is just the first goal of our ambition. China needs thirty to fifty years in order to become really developed and get close to, rather than surpass, developed countries.

---

42 Deng Xiaoping (1993). *How Deng Xiaoping Views Hong Kong Issue* (《鄧小平論香港問題》). Hong Kong: Joint Publishing. P38-39.

[…] Maintaining Hong Kong's prosperity meets China's vital interests, such that 'fifty years' was not put forward casually but it actually took into consideration China's actual situation and developmental needs."[43]

Lu Ping, Director of the Hong Kong and Macao Affairs Office of the State Council, made a similar remark: "I predict that 'one country, two systems' will not only remain unchanged for fifty years, but also remain unchanged even longer, because we need several hundred years to turn China into a strong socialist country. During the whole process, Hong Kong will play an important role as the bridge between China and the Western world."[44]

It is to be expected that China and Hong Kong will change constantly after Hong Kong's return, and China will change dramatically with the "reform and opening up" policy. In fact, Chinese leaders realise that "one country, two systems" will face continuous changes in Hong Kong and the mainland, but they remain optimistic and positive about the soundness of "one country, two systems". Deng Xiaoping said: "'Remaining unchanged for fifty years' is what we wrote on the Sino-British Joint Declaration, and it will definitely remain unchanged for fifty years. […] In fifty years, the mainland will be much stronger. At the end of fifty years, will the mainland deal with these issues in a narrow-minded way? Don't worry about change, and our policy towards Hong Kong will remain unchanged. Besides, change is not always bad. Change can be good, and what is important is what is changed. Hong Kong's return to China is also a change. So don't generally be afraid of any change. If anything changes, it must be for the better and will likely be beneficial to Hong Kong's prosperity and development, rather than impairing Hong Kong people's interests. Such change is worth welcoming. If someone says nothing will be changed, don't believe it. Is everything about Hong Kong's capitalism perfect? […] Guiding Hong Kong towards healthier development is also a change. Hong

---

43 Ibid, P28-29.
44 Speech delivered by Lu Ping, Director of Hong Kong and Macao Office of the State Council, to the luncheon party of Hong Kong five major chambers of commerce. See Zhao Rui & Zhang Mingyu (1997). *Chinese Leaders' Comments on Hong Kong* (《中國領導人談香港》). Hong Kong: Ming Pao Publications Ltd. P39-40.

Kong people will welcome such change and they will definitely ask for such change. The most important unchanged thing is socialism. "One country, two systems" is a great change, and China's rural policy is also a great change....The issue is whether it'll change for better or worse. Don't resist change. If changes are resisted, there will be no possibility to make progress."[45]

Deng Xiaoping also issued a sincere warning, stating, "If we really want to realise the goal of 'remaining unchanged for fifty years and more', we should keep the socialist system of the mainland unchanged."[46] Deng Xiaoping made an incisive explanation on this: "If we want to maintain Hong Kong's prosperity and stability for fifty years and even longer, we should maintain the socialist system under the leadership of CPC. Our socialist system is socialism with Chinese characteristics, one important characteristic of which is the way we deal with the issues of Hong Kong, Macao, and Taiwan, known as 'one country, two systems'. If China changes its socialist system and its socialism with Chinese characteristics under the leadership of the CPC, what will happen to Hong Kong then? Its prosperity and stability will vanish. If we really want to realise the goal of 'remaining unchanged for fifty years and more', we should keep the socialist system of the mainland unchanged."[47] In other words, "one country, two systems" and "remaining unchanged for fifty years" come in a special historical moment when the CPC and Hong Kong people shared the same interests. Only the CPC would like and be able to provide Hong Kong people with a set of policies highly envied by mainland compatriots. His admonition obviously was a response to some popular views in Hong Kong. Hong Kong's opposition thought that the best guarantee for Hong Kong's future was to end the "one-party regime", thereby placing Hong Kong people in opposition to the CPC. The opposition hoped to exploit the fear of many Hong Kong people with respect to the CPC in order to

---

45 Deng Xiaoping (1993). *How Deng Xiaoping Views Hong Kong Issue* (《鄧小平論香港問題》). Hong Kong: Joint Publishing. P12.

46 Ibid, P32-33.

47 Ibid, P33.

obtain political capital. In fact, the opposing camp set out its claims in ignorance of both the historical background and the current real situation.

In fact, it is neither possible nor sound to keep Hong Kong's original institutions and lifestyle "unchanged for fifty years". The Chinese Government tried its best to maintain Hong Kong's "current conditions" to the greatest degree possible, but it also knew that some changes were inevitable. On the other hand, it was nonetheless impossible to comprehensively and accurately predict and master those changes. Additionally, the Chinese Government didn't actively undertake research on the various changes that could transpire in Hong Kong after its return, let alone formulate plans for managing such changes.

Actually, making the transition from a British "colony" to a special administrative region of China already signifies a dramatic change in Hong Kong's political status. The wide range of political changes brought by that transition can't be underestimated. Those changes include: the end of the original colonial regime, the birth of the new SAR regime; a change in the source of Hong Kong governor's power; the unprecedented "grounding" of Hong Kong's common law system on the Constitution of the People's Republic of China and continental law system of the mainland; a reorganisation of the internal political forces of Hong Kong; changes in the balance of power among Hong Kong's political forces as well as the rise of some and fall of others; changes in the political mentality and mindset of Hong Kong people; and changes in the relationship between Hong Kong people and the Chinese nation, the Chinese Government as well as mainland compatriots. Even with the solemn commitment to "remaining unchanged for fifty years" and retaining the "executive-led" political system, the Chinese Government has still changed some elements of Hong Kong's political system. For example, less power was granted to the Chief Executive than to the governor of the "colony"; the executive became checked and balanced by the legislature and supervised by the judiciary; the Chief Executive and legislature was set up for eventual election through universal suffrage; a Court of Final Appeal was established;

Hong Kong's court was set up to enjoy the expanded power of interpreting the Basic Law; and the legislature was projected to have new power and functions. All these changes will have long-term, profound and immeasurable effects on the practice of "one country, two systems" in Hong Kong, and such effects have been occurring one after another following Hong Kong's return to China.

6. Hong Kong can't be "a base for subversion"

The central authority asks for "two systems, one country", to follow principles of mutual respect, and mutual nonaggression, a provision aimed mainly at Hong Kong people. As Ji Pengfei, Director of the Hong Kong and Macao Affairs Office of the State Council, said, "For a long time, the Chinese Government has taken policies beneficial to the economic prosperity and social stability of Hong Kong and Macao. The mainland provides Hong Kong and Macao with a large amount of staples and subsidiary foodstuffs, commodities, fresh water and industrial raw materials with concessionary prices, offering Hong Kong and Macao a reliable assurance for their economic development. This policy has remained unchanged for decades, regardless of various kinds of situations and difficulties. After the Third Plenary Session of the Eleventh Congress of CPC, China has entered into a new historical period. National unity is one of the three goals Chinese people have set for this new period. An important component of the grand cause of reunification is the reunification of Hong Kong and Macao and the resumption of sovereignty over Hong Kong and Macao. With the advent of 1997, the time was ripe to address Hong Kong issues. [...] the Chinese Government put forward a series of policies based on 'one country, two, systems' which in turn are based on the history and current situation of Hong Kong and Macao. Those policies can safeguard national sovereignty and ensure the continuous stability and prosperity of Hong Kong and Macao, accommodating the interests of all parties concerned. [...] Our policies will remain unchanged, because 'one country, two systems' is not an expedient but a great strategic policy. This policy can promote national unity peacefully and maintain the stability and prosperity of Hong Kong and Macao. With it, countries concerned can maintain and develop their reasonable interests in Hong Kong and Macao, which is beneficial to world peace. This policy has sufficiently taken interests of all concerned into account, and there is no reason to change it, as it is absolutely right and practical. As for the relationship between the mainland and Hong Kong and Macao, I hope for the following: economically, that the mainland, Hong Kong and Macao will bring their own advantages into play, cooperate with each other, and promote mutual benefit and mutual development; politically, that the mainland accepts the reality that Hong Kong and Macao will keep capitalism for an extended period, and that it will not apply the socialist system and policies to Hong Kong and Macao; in turn, Hong Kong and Macao should respect the socialist system of the mainland, and should not intervene in or attempt to change the socialist system of the mainland. No one will be allowed to make use of Hong Kong or Macao as a base for subversion. People from the mainland shall obey the laws of Hong Kong and Macao during their visits there; compatriots from Hong Kong and Macao shall also obey the national constitution and the laws of the mainland, and uphold national sovereignty and unity. In a word, the mainland and Hong Kong and Macao should seek common ground as parts of the same country and respect each other's characteristics so as to complement each other and develop together. The forty years of practice since liberation has demonstrated that Hong Kong and Macao have shared weal and woe with the mainland, and they have gone through ups and downs together. As with the "reform and opening up" as well as the enterprise of national development proceed, the cooperation between Hong Kong and Macao and the mainland will enjoy full-scale vigorous progress and further promote the stability and prosperity of Hong Kong and Macao."[48]

---

48   Zhao Rui & Zhang Mingyu (1997). *Chinese Leaders' Comments on Hong Kong* (《中國領導人談香港》). Hong Kong: Ming Pao Publications Ltd. P34-36.

To be frank, there are two "self-evident" requirements behind "one country, two systems": the first is that Hong Kong should contribute to China's economic modernisation instead of being a "burden" on the national economy; the second one is that Hong Kong can't become "a base for subversion", impairing national security and the security of the communist regime.

The first "requirement" expresses an important expectation of the Chinese Government regarding Hong Kong. As long as Hong Kong isn't a "burden" on the national economy, even if it suffers from economic recession, I believe "one country, two systems" will remain in effect, unless Hong Kong experiences severe political and social instability or turmoil as a result of serious economic difficulties. On the contrary, the second "requirement" is far more important. If this requirement can't be met, it will be difficult for "one country, two systems" to continue in Hong Kong. In the eyes of the Chinese Government, "maintaining the original institutions and lifestyle" also implies that Hong Kong should "continue" not to be "a base for subversion".

Undoubtedly, the most unfavourable factor for the long-term practice of "one country, two systems" is politics, which is whether Hong Kong under "one country, two systems" will be a threat to the national regime under the leadership of the CPC, or whether Hong Kong will be used by anti-communist and anti-China forces as "a base for subversion". Since the occupation by the British in 1841, Hong Kong has been an arena for activities of all kinds of political forces. The Qing government, the warlord regimes, the National Government, and the PRC Government were all provoked and threatened by revolutions, reactionary or anti-communist forces from Hong Kong. Insurgent elements sought the protection of the colonial government and attacked and attempted to change or overthrow the regime of the mainland. In order to avoid becoming an enemy of different governments of China and damaging the interests of Britain or Hong Kong, Britain restrained or suppressed anti-China government forces illegally settling in Hong Kong, and was rather successful, so the relationship between Britain and China was not one of mutual tension. However, if the colonial government was not resolved to eradicate those forces, they would not be eliminated. After all, political forces in Hong Kong reflected to a great degree the rising Chinese nationalism, and the anti-imperialism movement had a mass base in Hong Kong, so if the colonial government cracked down excessively, it might have given rise to anti-colonialist sentiments.

However, the Chinese Government was concerned about whether Hong Kong would possess the strength and effective policies to restrain all kinds of anti-communist and anti-Chinese forces after the withdrawal of the British, and whether Hong Kong could prevent those forces from making use of Hong Kong's freedom, human rights, laws, facilities, and conditions to connect with the international community, and also prevent them from threatening national security or overturning the Chinese Government. Particularly it's a fact that many Hong Kong people are "suspicious of the CPC", anti-CPC and "fearful of the CPC". Should the relationship between China and Western countries become tense, Western countries could use Hong Kong to attack China. Thus, Deng Xiaoping and other Chinese leaders thought it necessary to put forward solemn advice to Hong Kong people and ask them to behave properly, to refrain from making use of any advantage offered by "one country, two systems" in order to turn Hong Kong into an "anti-communist base" or "a base for subversion"; otherwise the Chinese Government would intervene and "one country, two systems" might be discarded.[49]

Deng Xiaoping once said, "However, is it possible that some events that happen in the HKSAR will also impair national fundamental interests? It might happen one day. In that case, should Beijing intervene? [...] The central authorities' policy is not to impair Hong Kong's interests, and it also hopes that there will be nothing that impairs

---

49 Hong Kong's opposition forces always consider matters from their point of view, and never admit their activities opposing the CPC constitute "subversion", and deny that they have the ability to overturn the CPC regime. They don't consider matters from the other's perspective, which is the main reason why they do what they think is right regardless of others' opinions and don't accept Chinese Government's criticism. That is also why they are not trusted by the central government.

national interests and Hong Kong's interests. […] After 1997, some Hong Kong people lash out at the CPC and China. We allow them to do that, but if they take actions and turn Hong Kong into an anti-mainland base under the banner of 'democracy', we must intervene. Hong Kong's executive authorities should first intervene, and it's not necessary to dispatch a garrison. A garrison will be dispatched only when there is turmoil or upheaval. But intervention is a must [when things are out of control]."[50]

At the beginning of the summer in 1989, the "Tiananmen Square Incident" transpired in Beijing, which shocked the Hong Kong people. It triggered several large-scale demonstrations with hundreds of thousands of people going to the streets, reproaching the CPC and some national leaders. However, the colonial government didn't stop those "anti-communist" and "anti-China" activities as was its custom. In 1989, instead of protecting itself politically by preventing the "anti-communist" and "anti-China" activities from devastating the Sino-U.K. relationship, the colonial government on the contrary tolerated and accommodated the protesters so as to maintain Britain's "international reputation" and avoid becoming a target of Hong Kong people. For a moment, the "anti-communist" and "fearing communist" sentiment among Hong Kong people was on the rise. The Chinese Government and the central government officials repeatedly warned and asked the Hong Kong people to observe the mutual non-aggression principle. Qian Qichen, the then Chinese Foreign Minister, reminded Hong Kong people that "The Chinese Government will not intervene in Hong Kong's current capitalist system, but we will not allow Hong Kong to be made use of as a base for opposing the Central People's Government and the socialist system of mainland."[51] Li Peng (李鵬), former Premier of the State Council, also said that "all Chinese people in the mainland and Hong Kong and Macao should respect each other, get along well with

---

50 Deng Xiaoping (1993). *How Deng Xiaoping Views Hong Kong Issue* (《鄧小平論香港問題》). Hong Kong: Joint Publishing. P36-37.

51 Zhao Rui & Zhang Mingyu (1997). *Chinese Leaders' Comments on Hong Kong* (《中國領導人談香港》). Hong Kong: Ming Pao Publications Ltd. P15.

each other, and respect the social system and lifestyle of each other. We should remain vigilant lest a small number of people with ulterior motives attempt to use Hong Kong as a base to subvert the central government and socialism."[52]

Zhou Nan, Director of the Hong Kong Branch of Xinhua News Agency, also stressed that "The mainland has never attempted to introduce, and will never introduce the socialist system, its lifestyle and values to Hong Kong. Likewise, Hong Kong people shouldn't attempt to introduce the capitalist system, its lifestyle and values to the mainland. Or else, they will be acting in opposition the will of 1.1 billion people in the mainland and also against the interests of Hong Kong residents. Creating and intensifying confrontations is not good for the harmonious relationship between the mainland and Hong Kong and will also impair Hong Kong's stability and prosperity as well as the fundamental interests of Hong Kong residents."[53] Former President Jiang Zemin said more bluntly and vividly, "After 1997, the mainland and Hong Kong will carry out 'one country, two systems'. Hong Kong has adopted a capitalist system, and the mainland has adopted a socialist system. The well water should not intrude on the river water and vice versa."[54]

During the "Tiananmen Square Incident", besides the large-scale mass movements against the Chinese Government and some national leaders that took place in Hong Kong, some Hong Kong people and organisations provided spiritual as well as material support for the anti-government protesters on the mainland, which reinforced the protesters' ability to sustain and prolong their struggles. Some anti-communist people in Hong Kong even collaborated with Western forces (including the U.S.'s CIA) to rescue the "pro-democracy activists" who were wanted in the mainland, a project called "Operation Yellow Bird"[55]. After the "Tiananmen Square Incident", some anti-communist

---

52 Ibid, P15.

53 Ibid, P16.

54 Ibid, P.31.

55 Operation Yellow Bird is the unofficial name of a secret activity after the "Tiananmen Square Incidents" to rescue the most important pro-democracy activists in the mainland. CIA played a key role there. "For a six-month period following the June 4 crackdown, a network

people in Hong Kong established the "Hong Kong Alliance in Support of Patriotic Democratic Movements of China" to support the political struggle against the Chinese Government and promote the "normalisation" and "long-term sustainment" of the work of "pro-democracy activists" at home and abroad. They demanded that "the democratic movement of 1989 be vindicated", "dissidents be released from prison", "those responsible for the June 4 massacre be held accountable", and that "the one-party regime be ended". Such anti-communist activities enabled them to enlarge the social base of the Hong Kong's opposition by mass mobilisation. This later gave rise to a long-term struggle with the Chinese Government and those Hong Kong "people who love mainland China and love Hong Kong".

With the anti-communist wave in Hong Kong caused by the "Tiananmen Square Incident" the Chinese Government suddenly realised that Hong Kong could be a threat and a hidden danger to the security of the nation and the communist regime, regarding which, the Chinese Government couldn't just stand by. Thus, at the latest stage of the Basic Law drafting, the Chinese Government added some articles aimed at safeguarding security of the nation and the regime. Meanwhile, in order to reassure the public, the Chinese Government delegated the power to legislate on national security, a prerogative of the central government, to Hong Kong, so that Hong Kong people could have the privilege of local legislation to safeguard national security based on Hong Kong's legal tradition, with an emphasis on human rights and freedom as well as the real situation in Hong Kong. The Chinese Government hoped for a win-win outcome. The Article 23 of the Basic Law aims precisely to promote such as win-win outcome: "The Hong Kong Special Administrative Region shall enact laws on its own to prohibit any act of treason, secession, sedition, subversion against the Central People's Government, or theft of state secrets, to prohibit

of dozens of the CIA's most valued agents in China, Hong Kong, and Macao provided a safe haven and a means of escape for the most important organizers of the pro-democracy movement." See Mark Perry (1992). *Eclipse: The Last Days of the CIA*. New York: William Morrow & Co. P246. "Two of the dissidents (who still cannot be identified) were among the most important CIA agents ever recruited in China." (Ibid, P250).

foreign political organizations or bodies from conducting political activities in the Region, and to prohibit political organizations or bodies of the Region from establishing ties with foreign political organizations or bodies." The aim of Article 23 is to safeguard national security in the traditional security areas. In actuality, the central authorities depended on Hong Kong to maintain national security, which was a basic duty of all Chinese nationals, expecting Hong Kong to shoulder voluntarily its responsibility to maintain national security. All in all, the key precondition for the long-term practice of "one country, two systems" is whether Hong Kong people will be able to prevent Hong Kong from becoming a potential threat to the security of the nation and the CPC regime.

Li Zhu (李竹), a mainland scholar specialising in national security laws made the following comment on Article 23 of the Basic Law: "One question that we will be faced with while practising 'one country, two systems' is how the central government and the SAR Government will jointly maintain 'one country's' national security under the current situation of 'two systems'?"[56] "Treason, secession, sedition, subversion against the Central People's Government and theft of state secrets are under Chapter I, 'Crime of Jeopardizing State Security,' of the Constitution of the People's Republic of China. Thus Article 23 of [the Hong Kong] Basic Law can be called a 'national-security provision', or 'treason provision'."[57] "In China, the relationship and powers between the central authorities and the SAR Government are different from those countries with federal systems. The central authority only has power over Hong Kong's foreign affairs, national defence, the appointment of the Chief Executive and judicial officials, along with determination of matters relating to sea and air transportation between the SAR and foreign countries. Those powers amount to rather little as compared with the absolute powers of the central authority in federal systems. The HKSAR has all the powers of legislation, administration and final adjudication except for those powers of the central authorities mentioned

---

56 Li Zhu (2006). *Study on the Legislation of National Security* (《國家安全立法研究》) Beijing: Peking University Press. P188.

57 Ibid, P187.

above. Thus, [the Hong Kong] SAR Government...must make laws on some matters relating to national security so as to jointly shoulder responsibilities for safeguarding national securities [with the mainland]. Or the country's national security will be weakened, as the aim of the Article is [protecting the] country as a whole. Article 23 of the Basic Law is a necessary requirement in order to ensure national security when there are two (or more) jurisdictions of law under a single national sovereignty, and it is a special provision safeguarding national security. It emphasises the HKSAR Government's responsibilities to a sovereign nation in safeguarding and maintaining national security. Meanwhile, the SAR Government is delegated the power to adopt a concrete method to undertake the legislative work by China's highest organ of state power, making it a part of its overall legislative power. That's why the expression used in the Basic Law is that the SAR 'shall' enact laws on its own. As an enabling clause, Article 23 of the Basic Law takes 'one country' into consideration, upholds national unity, maintains national sovereignty and territorial integrity, and maintains national security; at the same time, it sufficiently respects the HKSAR's legislative power under the principle a high degree of autonomy, respects the different social system and lifestyles of the HKSAR as implied in the meaning of 'two systems'. Article 23 represents 'one country' and 'two systems' simultaneously. It is a pioneering work in legislation and legislative science and there has been no precedent in the legislations of other countries."[58]

The central authorities gave up the power of legislating on national security for Hong Kong but allowed Hong Kong to legislate by itself, which is extremely rare. On the one hand, the central authorities worried that if they legislated for Hong Kong, it might cause fear and a loss of support among the Hong Kong people. On the other hand, it also reflected their confidence in Hong Kong's situation after reunification. The central authorities believed that "reunifying people's minds" was the trend and that Hong Kong people would be able to appreciate the

---

58   Ibid, P189.

intention of the central authorities and would show the courage to maintain national security.

"Ideal Type" of "One Country, Two Systems"

According to the strategic goals and core contents of "one country, two systems" put forward by the Chinese Government, we can imagine a picture of the successful practice or "ideal type" of "one country, two systems". That is to say, we can envisage what will happen if everything happens according to the central authorities' "blueprint". "One country, two systems" refers to a series of phenomena that will happen within Hong Kong, between the central authorities and the HKSAR, as well as between Hong Kong and the mainland. Of course, it only represents the "best", or the "ideal" situation, and there must necessarily be a discrepancy between it and the actual conditions. Some actual conditions may even run contrary to the "ideal" conditions. But considering the difficulties and complexity of Hong Kong's reunification, "approaching" the "ideal form" to a great degree is already quite a good result. An "ideal type" is basically an analytical concept used to analyse actual conditions. By contrasting "ideal conditions" with the actual conditions, we can find deviations and deficiencies in the practice of "one country, two systems", seek explanations for these deviations and explore measures to correct and remedy them, so as to achieve the practice of "one country, two systems" in accordance with the original strategic intention and consequently benefit the country and Hong Kong.

In my opinion, the "ideal type" of "one country, two systems" includes the following ten important aspects:

1. Compatriots of the mainland and Hong Kong both agree on the whole about the meaning and content and have confidence in the "one country, two systems" policy as formulated and interpreted by the Chinese Government. They believe that "one country, two systems" is beneficial to the country and Hong Kong. Likewise, the central author-

ity's understanding and interpretation of the Basic Law as a constitutional document for the practice of "one country, two systems" are agreed upon and accepted by the compatriots in the mainland and Hong Kong. This will prevent severe political disputes, clashes or polarisation from occurring as a result of different understandings or deliberate misinterpretation of "one country, two systems" and the Basic Law.

2. Even if there are different understandings of "one country, two systems" and the Basic Law, some institutions or organisations like the central authority and the Standing Committee of the National People's Congress can resolve the disputes by providing the authoritative and final interpretations, thus playing the role of "dispelling doubts and resolving disputes".

3. There is an effective and regular "separation" mechanism between Hong Kong and the mainland to prevent Hong Kong from becoming a political threat to mainland China and the central authority, to achieve mutual non-aggression and to safeguard the peace between "two systems". The British played a role in "separating" Hong Kong and the mainland to some extent when Hong Kong was a "colony" of Britain. The influence of other national or international organisations was limited in this regard. After the British withdrew from Hong Kong, such role would be performed competently and effectively by the Basic Law, the central authorities and the HKSAR Government.

4. Hong Kong maintains its prosperity and stability, enjoys continuous economic development, transforms and upgrades its industrial structure, achieves progress in information and technologies, and improves the standard of living of the people. Hong Kong continues to play a positive role in national development and makes new contributions to the country, instead of being reduced to an economic burden on China. The central authorities have no need to constrain the development of the cities, enterprises, and industries in mainland China in order to maintain Hong Kong's development, which thus helps prove the correctness and practical value of "one country, two systems" and strengthen mainland compatriots' confidence and faith in "one country, two systems".

5. The original political, social, and economic institutions, major public policies, lifestyle and values of the Hong Kong people as well as the belief systems behind the lifestyle remain unchanged on the whole and are tolerated by all sectors of Hong Kong continuously. Even if called for, reforms will be moderate adjustments to the status quo rather than momentous changes. The possibility of a demand for revision of "one country, two systems" and the basic law will remain slight. Even if both Hong Kong and mainland China undergo rapid changes, the relationship between the central authority and the HKSAR, and between Hong Kong and mainland China will only change for the better as opposed to being negatively affected. The changes and mutual influence of Hong Kong and mainland China will also not trigger severe contradictions or disputes between the compatriots of both sides.

6. Although the main idea of "one country, two systems" lies in maintaining the status quo of Hong Kong and doesn't require the Hong Kong people to love China and the CPC, Hong Kong people still take a broad perspective and are knowledgeable about the situation, standpoint, interests, needs, worries, and difficulties of China and the central authority in order that they refrain from doing harm to the country and the central authority, especially in terms of national sovereignty, security and development interests. Hong Kong people respect national interests and understand the functions and powers of the central authority. They do carry out actions to harm national interests, and not impede, challenge or weaken the functions and powers of the central authority; do not attempt to subvert the governance of the CPC or change the political system in mainland China or permit domestic or foreign anti-China forces to turn Hong Kong into "a base" for jeopardising national security. It will be best if Hong Kong people show "patriotism", and safeguard the interests of China and the central authority sincerely and voluntarily. Even without "patriotism", Hong Kong people should still have their own interests clearly in mind and avoid carrying out actions harmful to the nation and the central authorities, and refuse support for or work to oppose any domestic or external forces defined by the central authorities as anti-Communist forces, anti-China forces or threats to

national security and stability.

7. All political forces agree on the Basic Law and the "legitimacy" as well as "appropriateness" of the political system stipulated by the Basic Law, and are willing to conduct activities within the political system. The power and responsibilities of the central authority under "one country, two systems" are recognised and respected.

8. The Government of the HKSAR administers Hong Kong effectively and forcefully according to an "executive-led" system, ensures the practice of "one country, two systems" and the Basic Law, safeguards national security and interests, establishes positive relationship between the HKSAR and the central authorities, and limits the space available to the activities of forces based in Hong Kong opposed to the central authority. Thus, the central authority is not required to intervene in the internal political affairs of Hong Kong in order to ensure the practice of "one country, two systems" and the Basic Law. Therefore, it will not be condemned by Hong Kong and Western countries for "intervening" in the affairs of the HKSAR and undermining the "high degree of autonomy".

9. Political forces originally against the CPC start to accept the political reality step by step, give up their animosity against the central authorities, desist in their attempts to change the political situation in mainland China, respect the power and interests of the central authority, accept the central authorities' interpretation of "one country, two systems" and the Basic Law. Political forces originally against the CPC recognise the legitimacy of the HKSAR Government, and are willing to carry out activities within the political system as stipulated by the Basic Law, and finally become reformists with the system or within the pro-Beijing camp. Even if some people remain opposed to the CPC, they be diverted by successive internal conflicts and marginalised in politics and unable to gain the support of most Hong Kong people. Moreover, they lose their value to Western forces and their support accordingly, coming to a dead end in politics.

10. Hong Kong still enjoys careful attention from the West and gains support from it and still acts as a bridge connecting China and

the international community. External forces do not interfere in Hong Kong's internal affairs or support individuals or groups that are against the CPC or China or take advantage of Hong Kong to make trouble for China. In particular, these forces give up using Hong Kong as a tool to induce China to undergo "peaceful evolution".

CHAPTER THREE

The Contradictions, Difficulties and Insufficient Supporting Conditions for "One Country, Two Systems"

Since Hong Kong's reunification with China, the implementation of "one country, two systems" has been basically successful, though there is still a gap between the actual situation and the "ideal type" in some important aspects. We should carefully analyse the gap and reasons behind it, as it is of vital importance to the successful practice of the policy in Hong Kong. Only in this way can we take measures for improvement and set an example for the future in terms of the long-term development of Hong Kong and the relationship between Hong Kong and the mainland during the fifty-year period.

In the following section, I would make a comparison between the reality of the practice of "one country, two systems" and "the ideal type":

1. Hong Kong people basically accept the "one country, two systems" policy of the central authority, and agree with the facts that the policy has been by and large implemented after the reunification and that it is beneficial to safeguarding Hong Kong's interests and promoting Hong Kong's development. Nevertheless, there have been ups and downs in Hong Kong people's evaluations of the success of "one country, two systems". In the last few years, Hong Kong people didn't place firm confidence in the policy, and there was a somewhat downward trend in public confidence. Needless to say, this trend doesn't mean that Hong Kong people are ready to discard the policy

or turn to another solution with regard to "the issue of Hong Kong's future". The advocates for the so-called "nativism" or "Hong Kong independence", most of whom are young people who are frustrated with their own personal conditions and Hong Kong's situation, have drawn a lot of media attention lately but in fact don't have a big audience in society. People in the mainland also basically accept the practice of "one country, two systems" as it is, but some might doubt its practical value, particularly the value of Hong Kong to the mainland, and they are not certain whether Hong Kong might become an economic burden on the national economy or even a political threat. Even though the central authority's understanding and interpretation of the Basic Law as a constitutional document for the practice of "one country, two systems" is not challenged in the mainland, some scholars on the mainland still have their own alternative interpretation of its meanings. Nevertheless, these differences between scholars and the central government are peripheral and thus, not directly challenged.

The greater issue is that the interpretation of the central authority, which is in fact incontestable, is not what we see in Hong Kong. It can be said that the central authority has not yet had the final say in Hong Kong. Many Hong Kong people, especially the opposition, have an "alternative interpretation" of "one country, two systems" and the Basic Law, and they tend to see Hong Kong as an independent political entity and deny, or even try to extirpate, the central authority's powers in Hong Kong. Such ignorance or even denial of the central authority's interpretation hinders the implementation of "one country, two systems" in Hong Kong. In fact, since the return of Hong Kong, many severe conflicts can be attributed to the violent collision of the central authority's interpretation and the "alternative interpretation" of the opposition. Actually, the difference in interpretation is not only a matter of understanding but rather a political struggle between the central authority and Hong Kong's opposition forces as well as Western forces, which will in the end affect the practice of "one country, two systems" in Hong Kong.

2. After reunification, serious divergent views have emerged between the central authority and a cross-section of Hong Kong people,

especially the opposition, the legal circle and the judiciary, on the issues of the central authority's powers and obligations under "one country, two systems", the relationship between the central authority and the HKSAR, and the interpretation of some articles of the Basic Law. To resolve these divergences, the NPC Standing Committee has made some interpretations of several articles of the Basic Law, but major controversies on other articles remain. To avoid triggering future political disputes in Hong Kong, the central authority and the NPC Standing Committee are not willing to provide to them the final interpretations. Such disputes include the legislative powers of the Legislative Council, the powers of constitutional review and Basic Law interpretation of the Court of Final Appeal, and the definition of a permanent resident of Hong Kong. Due to the central authority's and the NPC Standing Committee's "self-restraint" on interpretation of the Basic Law, parts of the powers of the HKSAR Government and the central authorities have been eroded and Hong Kong's social order and Hong Kong people's livelihood are also affected. Also, since these disputes have not been resolved yet, political polarisation and struggle will likely continue in Hong Kong, and the opposition activists still have ample opportunities to agitate and stir up the general public.

3. Since Hong Kong's reunification, words and actions challenging the CPC and the central authority have emerged and continued to increase. There are even Hong Kong people working with anti-government groups in the mainland to engage in activities that endanger the political stability of the mainland as well as the security of the country and the CPC regime. Nevertheless, since local legislation for Article 23 cannot be accomplished in the foreseeable future, the function of the Basic Law to prevent Hong Kong from becoming a "base of subversion" cannot yet be fulfilled. The central authority and the HKSAR Government, hampered by the jittery state of public opinion in Hong Kong, have no other method but to handle these challenges passively and rather ineffectively. Though it is true that the detrimental influence from the "base of subversion" should not be overestimated, given the increasing national power of China, those rumours and activities

harmful for the stability of the mainland and the security of the central authority do impede the operation of "one country, two systems" and might foster misunderstandings and grudges among mainland people against the people of Hong Kong.

4. Hong Kong has achieved moderate economic growth after its return, which is positive especially when compared with developed countries. However, Hong Kong's economic performance is less impressive as compared with its major competitors. As an international financial centre, Hong Kong can't escape intermittent international or regional financial crises stemming from financial globalisation. Luckily, Hong Kong has a well-established financial framework, so in spite of difficulties, Hong Kong's economy still remains stable. Still, Hong Kong's industrial structure, exceedingly consigned to the large financial industry and real estate industry, is unsound in nature and unsustainable in the long run, and its international competitiveness in traditional industries is falling too. In past decades, the transformation to a more highly value-added and knowledge-based industrial structure has not made much progress. As a result, Hong Kong's sluggish economic growth and lack of a diversified industrial structure have resulted in an increasing gap between the rich and the poor and in worsening social and livelihood problems.

With the rapid development and transformation of the Chinese national economy, Hong Kong's overall economic importance to China will undoubtedly decrease, but its financial industry and professional services will continue to contribute to the national strategy of "going out". With the slowing economic growth of developed countries, China's economic development and vitality have become increasingly important to Hong Kong. Preferential policies for Hong Kong in service to the national interests are beneficial to Hong Kong's long-term development and will also promote the economic integration of Hong Kong and the mainland.

Yet, frequent interactions between Hong Kong and the mainland make some mainland compatriots and officials feel that Hong Kong lacks the necessary motivation to move ahead and instead relies too much on the mainland. According to their criticism, Hong Kong's contribution to national economic development is less than expected, leading to a concern that Hong Kong could potentially become a "burden" on the national economy, casting doubts on the effectiveness of "one country, two systems". Meanwhile, some Hong Kong people argue that the economic integration of the mainland and Hong Kong doesn't serve Hong Kong's core interests but will gradually impair Hong Kong's high degree of autonomy and lead to the "mainlandisation" of Hong Kong, so they also harbour some doubts about the usefulness of "one country, two systems". Other Hong Kong people are worried that the central authority's resolution to maintain "one country, two systems" will be shaken because of Hong Kong's decreasing economic value to China. Some have demurred based on the grounds that "one country, two systems" overly separates Hong Kong and the mainland, hindering the linking of the two systems, and since Hong Kong people can't get the "national treatment" in the mainland, Hong Kong is thereby inhibited from taking full advantage of opportunities engendered by national development.

5. "One country, two systems" promises to retain unchanged Hong Kong's original system and lifestyle for fifty years, which has greatly eased Hong Kong people's worries about Hong Kong's future. However, before Hong Kong's reunification, some Hong Kong people were in fact not satisfied with Hong Kong's status quo, thinking it was characterised by a great deal of unfair and unjust political, economic and social policies and phenomena. They believed that there must be moderate reforms of the existing conditions, and stressed that political reform was the mother of economic and social progress. Individualism, human rights, fairness and justice, environmental protection, child care, and post material values increasingly became parts of Hong Kong's core values, from which doubts about and opposition towards the status quo arose. Consequently, a plethora of economic, social and political contradictions emerged following Hong Kong's reunification, and currently, the call for reform is on the rise, particularly from the opposition and the young people. There is no strong demand for revising the Basic Law

yet, but its "conservatism" and "static thinking" do arouse reflection and weaken the authority of the Basic Law and "one country, two systems" to some degree. Since 1997, the situation of the world, the mainland and Hong Kong have changed sharply, and most of the changes that have happened couldn't have been imagined, let alone anticipated, before Hong Kong's reunification. These changes are interrelated and mutually affecting another. Some changes work to forge closer relationships between Hong Kong and the mainland, while other changes have brought about new contradictions between the central authority and the HKSAR as well as Hong Kong and the mainland. These contradictions and frictions would inevitably change the attitudes of the central authority as well as a certain number of people in both the mainland and Hong Kong people towards "one country, two systems".

6. After the reunification, Hong Kong people's attitudes towards China and the CPC have improved to some degree, because China has become stronger, more prosperous and in possession of a higher international status as well as increasing international influence, and the Chinese people are enjoying a rise in status in the international community. Nonetheless, Hong Kong people remain suspicious of the CPC, nor do they fully recognise the People's Republic of China as established by the CPC. Some even feel resistant to the CPC and the People's Republic of China. In general, Hong Kong people tend to equate the People's Republic of China with the CPC and equate national interests with the CPC's interests. They tend to accept only an "abstract" China (historical China, cultural China or ethnic China) that is related to them in blood, history, ethnicity and culture, but they extend a very weak sense of identity toward the "concrete" People's Republic of China and are not accustomed to being called "citizens of the PRC".

Generally speaking, Hong Kong people don't understand and are not willing to think proactively or voluntarily in order to address the interests and needs of the nation and the central authority, let alone stepping forward and serving the country when national and central security are under threat. Some Hong Kong people even sympathise with and support those political forces in Hong Kong that oppose the CPC, voting for leaders of oppositional political forces during the Legislative and District Council elections, so that activist leaders can then use their positions and power to "represent" the interests of those people, challenging Beijing and handicapping the HKSAR Government. As for the opposition's deliberate words and actions challenging and threatening the CPC, many Hong Kong people adopt a sympathetic and tolerant, and even encouraging stance. Even a fair number of civil servants, due to the long-term political influence of their former colonial masters, oppose the CPC or are disloyal to it.

Indeed, based on their own interests and practical circumstances, Hong Kong people should show a diminished motivation to confront the central authority after Hong Kong's return. To a certain extent this is in fact the case, as the opposition and the anti-communist forces have restrained themselves to a great extent. However, Hong Kong people still place top priority upon Hong Kong's own interests and have a rather limited understanding of and concern for national interests and security. In such a context, Article 23 of the Basic Law is still a thorny issue. In the eyes of mainland compatriots, Hong Kong people have never been grateful to the central authority and the mainland compatriots and have failed to fulfil its responsibility to the country even though Hong Kong has been returned to the motherland for almost twenty years. In a sense, Hong Kong is still a potential threat as a "base for subversion" or an "anti-communist base". Such a potential threat is not conducive to a positive relationship between the central authority and the HKSAR, or a cordial relationship between compatriots of the mainland and Hong Kong.

7. Based on their adherence to Western political values and deep-rooted anti-communist beliefs, various kinds of opposition forces in Hong Kong have been resisting Hong Kong's reunification with China right from the very beginning, and they are unwilling to accept the arrangement of "one country, two systems". Not having been able to impede reunification, they turned to making "alternative interpretations" of "one country, two systems" and the Basic Law, rooting their protestations in a paradigm that saw Hong Kong as an "independent political

entity". After Hong Kong's return, these opposition forces have continuously challenged and denied the central authority's powers under "one country, two systems"; doubted the applicability of the Constitution of People's Republic of China in Hong Kong; slandered NPC Standing Committee's interpretations of the Basic Law; and sought to unilaterally determine the process of political reform in Hong Kong.

Although the opposition takes part in the elections of the chief executive, the Legislative Council, and the District Councils and accepts appointments to various advisory committees and statutory bodies as delivered by the Hong Kong SAR Government, the opposition is still essentially "the opposition outside the system", which doesn't recognise the "legality" and "legitimacy" of the current political system and sees its political task as overturning the current system. The opposition's "alternative interpretation" of "one country, two systems" has thus far been successful in that their influence on crucial issues relating to "one country, two systems" and the Basic Law is perhaps more profound than either the central authority and the Hong Kong SAR Government. Even in the legal circles and the judiciary, many people hold opinions about "one country, two systems" and the Basic Law that differ from the stance of the central authority. The pronounced influence of this "alternative interpretation" is the source of the severe political rift between different political forces in post-1997 Hong Kong and also accounts largely for the difficulties and deviations in the practice of "one country, two systems" after Hong Kong's return.

8. Since Hong Kong's return, the Hong Kong SAR Government has been saddled with ineffective governance, and it has not been able to establish a new regime to deal with the complicated political situation and conduct effective governance. All these are reflected in such aspects as its lack of political prestige, incapable leadership, lack of solidarity in the governing coalition, discord between the executive and the legislature, obstruction of the judiciary, weak public support, frequent collective protests, and difficulties in implementing new policies. The "executive-led system" therefore remains in name only, and a weakened Hong Kong SAR Government has limited ability to protect national

security and interests, to implement "one country, two systems" and the Basic Law, and to instil an authoritative interpretation of "one country, two systems" and the Basic Law among the Hong Kong people.

What is worse, for fear of bringing about public dissatisfaction, of attacks from the opposition activists as well as external forces, and in particular of triggering interpretations of the Basic Law by the NPC Standing Committee, the Hong Kong SAR Government, which has been in a difficult position for a long time, lacks also the political courage to correct elements not in conformity with "one country, two systems" and the Basic Law. And, of course, the reason why the HKSAR Government does not dare to correct things that are not in conformity with "one country, two systems" and the Basic Law also lies in the fact that a portion of the government officials themselves, particularly officials in charge of legal affairs, have views about "one country, two systems" and the Basic Law that are similar to those of the opposition activists.

Thus, on some major issues, the central authority is forced to take actions to "impose order on chaos", but doing so will inevitably lead to frictions between the central authority and a portion of the Hong Kong people, resentment from some Hong Kong SAR Government officials, in addition to causing some Hong Kong people along with external forces to question the central authority's determination in implementing "one country, two systems".

9. The opposition forces that challenge the CPC and the central authority and held "different interpretations" of "one country, two systems" and the Basic Law will remain a considerable political force in the foreseeable future, with a further tendency towards radicalism and violence, even though at the same time the opposition forces have seen a gradual decline since Hong Kong's return. Admittedly, some of the opposition activists will stop confronting the central authority after analysing the situation and gradually transform themselves into "loyal opposition activists"; but these people will only be a minority in the opposition camp in the foreseeable future. Thus, the fundamental political landscape in Hong Kong will not change significantly in the foreseeable future. Accordingly, a few years ago, the central authority

had started to change the way it dealt with Hong Kong. The central authority is using, as far as possible, their powers under "one country, two systems" to curb the opposition ideologically and behaviourally. The purpose is to restrict the political space, weaken the strength, and reduce the ideological influence of the opposition in Hong Kong.

10. The attitude of the West towards Hong Kong after Hong Kong's return is quite ambiguous and complex. On the one hand, the West still approve of Hong Kong's capitalist system, legal system, protection of freedom and human rights, the people's "anti-communist" tendencies, and Hong Kong's extensive contact with the West. On the other hand, the West also recognises that Hong Kong no longer belongs to "the Western camp" politically due to the fact that Hong Kong can only fall in line with the Chinese Government on major political issues and that Hong Kong's economic integration with mainland China will continue unabated. Therefore, the West is losing confidence in the possibility of Hong Kong playing a significant role in promoting China's "peaceful evolution". It is possible that eventually the West will totally lose hope in Hong Kong on that matter.

Furthermore, in recent decades, China has worked out a development path different from that of the West. It was commonly called "the path of socialism with Chinese characteristics" or the "Beijing Consensus" over the past decade. China has been able to put forward some creative theoretical ideas and draw attention from other countries, even prompting some of them to borrow from China's experience. As such, to some extent there has been a weakening of the worldwide appeal and discourse power of the "Western model" and the "Washington Consensus". In this way, Hong Kong is not only unable to induce China to follow the path of "peaceful evolution", but on the contrary had made some contributions to China's rise. Consequently, whether Hong Kong's continuous prosperity and stability is in the interests of the West, particularly the U.S., will inevitably become a major concern of many Western strategists who see China's rise as a mortal threat to Western interests and security. What is more, the strategic battles between China and the U.S. in East Asia and in broader areas do raise

Hong Kong's strategic importance in the strategy of the U.S. to contain China, thus increasing the instrumental value of Hong Kong's opposition forces to U.S. strategists. Intervention in Hong Kong affairs by outside forces has endowed the conflicts between the central authority and Hong Kong with the broader implications of geopolitical rivalries among great powers.

In the future, it is an open question as to whether Hong Kong will continue to enjoy the full support of the West. It is also an open question whether the West will "create chaos" in Hong Kong so as to curb China's rise. In any case, these questions will inevitably arouse the attention of the central authority and the people in Hong Kong and the mainland. Nobody has definite answers to these questions today. In any event, the lack of mutual trust between great powers is a common scenario in international politics. Contests between China and the U.S. over Hong Kong will cause increasing concern on the part of the central authority as to whether Hong Kong constitutes a security threat to the country and the CPC regime, and whether such concerns are likely to lead the central authority to think actively about how to deal with Hong Kong to forestall any trouble coming from there, thus making the relationship between the central authority and the HKSAR more complicated and paradoxical.

Obstacles in Implementing "One Country, Two Systems"

"One country, two systems" as the means to deal with "Hong Kong's future" was not only the best arrangement against the historical backdrop of the time in the 1980's, but remains the best arrangement even from today's perspective, because in actuality no better ideas have been proposed so far. Today, there are always people claiming that improvements can be made to various aspects of "one country, two systems" and the Basic Law, some of whom even think that "one country, two systems" is not the best alternative. Today, an extreme minority of Hong Kong people advocating "self-determination" or "Hong Kong independence" are saying that Hong Kong will lose its unique and valuable attributes and forfeit its development because of "one country, two systems" and that it is better for Hong Kong to detach itself or even separate itself from China. Nevertheless, in evaluating a policy formulated three decades ago primarily from today's perspective and without considering all the relevant factors and issues is neither fair nor reasonable. It can be said in all fairness that "one country, two systems" is a difficult but wise arrangement. It provides for the smooth and peaceful return of Hong Kong to China as well as the preservation to a great extent of the prosperity, stability, freedom, justice, efficient administration, clean government and good governance after its return notwithstanding all the troubles and difficulties. The problems encoun- tered after Hong Kong's return such as administrative difficulties, polit- ical chaos, conflicts between the central authority and the opposition, friction between compatriots from mainland China and Hong Kong are less serious than what people expected before Hong Kong's return. As a matter of fact, a large number of people bore negative expectations or pessimistic attitudes towards Hong Kong's future before 1997. Indisput- ably, the post-return conditions have left many of those people "joyfully surprised".

However, we also need to understand that there is no perfect policy in the world, and that no policies can be exempt from the test of new problems presented by drastic social changes. As a complex major national policy which involves different interests and standpoints and was designed to be applied for fifty years or more, it is more than natural for the "one country, two systems" policy to encounter some problems and difficulties during its implementation process, some of which can nevertheless be predicted beforehand. To put it simply, as a policy or an arrangement, it contains some lurking inherent contradictions and difficulties and also lacks some "compatible" conditions. The so-called contradictions mainly refer to the internal contradictions inherent in "one country, two systems", such as the contradiction between preserving capitalism with Hong Kong characteristics and promoting democracy in

Hong Kong step by step; the contradiction between opposition forces' interference in and distortion of "one country, two systems" and the central authority's "non-interference" policy; as well as the contradiction between remaining an "executive-led" system and the counterbalance between legislative power and judicial power. The difficulties mentioned above involve the slow process of winning back people's support, the hardships in establishing a new regime for the HKSAR, obstructions from all kinds of opposition forces, a lack of common understanding of "one country, two systems" and the Basic Law, etc. The "lack of compatible conditions" mentioned above basically refers to a serious shortage in talents capable of administering Hong Kong, most of whom lack a sense of solidarity, labour under an ineffective interpretation concerning "one country, two systems" and the Basic Law, and lack a comprehensive strategy for preventing Hong Kong from becoming a "base for subversion", etc.

Continuous changes in the international order, the domestic situation, and the circumstances in Hong Kong not only intensified part of the lurking contradictions and difficulties, further highlighting the dilemma of insufficient "compatible" conditions, but also brought about new challenges in implementing "one country, two systems". A proper understanding of these contradictions, difficulties, and "compatible" conditions would be conducive to better implementing "one country, two systems" and could also provide a useful reference regarding how to plan Hong Kong's future after 50 years of the policy's implementation.

The Tendency Towards Compromise in "One Country, Two Systems"

Twenty years of experience since Hong Kong's return demonstrates that major contradictions, difficulties, and "compatible" issues in implementing "one country, two systems" stem from several aspects. The first one has to do with the tendency towards compromise in "one country, two systems". In order to achieve a peaceful resolution of the Hong Kong issue that was left over by history, as well as to ensure a smooth and steady transition and a prosperous and stable Hong Kong after its return, "one country, two systems" has to take care of and coordinate the interests, concerns, and thoughts of many different parties. These different parties mainly refer to China, the U.K., the Western world, the Hong Kong business community, the opposition, and the general public of Hong Kong. Indeed, taking care of different parties does not mean that each party is to be treated "equally", but rather that we need to measure the relative importance of different parties in implementing "one country, two systems" and the significance each party bears in achieving the strategic objectives of "one country, two systems". From these perspectives, the interests of China, the U.K., their political allies, as well as investors in Hong Kong have the highest priority.

For China, "one country, two systems" is a major national policy concerning national sovereignty, security and development interests, and it is natural for China to put its objectives and needs first; only under this premise can China consider and try to satisfy the needs of other parties and make necessary "concessions". For the U.K. and the West, significant "concessions" include maintaining the interests of the U.K. and the West in Hong Kong, trying to enable political elites supported by the colonial government, maintaining all civil servants' positions after Hong Kong's return, preserving the concept of conservative governance, retaining the basic administrative direction and major public policies of the colonial period, making a commitment to promoting democratic reform in Hong Kong, accrediting the contributions that the U.K. made to Hong Kong's development, approving the original system in Hong Kong, agreeing not to engage in ideological "decolonisation", trying to "respect" the administrative "autonomy" that the colonial government enjoyed during the transitional period, and above all, "reluctantly" tolerating and accepting various political and administrative reforms initiated by the U.K. without China's input or in spite of China's opposition during the transitional period, absorbing the recommendations put forward by British people during the drafting process of the Basic Law, etc. Indubitably, the "concessions" that were made to the U.K. have heavily influenced and presented great difficulties for the post-return political landscape, administration pattern, and development path, and hindered the formation of "compatible" conditions; but under the circumstances at the time, securing the cooperation of the U.K., in particular that it refrained from conducting any sabotage, was of great significance. This combined with a UK-dominant transitional period of as long as fifteen years meant that the U.K. surely enjoyed a high "bidding" status.

The "concessions" that the Chinese Government made to investors (commercial and industrial circles, zaibatsu in particular) were mainly reflected at maintaining Hong Kong's capitalist system characterised by its original free, laissez-faire, liberal business environment, limited government regulation, simple and low tax regime, prudent financial management system, and limited welfare system, along with building up a political system where investors could exert a great influence, the latter aspect being particularly important. The reason for these decisions was because investors have already enjoyed a favourable economic system and a fiscal policy, and "one country, two systems" additionally expanded their political rights and power. Under the colonial administration, business organisations had exerted considerable influence on government administration, but it was hard for them to have a say in the formation process, government policy objectives, and personnel deployment of the British government and the colonial government. In consideration of the concerns over investment withdrawal from the business sector, plus the fact that the business sector lacked any political organisations or representatives of its own, as well as the lack of organisation and the intention and confidence to participate in politics, "one country, two systems" therefore provided investors with special care, granting them more power than other social groups in the election of the chief executive and Legislative Council.

For Hong Kong people, the primary "concession" is not to prac-tise socialism in Hong Kong, along with granting them more rights and preferential treatment than their mainland compatriots. For instance, Hong Kong people do not pay taxes to the country or participate in mili-tary service; they face no restrictions on child-bearing and finance and no military spending. Furthermore, Hong Kong can conduct national security through local legislation on behalf of the central authority in accordance with the specific circumstances in Hong Kong, and the Communist Party of China as the country's ruling party does not carry out any public activities in Hong Kong (including refraining from participating in any kind of elections) and is committed to not inter-vening in Hong Kong affairs. In addition, Hong Kong's mainstream forces are allowed to continue to administer Hong Kong after its return, without being required to undergo a process of "decolonisation", and Hong Kong can work toward achieving the ultimate goal of selecting the chief executive and electing the Legislative Council through universal suffrage step by step. In fact, since China still sees "grand unifica-tion" as its ultimate end and socialism as its noblest ideal, allowing the continuous existence of the capitalist system in Hong Kong, which is unacceptable for the Communist Party of China at the ideological level, is the biggest "concession" for the Chinese Government. In addition, we cannot but admit that the most significant "concession" for Hong Kong people and the opposition forces in particular is the Chinese Govern-ment's permission for Hong Kong to gradually bring in a series of elec-toral arrangements with Western characteristics, of which the contents, meanings, and risks are hard to understand for people on the mainland. Furthermore, in order to reduce or avoid immediate controversies and prevent them from hindering efforts to reassure the public, the Chinese Government would rather leave some problems to be resolved in the future, though these problems may involve fundamental issues such as the central authority's power, responsibilities, and functions.

The British Government's "concessions" include the waiver of three unequal treaties, agreeing to return Hong Kong Island and Kowloon Peninsula along with the New Territories back to China in

1997, admitting that the Hong Kong issue is a shared issue between China and the U.K., giving up the intention of involving a "third party", avoiding "over-internationlisation" of the Hong Kong issue, agreeing to maintain effective administration and remain committed to maintaining the prosperity and stability of Hong Kong before its return.

Hong Kong people's "concessions" include acknowledging the central authority's sovereignty over Hong Kong, acknowledging "one country, two systems" and the Basic Law, continuing to work and invest in Hong Kong, on the whole, not requiring the U.K. to fulfil political and moral duties to Hong Kong people even if they are considered U.K. "citizens", not extending appeals to the international community and organisations, avoiding actions that are detrimental to national sovereignty and regime security and so on.

Since "one country, two systems" is the result of contests among different forces, it is reasonable for it to contain some potentially controversial aspects; it is not surprising at all to encounter difficulties and insufficient "compatible" conditions at the beginning of its implementation.

Multiple Attributes of the Basic Law

The multiple attributes of the Basic Law can lead to different interpretations of its spirit and of its separate articles, resulting in different perceptions about "one country, two systems", thus affecting its successful implementation in Hong Kong. Different observers will interpret the Basic Law based on the attributes they favour and promote an understanding of "one country, two systems" that is most in line with their real interests, values, political inclinations and legal interpretations. The multiple attributes of the Basic Law are reflected in that it can be classified as both national law and "constitutional" document at the local level; it is not only a legal document, but also a policy document; it belongs to a socialist civil law system and common law system at the same time.

Generally speaking, experts and scholars on the part of the central authority and mainland China deal with problems related to the Basic Law from the perspectives of national law, the central authority's policies towards Hong Kong, legislative intent, and civil law; whereas Hong Kong's opposition forces, many judges, legal professionals, scholars, and the media prefer to look at the problems from the perspectives of local law, common law, and simple legal text. In saying this, I do not mean that the two sides ignore each other's opinions completely, but rather, simply want to highlight the main differences between the two sides. Different attributes of the Basic Law bring to bear different perspectives in the interpretation of the Basic Law and "one country, two systems", thus leading to disputes that are hard to reconcile. Controversies caused by the multiple attributes of the Basic Law can be clearly seen in the right of abode, disputes over the legislation of Article 23 of the Basic Law, and constitutional reforms. Experts and scholars on the part of the central authority and mainland China tend to start from the perspectives of the central authority's policies towards Hong Kong and its legislative intent, stressing that the Basic Law is the means to implement "one country, two systems". In fact, the preamble of the Basic Law states, "In accordance with the Constitution of the People's Republic of China, the National People's Congress hereby enacts the Basic Law of the Hong Kong Special Administrative Region of the People's Republic of China, prescribing the systems to be practised in the Hong Kong Special Administrative Region, in order to ensure the implementation of the basic policies of the People's Republic of China regarding Hong Kong."

To put it simply, legal scholars on the part of the central authority and mainland China and Hong Kong people who "love mainland China and Hong Kong" tend to look at the Basic Law and understand "one country, two systems" from the perspectives of national interests, civil law, and the central authority's policies towards Hong Kong; whereas people from Hong Kong's opposition forces, academia, legal community, judicial world, and some media prefer to understand the Basic

Law and understand "one country, two systems" from the perspectives of Hong Kong's status as a Special Administrative Region, individual rights, and legal provisions. In terms of the issue of the right of abode, the former party emphasises that the purpose of the policy is to continue the policy of the right of abode proposed by the former colonial government, aimed at eliminating Hong Kong people's political concerns by prescribing strict restrictions on conditions under which people from mainland China can enjoy right of abode in Hong Kong, so that Hong Kong people need not worry about mass immigration of people from the mainland to Hong Kong. Whether relevant articles in the Basic Law are clear enough or not, due to the clarity of the purpose of the central authority's policy on the right of abode, the former part holds that we must understand these articles from the perspective of both national and Hong Kong interests and give priority to the legislative intent of the Basic Law. On the contrary, the latter party highlights individual rights and welfare according to relevant articles of the Basic Law and deal with problems concerning who enjoys the right of abode by referring to international (mainly Western) human rights conventions. Many issues relevant to the right of abode still remain unresolved and the problem of the right of abode continues to trouble Hong Kong, though some of the problems were solved through the NPC Standing Committee's interpretations of the Basic Law.

Dong Likun, a legal scholar from mainland China, pointed out, "It seems that there are not any explicit articles that declare what methods to use in interpreting laws; however, in practice and theory, the method of analysing legislative intent is most commonly used."[1] He further proposed that the courts in Hong Kong should also use legislative intent as their guideline to interpret the Basic Law. "The Basic Law is Hong Kong's constitutional law, Hong Kong's court even refers to it as the constitution of Hong Kong; Hong Kong's court should increase their use of purpose-oriented interpretation methods, which

---

1　Dong Likun (2014). *Relationship between the Governance Power of the Central Authorities and the High Degree of Autonomy of Hong Kong Special Administrative Region* (《中央管治權和香港高度自治權》). Beijing: Law Press. China. P81.

means to take into serious consideration the legislator's purpose and aim in enacting the law, in other words, should consider the legislative intent of the law to be interpreted. Perplexingly enough, Hong Kong's court sets literal interpretation and purpose-oriented interpretation in opposition and over-emphasises literal interpretation, regarding as heresy the idea proposed by the NPC Standing Committee that interpretation of the Basic Law should be based on legislative intent."[2] The criterion for deciding whether to use the common law method of legal interpretation or the method of analysing legislative intent to interpret the Basic Law lies in whether the interpretation method chosen is in accordance with the provisions of the Basic Law and the NPC Standing Committee. According to relevant provisions of the Basic Law and the NPC Standing Committee, certain articles of the Basic Law should be interpreted according to the legislative intent of the Basic Law. First, the Basic Law is a national law; we should interpret it in accordance with China's method of legal interpretation instead of "modifying" the Basic Law through the use of common law interpretation method by Hong Kong's court.[...] Since the NPC Standing Committee exercises the right to interpret the Basic Law, there is no doubt that we can only interpret the relevant provisions of the Basic Law in accordance with China's interpretation method, in order that the interpretation of the Basic Law is consistent with its legislative intent.[...] Second, by formulating provisions on the interpretation of the Basic Law, the NPC Standing Committee defines the interpretation method that should be used to interpret the Basic Law; Hong Kong's court shall exercise the right to interpret the Basic Law unconditionally in accordance with the provisions of the NPC Standing Committee."[3]

The painful experience of failing to legislate Article 23 of the Basic Law also reflects the great differences between these two stances. The former sets safeguarding national security and interests as its top

---

2   Dong Likun (2014). *Relationship between the Governance Power of the Central Authorities and the High Degree of Autonomy of Hong Kong Special Administrative Region* (《中央管治權和香港高度自治權》). Beijing: Law Press. P83.

3   Ibid, P83-84.

priority, and exercises certain restrictions on individual rights, while the latter considers guaranteeing individual rights and freedom to be the main consideration, and is willing to protect national security only under this premise. In places other than Hong Kong, laws related to national security are usually enacted by the central authority, rather than by local institutions. However, in order to consolidate Hong Kong people's confidence in the future of Hong Kong and show respect for the Hong Kong people, the central authority handed over the task of legislating laws safeguarding national security to the HKSAR, so that laws concerning national security legislated by Hong Kong can effectively safeguard national security while ensuring the pursuit of human rights, freedom and justice. However, Hong Kong's opposition forces, most legal professionals, and the media all focused on the issue of whether human rights, freedom and justice would be impaired by Article 23 of the Basic Law, showing that they did not really care about national security. Some people even referred to Article 23 of the Basic Law as an "evil law", and they opposed any work related to local legislation, even though the bill proposed by the HKSAR was much looser than the national security laws in some Western countries. With ever more widespread, complicated, and severe security risks facing China, the content covered by Article 23 of the Basic Law cannot meet the country's needs to safeguard national security. Therefore, the divergences over the issue of national security among various parties in Hong Kong will become more and more severe.

Different viewpoints reflected in the controversies over political reform are particularly evident. Experts and scholars taking up the part of the central authority and mainland China stress that Hong Kong is a Special Administrative Region of the People's Republic of China, and the "high degree of autonomy" that it enjoys is granted by the central authority; the central authority has dominant and decision-making power over political reform issues; the political system (including the electoral system) must be in line with the implementation of "one country, two systems" and serve its strategic objectives. Therefore, we cannot regard the development of democracy in Hong Kong as an isolated process. Moreover, democratic reform is not the most important objective of implementing "one country, two systems", and other more important objectives include defending national sovereignty, security, and development interests, maintaining prosperity and stability in Hong Kong, preserving the original capitalist system, establishing a good relationship between the central authority and the HKSAR, and caring for the interests of all sectors and in all aspects. Basing their perspective on Hong Kong as the point of departure, the opposition forces and their supporters blindly adhere to a concept of Western democracy, firmly believing that this is the only way to achieve effective administration, demoting national security and interests to a second priority.

Potential Contradictions in "One Country, Two Systems"

Since "one country, two systems" is characterised by concessions and static thinking, it is very natural that it should encounter internal contradictions. The problem is that not all these contradictions can be dealt with properly through mechanisms established by "one country, two systems" and the Basic Law; some problems can only be dealt with through means other than law, among which political means are particularly important. However, the use of political means will bring risks, and political means are likely to conflict with the law. In the following section, I would like to focus on a number of important internal contradictions.

1. The contradiction of "economic integration" and "political separation" between Hong Kong and the mainland

"One country, two systems" assumed that the economic relationship between the two sides would grow closer after Hong Kong's return, with increasing common interests and interdependence. However, the guiding ideology at the political level was to ensure that "the well water should not intrude on the river water and vice versa". However, as more and more mainland enterprises enter Hong Kong and more investments from mainland enterprises are made in Hong Kong, an increasing number of mainland talents choose to go to Hong Kong for further development; due to the fact that Hong Kong's strategic position as China's RMB offshore centre is becoming more crucial to China's financial strategies, in addition to the fact that Hong Kong plays the role of the platform for China to "go global", Hong Kong is invested with a greater part of the national interest and plays a more important role. Some of the policies and laws enacted by the Government of the HKSAR, especially those related to investment, taxation, currency, financial business supervision, fair competition, information disclosure, labour welfare, business environment, industrial structure, environmental protection, "immigration" policy, etc., all have a significant impact on enterprises and talents from mainland China. Issues such as whether Hong Kong can conduct effective administration, whether policies of the HKSAR are stable and appropriate, whether Hong Kong's political situation is stable, whether external forces interfere in Hong Kong's affairs, and whether Hong Kong's investment environment is attractive are more closely related to national security.

Investors and talents from other countries and regions often lobby the Government of the HKSAR for their own interests through the government of their own countries and chambers of commerce. On the other hand, as for enterprises and talents from mainland China, it is not unusual for them to promote their own interests by influencing policies of the HKSAR; the central authority cannot eliminate the possibility that such sources may put pressure on the Government of the HKSAR by "secretly" lobbying the central authority and some local governments. Prior to the return of Hong Kong, when the Preparatory Committee for the HKSAR was working on the method for electing the first Legislative Council, some central officials suggested making mainland-funded enterprises in Hong Kong a functional group, so that they could protect their own interests by sending representatives to the Legislative

Council. This suggestion was not recognised by other central officials, and eventually came to nothing. The results reflected that the central authority did not want to see people use this as an excuse to criticise the central authority's role and to sabotage the "high degree of autonomy" Hong Kong enjoys. However, as Hong Kong bears more and more mainland interests, and as these interests increasingly relate to national security, there are more incentives for the central authority to interfere in the economic policies of the HKSAR Government.

With Hong Kong's active involvement in China's Five-year Plan, the process of economic integration between the two places is bound to accelerate. The integration of the mainland's and Hong Kong's development strategies requires both input by the central authority concerning Hong Kong's economic development and cooperation between the two places along with the development of industrial policy of the HKSAR Government; the central authority can thereby assist Hong Kong and safeguard national security at the same time. There is no doubt that policies related to the economy spill out of the economic sphere and have an impact on other public policies. In short, the central authority and the Government of the HKSAR should conduct more frequent economic interactions, thus strengthening the political cooperation on all sides, but in this way the original intention to "separate economy with politics" must be adjusted.

Similarly, the HKSAR Government also has incentives to influence and lobby for the central authority's economic policies towards Hong Kong, focusing on striving for more HK-preferential policies and measures from the central authority; but due to the disparity in political strength, Hong Kong's "interference" in mainland politics should have no big influence on the political situation in the mainland.

2. The contradiction between "maintaining the status quo" and "changing the status quo"

The Chinese Government was committed to ensuring the status quo of Hong Kong to "remain unchanged for 50 years" after its return, which surely gave Hong Kong people a sense of "reassurance". The "status quo" refers to the status of the late 1980s. This commitment assumed that most Hong Kong people were satisfied with the "status quo" of that time and that they were extremely worried that they might lose the "status quo" or that the "status quo" might be changed completely. To assuage their worries, the Basic Law fixes or "freezes" Hong Kong's status quo of that time by way of law and power. There-fore, as a constitutional document, one of the major features of the Basic Law is that there are a great many articles specifying some public policies and preserving some aspects of the "current status".

For example, Article 107 stipulates, "The Hong Kong Special Administrative Region in drawing up its budget shall follow the prin-ciple of keeping expenditures within the limits of revenues in drawing up its budget, and strive to achieve a fiscal balance, shall avoid deficits and keep the budget commensurate with the growth rate of its gross domestic product." Article 108 requires Hong Kong to practise a low tax policy. Article 141 states,"Religious organisations shall, according to their previous practice, continue to run seminaries and other schools, hospitals and welfare institutions and to provide other social services." Article 145 prescribes, "On the basis of the previous social welfare system, the Government of the Hong Kong Special Administrative Region shall, on its own, formulate policies on the development and improvement of this system in the light of the economic conditions and social needs." Of course, there are a few countries such as India whose institutions also include a series of provisions on social and economic policies; but these provisions usually concern people's desires and goals, and they are not directly linked to articles and provisions prescribing a process of legal investigation by means of litigation in a court of law. In contrast, the policy provisions in the Basic Law are justiciable, so they are legally binding, and their "legal rigidity" is very strong. Admittedly, there is certain flexibility in those policy-describing provisions and provisions pertaining to maintaining the "current status", enabling the Government of the HKSAR to adjust those provisions according to the development and administrative needs of Hong Kong;

but the room for adjustment is very limited, otherwise the provisions might lose their power to maintain the "current status" and become meaningless words.[4]

However, even in the 1980's, many people, the lower class and a portion of the middle class in particular, felt unsatisfied with the "status quo" of Hong Kong. They looked forward to a certain degree of social reform, lamenting the widening gap between the wealthy and the poor along with various inegalitarian political and economic phenomena. Hong Kong was faced with a great deal of pressure from public, social welfare, religious, and political institutions to promote various forms of reform. Therefore, not all people are satisfied with the Basic Law's legal power to "freeze" policies and in turn regiment social status, because, in some people's eyes, such power preserves unfair and unjust social conditions.

Since Hong Kong's return, difficulties and contradictions of all kinds have flooded in; these have been reflected in many ways, such as the widening gap between the wealthy and the poor, the lack of vitality in economic growth, a devastated middle class, the lack of diversification in the industrial structure, the insufficient development opportunities for young people, a rapidly aging population, the rise of severe poverty, the incidence of high prices and insufficient supply of housing, the emergence of "development concepts" of "post-modernism" and "post-materialism", etc.[5] An assortment of deeply-rooted problems arose one after another, overwhelming the Government of the HKSAR.[6]

---

4   Hong Kong was hit by the Asian Financial Crisis soon after its return; The SAR Government's budget was tight, and there was a deficit over a few consecutive years. However, since the various parties involved understood about the reason, it did not lead to litigations.

5   Refer to Lau Siu-kai (1997). The Fraying of the Socio-economic Fabric of Hong Kong. *The Pacific Review*, Vol. 10, No. 3, P426-441; Chan Koon-chung (2005). *My Generation of Hong Kong People* (《我們這一代香港人》). Hong Kong: Oxford University Press; Lui Tai-lok (2007). *Four Generations in Hong Kong* (《香港四代人》). Hong Kong: Step Forward Multi Media; Lau Siu-kai (2013). The Divergence between Different Opinions and Policy Consensus (分歧與政策共識的剝落). In *Hong Kong's Politics after Its Return* (《回歸後的香港政治》). Hong Kong: The Commercial Press. P279-307.

6   Wong Yue-chim (2012). *Hong Kong's Deep-Rooted Contradictions* (《香港深層次矛盾》). Hong Kong: Chung Hwa Book Company; Chen Li-jun (2015). *A Study on Hong Kong's Social Relations and Contradictions* (《香港社會關係與矛盾變化研究》). Hong Kong: Chung

Hong Kong scholar Lui Tai-lok (呂大樂) even asserted that the "Hong Kong Model" on which Hong Kong's success was based in the past was outdated, requiring a new approach in order to move forward. He specifically pointed out the problems that Hong Kong was faced with at that time—that the SMEs were faced with decreasing development potential and space; that the "positive non-intervention" policy was outdated; that the administration-oriented government was not able to handle the increasingly complex political situation any longer; that "consultative politics" failed to coordinate and integrate the interests of all parties and meet people's desire to participate in politics; that the social order had begun to collapse, and the "new town" development strategy could not stand on its own any longer, etc.[7]

In such an atmosphere of dissatisfaction and anxiety, there were various appeals and suggestions to change the "status quo". But that does not mean that Hong Kong people wanted to give up on or completely change the "status quo", they were just stuck in the dilemma of whether to "maintain the status quo" or "to change the status quo"; they felt anxious and at a loss as to what to do or what choices to make. People felt less confident about Hong Kong's political system and policies that had proven effective in the past, but they did not dare to give up on the previoun "way to success". They lacked consensus on how to conduct reform and felt unsure and anxious about which possible changes to enact. Controversies on whether to "maintain the status quo" or "change the status quo" would not only be detrimental to effective governance, but also hinder the proposition and implementation of new policies.

Debates from all aspects grew fierce, and they were generally related to several aspects: whether Hong Kong should make economic development its top priority or also pursue some "non-material" goals such as environmental protection, cultural and social conservation, community restoration and preservation, as well as humanitarian

---

Hwa Book Co.

7   Lui Tai-lok (2015). *Hong Kong Model: from Present Tense to Past Tense* (《香港模式：從現在式到過去式》). Hong Kong: Chung Hwa Book Company.

concerns; whether economic development strategy should focus on promoting economic integration with the mainland or strengthen ties with Western countries; whether to broaden and strengthen the SAR Government's roles and functions in economic and social development; whether the SAR Government should actively support new high-value-added industries; whether to make a major change in Hong Kong's fiscal policy; whether serious action should be taken to deal with problems related to Hong Kong's low tax policy and narrow tax base; whether to greatly increase social welfare in Hong Kong; whether to make social welfare a universal right; whether the SAR Government should play a significant role in narrowing the gap between the rich and the poor; how to change the relationship between the government and business consortiums; whether to improve Hong Kong's anti-monopoly policies; whether it was necessary to conduct a radical reform in its educational system; whether to increase labour security; how to divide work between the government and society when dealing with various kinds of social problems, etc. We can say that the greater part of the "consensus" reached in the past pertaining to the political system, policies, development direction and social form has ceased to exist. With increasing concerns with regard to Hong Kong's economic future would come greater disagreements over the direction of Hong Kong's development along with the role of the government and public policy. Admittedly, most people still agreed with the way things were originally, but there tended to be more people questioning or opposing the "status quo", even though they did not have a unified idea on how to construct another development "model" for Hong Kong.

Even so, more and more people grew resistant to the provisions in the Basic Law that were meant to maintain the "status quo", believing that "one country, two systems" limited the ability and space for Hong Kong to solve problems on its own. For now, provisions of the Basic Law still allow for some elasticity in changing Hong Kong's "status quo", for instance, fiscal prudence does not mean that Hong Kong cannot have a deficit or borrow money to cover government spending, and a low tax policy does not necessarily mean that it should maintain the current excessively low tax rate.[8] The Basic Law also did not create insurmountable difficulties for the SAR Government with regard to coordinating the interests of different classes. As long as a social "consensus" was reached to change the "status quo", and the central authority was not against it, it was hard to imagine that Hong Kong's court would accept lawsuits seeking to overthrow the "consensus". Even so, as Hong Kong's social divides continued to grow deeper as people grew more dissatisfied with the "status quo", and consequently, "one country, two systems" would face severer challenges because of its "conservativeness" and "static thinking".

3. The contradiction between "preserving the original capitalism" and "promoting democracy step by step"

Hong Kong's previous capitalist system and the colonial governance with Hong Kong characteristics were a pair of twin brothers, rare in human history, Chinese history, and even in colonial history. Notable features of the original Hong Kong capitalism included "small government" (less government involvement and intervention in economic activities), a low and simple tax regime, relatively fair market competition, limited welfare, the government and capitalists working together to suppress public demands, the government's freedom from the control of the bourgeoisie, the bourgeoisie's freedom from government manipulation, and the role of big capitalists, especially British corporations, as leaders of the bourgeoisie class.

From the theoretical point of view, this series of features should only appear under an autocratic or authoritarian regime, but the regime of Hong Kong was willing to respect market competition,

---

8 When he was interviewed by a newspaper, Chief Secretary David Akers-Jones of the former British Hong Kong Government made the criticism that some provisions in the Basic Law had placed too many and too stiff restrictions on the SAR Government's governance, for example, it could not borrow money when stuck in financial difficulties, so it could not respond effectively to the series of serious social problems facing Hong Kong after its return, such as the aging population. See *South China Morning Post*, 7 September, 2015, pC3.

limit economic and social functions of the government, pursue prudent fiscal policy, and commit itself to protecting individual freedom and rights. Undoubtedly, all colonial regimes are autocratic or authoritarian regimes, but the colonial government of Hong Kong was a rare exception in that it was willing to act as a "limited government" or to conduct "benign governance". The primary reason lies in the fact that the British needed to attract capital, enterprises, talents, and labour to Hong Kong through conciliatory methods after its landing on Hong Kong, "the barren island", so as to turn Hong Kong into a commercial port that would enable the U.K. to conduct business and trade activities in the "Far East".[9] Throughout the "colonial" history of Hong Kong, the administrative mode of conciliation gave birth to the capitalism with Hong Kong characteristics.

Since the capitalism with Hong Kong characteristics was based on a non-democratic (non-elected) colonial regime, which had sufficient autonomy and independence to limit its administrative functions, the introduction of democratic elections of any kind would do damage to the capitalism with Hong Kong characteristics. On the one hand, "one country, two systems" was designed to resolutely preserve capitalism with Hong Kong characteristics; on the other hand, it was intended to promote democracy in Hong Kong step by step; with the emergence of democratisation in Hong Kong, capitalism with Hong Kong characteristics would inevitably undergo transformation or even decline and decay over time.

Given that Hong Kong's democratisation can only be realised step by step, Hong Kong is yet categorised as a partial democracy twenty years after its return (some people still consider Hong Kong to be an authoritarian regime) since its chief executive and the Legislative Council have not yet been elected through universal suffrage. Nevertheless, "capitalism with Hong Kong characteristics" has been significantly affected by democratisation, with a number of subtle changes emerging.

---

9   See Lau Siu-kai (2014). *Hong Kong's Unique Road towards Democracy* (《香港的獨特民主路》). Hong Kong: The Commercial Press. P1-28.

First of all, the chief executive was no longer appointed by the "suzerain"; instead, the chief executive was elected by an election committee, and then the chief executive was officially appointed by the central authority. At the same time, Hong Kong's business circle and professional elites could influence the election results, since they accounted for a large share in the election council. The political objective of this election method is to grant Hong Kong capitalists and their supporters' political power that was unattainable in the "colonial" period. Moreover, big capitalists can lobby directly the central authority and further enhance their influence on the chief executive and the SAR Government since the central authority has attached great importance to Hong Kong's bourgeoisie class. Since the chief executive relied more heavily on the support from people with common interests along with the capitalists, the policies of the HKSAR obviously favoured the bourgeoisie. Against the background of the deteriorating situation with respect to the widening gap between the wealthy and poor, the SAR Government's subordination to the bourgeoisie class would lead to nothing but greater numbers of social issues, as well as growing dissatisfaction with the bourgeoisie class and the status quo among Hong Kong people.

Second, the composition of the bourgeoisie class had undergone significant transformation after Hong Kong's return; on the one hand, the composition had become increasingly complex, reflected in a diversified capital structure, the coexistence of mainland capital, local Chinese capital, and all kinds of foreign capital, as well as fierce competition among all parties. The SAR Government, having lost independence from the bourgeoisie class, found it difficult to coordinate the interests of different capitalists, resulting in decreasing support for the SAR Government from the bourgeoisie class, and meanwhile, capitalists had growing complaints about the SAR Government's unfairness in handling affairs.[10]

---

10 Lui Tai-lok & Stephen W.K.Chiu (2007). "Governance Crisis in Post-1997 Hong Kong: A Political Economy Perspective". *The China Review*, Vol. 7, No. 2 P1-34; Fong Chi-hang (2014). The Partnership between the Chinese Government and Hong Kong's Capitalist Class:

Third, the position and strength of the British-owned consortia were not as prominent as before, rendering them unable to take the lead among all the capitalists. Rising Chinese businessmen were in want of a leader and were troubled by severe internal controversies. Mainland corporations generally complied with the central authority and did not obey the authority of others. Other foreign corporations were incapable of acting as leaders. In addition, the lack of solidarity in the bourgeoisie not only weakened their political strength and increased their concerns regarding the rise of democratic politics, but also intensified their incentives to act out of their own will and pursue their individual interests. From another perspective, it was hard for a bourgeoisie that lacked a sense of solidarity to adopt a long-term perspective, to actively participate in political activities, to proactively take measures to reduce social conflicts, or to strategically make limited "concessions" to improve its relationship with other classes so as to effectively defend Hong Kong's capitalist system.[11]

Fourth, populist politics had already been on the rise despite the fact that Hong Kong had not yet implemented the practice of selecting the chief executive and the Legislative Council through universal suffrage. In order to secure greater "acceptability", win public support, reduce difficulties within the administration, and deal with pressure from the Legislative Council and political parties, every chief executive would undertake some policies to please the public, among which increasing social welfare and addressing public grievances were inevitably designated. As a result, it was more difficult for the government to perform as a "limited function" government and stick to principles of fiscal prudence. While the previous welfare and labour policies were

---

Implications for HKSAR Governance 1997-2012. *The China Quarterly*, Vol. 217, P195-220; Fong Chi-hang (2014). *Hong Kong's Governance under Chinese Sovereignty: The Failure of the State-Business Alliance after 1997.* London: Routledge; or refer to Lui Tai-lok (2015). *Hong Kong Model: from Present Tense to Past Tense* (《香港模式：從現在式到過去式》). Hong Kong: Chung Hwa Book Company. P49-61.

11 American bourgeois class was faced with the same situation. Internal differentiation led to the decline in their social responsibility and investment in public affairs. Refer to Mark S. Mizruchi (2013). *The Fracturing of the American Corporate Elite.* Cambridge, MA: Harvard University Press.

relatively "moderate", the accumulated financial burden could not be overlooked.

With the deepening democratic development in Hong Kong and the advent of universal suffrage in particular, it was nearly impossible to maintain "capitalism with Hong Kong characteristics" in an environment where a variety of social and livelihood issues converged. Demands for welfare and populist politics arose one after another, but once these demands were met, more would arise in their place. Even though the Basic Law provides legal protection for "capitalism with Hong Kong characteristics", more and more Hong Kong people started to question the Basic Law. Since the government and society could not respond to public appeal through reforms because of the Basic Law, people's confidence in "one country, two systems" and the Basic Law was compromised.[12]

4. The contradiction between "small government" and its economic integration with the mainland

Against the background of "capitalism with Hong Kong characteristics", Hong Kong's "small government" policy did not encourage government intervention in economic affairs. Under the principle of "small government", the government tried its best not to interfere in the operation of the market. Even in the event of a market monopoly, the government lacked the willingness and means to make any strong response. The government basically did not own or run businesses at all. In addition, the government usually did not make long-term policy plans except to enact laws and maintain necessary facilities and services such as land, roads, housing, welfare, and education. Even in the policy areas mentioned above, long-term plans were not made on a frequent

---

12 When drafting the Macao Basic Law, the central authorities applied lessons learned from the Hong Kong Basic Law. As a result, Macao's Basic Law did not incorporate the article prescribing the production of the governor and Legislative Council through universal suffrage.

basis. Economic development plans were even considered preposterous because many government officials regarded economic plans as "socialist" measures, which they thought were incompatible with the established market principles. Moreover, rarely did the government take the initiative to support certain sectors or industries. Since the government was only willing to deal with administrative affairs, the government was short of experts who were skilled at planning economic and technological development; even if the government intended to increase its intervention in economic affairs, the lack of personnel was a formidable obstacle that was hard to overcome.

The increased economic exchanges between Hong Kong and the mainland were anticipated and stimulated by "one country, two systems"; without "one country, two systems", Hong Kong could not have served as the impetus to propel the development of the mainland. After its return, Hong Kong encountered many difficulties, and with the continuous global financial and economic fluctuations and obstacles encountered in trade liberalisation, Hong Kong relied more and more on China, which was on an economic rise. In 2003, a large-scale anti-government parade took place in Hong Kong. The legislation of Article 23 was an important catalyst, and economic downturn in Hong Kong along with Hong Kong people's pessimism with regard to its economic outlook were also important factors. In order to boost Hong Kong's economy and consolidate the SAR Government's rule, the central authority introduced a series of economic and financial policies that were beneficial to both Hong Kong and the country at large, among the most important of which included: the Closer Economic Partnership Arrangement (CEPA); permitting mainland compatriots to visit Hong Kong on an "individual visit" visa; permitting mainland enterprises to go public and raise funds in Hong Kong; increasing economic cooperation between Hong Kong and Guangdong Province as well as between Hong Kong and Shenzhen; expanding RMB trade, turning Hong Kong into an RMB offshore centre; and permitting mainland enterprises and funds to "go global" via Hong Kong. In particular, what was of landmark and long-term significance was Hong Kong's active participation in the country's Five-year Plans, which meant that Hong Kong's economic development strategy should be coordinated with the national development strategy.

Because the mainland practised a socialist market economy with Chinese characteristics, which was very different from the laissez faire capitalism practised by Hong Kong, Hong Kong needed to take great advantage of national development to promote its own development– while also playing a positive role in the country's development. In this way, it was inevitable for the SAR Government, Hong Kong's entrepreneurs, and professional elites to adjust their economic and business ideas. Hong Kong needed to understand the country's major economic and social policies, the country's planning process and the content of those plans, as well as the opportunities and challenges brought about by national development. In order to integrate with national development, Hong Kong also needed to make some projections concerning Hong Kong's future economic development, make "plans", and invest resources. At the same time, the central authority also needed to know the "long-term" economic policies and arrangements of the HKSAR Government so that it could come up with some guiding principles, policies, and measures that were conducive to Hong Kong's development. The central authority and the Hong Kong Government needed to constantly exchange information and conduct policy consultations so as to coordinate each other's actions.

Along with Hong Kong's business sector and professional circle, the Hong Kong Government needed to plan for Hong Kong's future, lobby for favourable policies and arrangements from the central authority and the mainland, and train talents that could promote economic exchanges between the two sides, thus providing coordination and assistance for their development on the mainland. The "new tasks" mentioned above were no simple task for the Hong Kong government officials and business and professional elites who had been accustomed to short-term thinking and micro management. They had to operate from a long-term perspective; otherwise the future productivity and economic development of Hong Kong would face crises.

In conclusion, with the trend of economic integration between the mainland and Hong Kong, the image and operational model of "small government" would gradually but inevitably change to one in which it were more involved in economic affairs. With time, the gap between Hong Kong's economic system and the mainland's more market-oriented and liberalised economic system would narrow.

5. The contradiction between an "executive-led" system and the checks and balances between the legislative and judicial branches

In the "colonial" period, Hong Kong's "executive-led" system was built upon a basis of rather limited checks and balances on executive power. However, the colonial government was under the supervision of elites, public opinion and media functioning outside the political system. In the "colonial" political system, until Hong Kong's return, members of the Legislative Council were appointed by the governor of Hong Kong. The duty of the Legislative Council was to make suggestions and constructive critiques that were helpful for the government's administration, thus winning people's support for the colonial administration and consolidating the legitimacy of colonial rule. Anyone trying to oppose and challenge the colonial rule would not be appointed, and those in opposition already appointed as members of the council would be removed. Therefore, during the "colonial" period, the legislative branch, being too "obedient" to provide a check on government, was naturally affiliated with the executive branch.

After Hong Kong's return, the Legislative Council gained independent power and a base of social support; the method for electing members of the council was different from that of electing the chief executive; and the main role of the council was to monitor and check the executive branch. Despite the fact that the pro-establishment camp had occupied more than half of the Legislative Council's seats due to its method of election, members belonging to the pro-establishment camp did not belong to the same "political community" as the chief executive. In other words, since Hong Kong was not a parliamentary system where government is elected by the parliament, and the chief executive is not allowed to have political affiliations, they did not "share the same skin". In this way, if the chief executive enjoyed high popularity among the public and conducted effective governance, members belonging to the pro-establishment camp would usually be willing to support him; but it was hard for them to share his or her political honour, since the honour would be awarded to the chief executive himself or herself only. On the contrary, if the chief executive was not popular among the public and conducted bad governance, the pro-establishment members would not support him or her for fear that they might lose the election.

Since Hong Kong's return, the chief executive had been stuck in the dilemma of lacking public support and facing difficulties in administration, for example, when the opposition activists in the Legislative Council opposed him altogether or when the pro-establishment members dared not support him for fear that the electorate might change their stance. Without intervention and mediation from the central authority, along with their efforts to urge pro-establishment members to support Hong Kong's chief executive, it would have been even harder for the chief executive to overcome such difficulties in his administration. To put it simply, in the absence of the "ruling party" or a solid and powerful "governing coalition", the chief executive would lack stable, consolidated, and reliable support from Legislative Council members, making his position an uphill battle.

Similarly, the judiciary branch of the "colony" claimed itself to have "judicial independence", with the governor of Hong Kong rarely getting involved in the appointment and removal of judges. However, Hong Kong's judges highly respected the policies and decisions made by the governor and the colonial government, and they were willing to admit that the government was better equipped than the court to make and implement policies; thus they rarely intervened in or changed the government's decisions through litigation. Even if there were judicial review procedures, Hong Kong's judges tended to deal with this with discretion, lest they might give the impression of judicial power abuse or judicial trespassing. The behaviours of the judges from the "colony" were similar to those of the judges from the U.K., who generally would not question the decisions and actions of the government.

The behaviour of Hong Kong's judiciary branch has undergone some significant changes after Hong Kong's return. Firstly, the court can challenge the government's policies, decisions, and actions based on the interpretation of the Basic Law, in particular, the articles related to human rights and policies. And secondly, the court regards itself as an important force in maintaining Hong Kong's high degree of autonomy, safeguarding human rights and freedom, and preventing the administrative branch's abuse of power; therefore it assigns itself a major task. Thirdly, with a rise in "judicial vitality" and "judicial activism", the court is more likely to promote social reform through trial. Changes in this aspect have mainly been "contracted" from the West, in particular from the U.S. court. Under the pretext of defending human rights and preventing administrative branch's power abuse after Hong Kong's return, questions and challenges from the judiciary branch on the administrative branch's decisions and actions put a lot of constraints on the administrative branch, thus weakening the operation of the "executive-led" system.

6. Central authority's contradiction between "intervention" and "non-intervention"

Under the principles of "one country, two systems" and "Hong Kong people administering Hong Kong with a high degree of autonomy", the central authority had made a promise not to "intervene in" Hong Kong's "internal affairs". At the same time, the central authority had to perform important functions such as national defence, diplomacy, etc., and was authorised to exercise a wide range of power prescribed by the Chinese Constitution and the Basic Law. More importantly, the fact that bore most heavily on national sovereignty, security and development interests was whether "one country, two systems" was actually successfully implemented; therefore, the central authority took on more responsibilities than the HKSAR in implementing "one country, two systems" and also assumed the greatest liability. Moreover, "one country, two systems" required strengthened economic, trade, and financial cooperation between Hong Kong and the mainland; as the "integration" of the two sides became an irreversible trend, Hong Kong became more and more involved in the country's Five-year Plans. All of these factors meant that the state's major policies and measures would have a direct or indirect impact on Hong Kong. Therefore, it remains difficult to tell which part of the central authority's actions towards Hong Kong were "interventions" in violation of "a high degree of autonomy", and which part represented the central authority's "care" and "support" for Hong Kong. Past experience shows that Hong Kong people would not consider policies that are beneficial to Hong Kong "interventions", especially those related to the economy and people's well-being; instead, they would regard them as the central authority's "Hong Kong-preferential" measures and be grateful to the central authority. But as for those opposed to the "integration" of Hong Kong and the mainland, they were bound not to acknowledge the central authority's "Hong Kong-preferential" measures, but neither would they regard them as actions damaging to "one country, two systems".

On the grounds that the central authority should not interfere politically in Hong Kong's "internal affairs" certain voices criticised the central authority for the establishment of the Liaison Office of the Central People's Government in Hong Kong, which was seen as a clear proof of its "intervention" in Hong Kong affairs, especially when there was no mention of the establishment of the Liaison Office of the Central People's Government in the Basic Law. The central authority's deliberate cultivation of people who "love mainland China and Hong Kong", and directly or indirectly assisting organisations and candidates who love mainland China and Hong Kong, especially if such assistance bore on Hong Kong's elections, was also regarded by many people, the opposition activists in particular, as "blatant" and "rude" "intervention" by the central authority in Hong Kong's "internal affairs".

Some people did not agree with the central authority's putting political and economic pressure on Hong Kong's opposition activists and anti-Communist media. Meanwhile, controversies and critiques would arise when national leaders, central officials, and people from the mainland gave their opinions or suggestions regarding Hong Kong affairs. Taking into account all possible political consequences and repercussions, in normal circumstances, the central authority should be very reluctant to intervene in the affairs of Hong Kong. In fact, besides continued efforts to foster and support the forces that "love mainland China and Hong Kong", the central authority as well as people from the mainland, including the central authority's organisations and personnel in Hong Kong, had been trying to avoid getting involved in the affairs of Hong Kong, thus examples of commentary on Hong Kong's affairs were few and far between.

However, with a growing trend toward ever more frequent challenges to the central power and attacks on the Hong Kong Government, the central authority began to worry about whether "one country, two systems" could be successfully implemented, whether the administrative power of the central authority and the Hong Kong Government would be seized or overhauled by the opposition and pro-Western forces, and whether Hong Kong would become a "base of subversion". Because of this, the central authority increased its participation in Hong Kong's political affairs especially after the anti-Article 23 Legislation demonstration, the main purpose of which was to grant the central authority a greater say in explaining "one country, two systems", secure the central authority's power and authority, strengthen the Hong Kong Government's administrative ability, contain the opposition, and strengthen and consolidate the power of those who "love mainland China and Hong Kong". I will further elaborate on these developments in the following section.

In short, the central authority had been stuck in the dilemma of choosing between "intervention" and "non-intervention" in Hong Kong affairs since Hong Kong's return. The choice of "non-intervention" would surely be well-received by certain parties, who would think that the central authority had kept its promises and respected the principle of "Hong Kong people administering Hong Kong with a high degree of autonomy". However, with unfavourable conditions and obstacles and unreasoned resistance from various opposition forces standing in the way of the successful implementation of "one country, two systems", the central authority believed that it had no choice but to make some difficult and painful choices, hoping to create favourable conditions and atmosphere and bring the development of "one country, two systems" back on the right track through conducting "intervention" for a period; the central authority would not withdraw from Hong Kong affairs until people who love mainland China and Hong Kong commenced conducting efficient governance in Hong Kong, thus ensuring the completion of the "historic and strategic task" of implementing "one country, two systems".

The Transition Period Was Too Long, and the British Had Ulterior Motives

Historical experience showed that, generally, when the U.K. decided to retreat from or give up a colony or mandated territory, whether it had made careful plans beforehand or was fleeing helter skelter, it would not be long before the independence of the colony or mandated territory was secured, usually after a few years. In this transitional period, regardless of the eventual success or failure, the British would try to establish or consolidate the systems they were proud of and that would constitute a "glorious retreat", the aspects of which included the Westminster parliamentary system, the selection of representatives through universal suffrage, "political neutrality" in the civil service system, a "relatively independent" judiciary system, and a multi-party system with rotating leadership. It was also very important for the

British to choose a political successor that satisfied them; it would be best if the political successor had received a Western education, was steeped in Western political values, admired British people and culture, embraced an anti-communist attitude, and had cooperated with the British in political affairs before. For those who were leaders of anti-imperialist and anti-colonialism movements, as long as they opposed communism and were willing to protect Westerners' local interests after its independence, they were considered acceptable political allies for the British, with Lee Kwan Yew from Singapore being an outstanding representative. In fact, the British retreat plan in most cases was only a mixed success. Some of U.K.'s former colonies chose the socialist path and established authoritarian governments of various kinds (e.g. one-party dictatorships, personal dictatorships, and military regimes) after their independence. The relationship between these new regimes and the U.K. was very complicated and subtle.

China and the U.K.'s peaceful diplomatic negotiations were conducive to the U.K.'s willingness to return the entirety of Hong Kong to China. On the one hand, China promised to maintain Hong Kong's "status quo" and the continuity of Hong Kong's system and also agreed to retain and rely on the Chinese civil servants and political elites who had already engaged in assisting and supporting the colonial government. On the other hand, the U.K. promised to maintain Hong Kong's prosperity and stability in the transitional period prior to Hong Kong's return and to coordinate with the Chinese Government's policies towards Hong Kong, clearing the ground for the implementation of "one country, two systems" and ensuring Hong Kong's smooth transition into China's Special Administrative Region. The manner in which the U.K. withdrew from Hong Kong differed from the manner in which it had retreated from other colonies is that Hong Kong did not end up gaining independence; instead, it transitioned from a British "colony" to a Special Administrative Region of China; the basic arrangements for Hong Kong after its return were mainly decided by mainland China, so the U.K. could only engage in "decolonisation without independence" in Hong Kong.

From another point of view, as Hong Kong did not return to China or start to implement "one country, two systems" until the year 1997, the future successful operation of "one country, two systems" relied, to a great extent, on Britain's positive and close cooperation in the lengthy transitional period. The U.K.'s willingness to implement "one country, two systems" according to China's principles after Hong Kong's return had a great influence on whether this policy had a solid foundation and a positive start. Therefore, what the U.K. did in the transitional period has had a far-reaching influence on the successful implementation of "one country, two systems", which was why the U.K. reserved some bargaining chips on the Hong Kong issue and was able to force China to make more "concessions" to its demands and interests.

As is known to all, the British were very reluctant to leave Hong Kong; they were not confident enough in China's promises, and they did not really trust the Chinese Government; they hoped that Hong Kong would practise "one country, two systems" as defined by the U.K., even if their definition conflicted with the Chinese Government's definition of "one country, two systems". The British longed for a "glorious retreat", the core of which involved working to establish some policies and measures that had been implemented in some of their former British colonies, including setting in place a representative political system characterised by "governance by people" with the legislature playing a central role; strengthening human rights protection; promoting judicial "independence"; introducing and expanding parliamentary election components at all levels; and supporting all kinds of opposition forces, including anti-communist and pro-British forces. The British were particularly concerned about whether the colonial government could maintain governing authority and effective governance in the transitional period; the main reason was that it was inevitable for a colonial government on its way out to face a rapid decline in its governing authority and a loss of support from its sympathisers.

Furthermore, since Hong Kong's pre-return, transitional period (from 1982 to 1997) was very long, maintaining governing authority and effective governance during that period, even for the British people, who touted their political wit, was a major challenge. At the same time, with the colonial governance coming to an end, all kinds of hidden conflicts were exposed and quickly intensified. Those conflicts mainly occurred between the traditional pro-Beijing patriotic forces, the political elites occupying important positions in the "colony", the emerging "democratic" forces, and the conservative colonial elites, as well as different social classes. There appeared divergences among the original pro-colonial Chinese elites, because the U.K. was to retreat very soon, with many business elites leaning toward the Chinese Government in seeking their own interests, thus shaking the foundation of the colonial governance. More troublesome was Hong Kong people's concern for Hong Kong's future and antipathy towards the Chinese Government triggered by the Tiananmen Square Incident that took place in Beijing in 1989. Hong Kong people's rising anxiety, fear, restlessness and anti-communist sentiments not only added to the difficulties in colonial governance, but also posed obstacles for the cooperation between the Chinese Government and the U.K. Government on the Hong Kong issue. In addition to all of this, the British had been working hard to find a way to obtain maximum economic and other substantial interests for the U.K. during the transitional period.

In view of the above circumstances and factors, the way that the U.K. withdrew from Hong Kong was unique, with national interests and national honour being their priorities; the British people were navigating an unprecedented exit during the transitional period. At that time, cooperation between the central authority and Hong Kong on many practical issues was generally adequate, though not always smooth, especially when British economic interests were involved. According to Chen Zuoer's (陳佐洱) recollections, a former official of the Hong Kong and Macao Affairs Office of the State Council, the British were more willing to cooperate on practical issues such as the designation of land for the use of military purposes, budget preparation, construction of the new airport, the right of abode and the SAR passport, the Hong

Kong and Kowloon sewage project, mobile phones, etc.[13] However, the British maintained other political intentions, including preventing the weakening of the governing authority, bending "one country, two systems" according to their own vision, maximising the interests of the U.K., and achieving "an honourable retreat". Therefore, the U.K. did not fully cooperate with China. Instead, it tended to deliberately scuttle cooperation in a number of aspects, with the intention of forging a fait accompli and thereby forcing China to make concessions on a number of points.

With a lengthy transitional period and a lack of sufficient knowledge and awareness of the U.K.'s purposes and actions on the Chinese Government's part, the U.K.'s political deployment in the transitional period was very successful; even if the Chinese Government had been aware of their purposes and actions, the central authority would have had found it difficult to deal with under the circumstances at that time; the result was that when Hong Kong was returned to China in 1997, political forces, institutions, procedures and arrangements inconsistent and incompatible with the underlying concept of "one country, two systems" were already "rooted" in many areas in Hong Kong to some extent, posing obstacles and provoking difficulties for the implementation of "one country, two systems" in Hong Kong.

Chinese leaders were not completely blind to the possibility that the British might cause problems for Hong Kong's smooth return to China; after all, China and the U.K. had experienced tough negotiations, and China was aware that the British were not resigned to withdraw from Hong Kong. Deng Xiaoping's experience in interacting with the British had made him ever doubtful about their intentions and sincerity during the transitional period.

On July 31st, 1984, when he met British Foreign Secretary Geoffrey Howe, Deng Xiaoping made it clear to the U.K., "To be frank, as for the Hong Kong issue, we are paying close attention to the 13-year tran-

13 Chen Zuoer (2012). *The Negotiation on Hong Kong's Return that I Witnessed* (《我親歷的香港回歸談判》). Hong Kong: Phoenix Books Culture Publishing Company Limited.

sitional period; as long as the transitional period is well-planned, we are not worried about things after 1997. We hope that the following situations won't happen in the transitional period. First, we hope the position of Hong Kong dollar will not be shaken. How many Hong Kong dollars should be put into circulation remains a question. Hong Kong dollars have always enjoyed good credibility because of abundant reserves, which is in excess of the circulation; we cannot change the current status. Second, we agree to authorise land leases within fifty years after 1997, and agree that the British Government of Hong Kong may use revenues from land sales; but we hope that the revenues are used for infrastructure and land development rather than administrative expenses. Third, we hope the British Government of Hong Kong will not recruit more staff or increase salaries and pensions at will, which will increase the burden on the future SAR Government. Fourth, we hope that the British Government of Hong Kong will not independently form a group charged with leadership, to avoid imposing them on the SAR Government in the future. Fifth, we hope the British Government of Hong Kong can persuade British-funded enterprises not to withdraw their funds. We hope no problem will crop up during the transitional period, but we must be prepared for any that may arise despite our wishes. We look forward to improving Sino-British cooperation in the future."[14]

Deng Xiaoping noted these opinions again in 1984 when he met with the Hong Kong and Macao Compatriots Group of the National Day, "I mentioned a few problems that we hope will not appear in the transitional period during our dialogue with the British. One is the withdrawal and outflow of British funds and the other one is the violent fluctuation of the Hong Kong dollar. If the currency reserves were exhausted, the Hong Kong dollar would be depreciated, and unrest will occur. Is it OK if we do not keep an eye on the currency reserves during the transitional period? Moreover, we have a land problem; if

14 Deng Xiaoping (1993). *How Deng Xiaoping Views Hong Kong Issue* (《鄧小平論香港問題》). Hong Kong: Joint Publishing. P9-10.

we sold all of our land and used the revenues for covering administrative expenses and handed the burden over to Hong Kong's Government after 1997, is it OK if we do not intervene? I have made five points, and the British have expressed their willingness to cooperate with us, [...] As for the Sino-British Joint Declaration, we believe that not only we ourselves will abide by it; we are convinced that the British will too, and we are even surer about Hong Kong people's commitment to it. But it should be taken into consideration that there might always be some people who never intended to fully implement it. There will be some restless factors, disruptive factors, and unstable factors. Honestly, these factors will not come from Beijing, but we cannot exclude the possibility that they may be found in Hong Kong or among certain international forces."[15]

After the "Tiananmen Square Protests" in 1989, the British had estimated based on their observation and study that there was not much time left for the Chinese Communist Party; therefore, they decided to change the policy towards Hong Kong and China, which meant replacing cooperation with confrontation, trying to overturn Hong Kong's future arrangements, and enabling the U.K. to have another chance at governing Hong Kong. In response to the treachery of the British, Deng Xiaoping decided to adopt a tough stance, ensuring the smooth return of Hong Kong by engaging in tit for tat struggles with them, with China standing on its own two feet, guided by "Start All Over Again" diplomatic strategic theory and practice.

Former director of the Hong Kong and Macao Affairs Office of the State Council Lu Ping also expressed his distrust of the British. He recalled and said, "I've learned something about the British: they are astute indeed; they will look two steps forward before they take the first step; therefore, we need to look three steps ahead in order to deal with them. We have to fight them in a rational, advantageous, and decent way; we can achieve success only if we know ourselves as well as our

---

15 Ibid, P13-14.

enemies."[16]

However, despite the fact that the Chinese Government never relaxed its vigilance nor expressed distrust towards the British, they were successful in preventing or ameliorating certain plans initiated by the British. For their part, the ever-astute British remained to a great extent still capable of carrying out their political deployments according to their political intentions.

To put it simply, in the lengthy transitional period, British people made ten crucial political deployments. They are as follows:

First and most crucial stands the reform of the political system; namely, the British introduced "mass politics" and "public-opinion politics", and they took advantage of public opinion to defend their own position. Although the colonial government's "representative govern-ment" reform did not achieve its ultimate goal, through the implementa-tion of parliamentary elections, direct elections in the regions in partic-ular, they could increase the public's political influence dramatically, providing a broad space for development of various opposition and anti-Communist forces. The British also started to follow a policy direc-tion that was "based on public opinion" in order to make Hong Kong people feel that they were able to influence the governmental adminis-tration. In the latter part of the transitional period, governor Chris Patten appeared as the "populist governor"; disguised as the leader for "demo-cratic election" politics, he strongly encouraged confrontation between the public and the side represented by the central authority together with those who "loved mainland China and Hong Kong". From then on, Hong Kong's political environment witnessed irreversible changes. One major change was the violent collision between the original "elite poli-tics" and emerging "mass politics", which directly resulted in declining political authority and increasing political conflicts.

Second, the British no longer complied with the promises that they made to the Chinese Government. They did not contain anti-Com-

---

16 Lu Ping (2009). *Lu Ping's Verbal Recollections of the Reunification of Hong Kong* (《魯平口 述香港回歸》). Hong Kong: Joint Publishing. P79.

munist and anti-China behaviours; what was worse, they even encouraged them either overtly or covertly. It seems certain that the British knew that by doing so they would make relations with the Chinese Government even more awkward, damaging Sino-British cooperation. Still, with the decline of the political authority of the colonial government due to the ending of the "colony", the colonial government was increasingly incapable of restricting those forces. Meanwhile, those forces hoped to strengthen themselves at a time when Hong Kong was still under the colonial governance, in order to have a stronger hand in order to deal with the Chinese Government led by the Communist Party of China after Hong Kong's return. Therefore, encouraged by the British, dialogue and actions against the Chinese Government increased dramatically; this was quite a departure from the time when these forces were contained by the British.

Third, the British fostered anti-Communist democrats in order to establish a "half-establishment camp" or a "half political ally", and furthermore, directly supporting opposition and anti-Communist forces. In fact, the British had political incentives for providing a foothold for opposition and anti-Communist forces within the colonial government, namely, defending the U.K.'s interests in Hong Kong and achieving the objective of an "honourable retreat". These objectives were at risk, as a portion of the Chinese elites formerly affiliated with the British, who were now losing their mandate, turned to the Chinese Government, and some even defied the colonial government, shaking the foundation of colonial governance.

On the one hand, the British met the demands of the opposition and weakened challenges to colonial governance by implementing political and constitutional reforms; on the other hand, the British contained the Chinese Government, "pro-Beijing" forces, and formerly "pro-British" elites by capitalising upon anti-communist forces. To a certain extent, the opposition forces were no longer repressed or under surveillance; instead, they had suddenly transformed into a half "ally" of the colonial government. British people's new political deployments and administrative policies also changed the previous colonial governance model, bringing part of the opposition forces and the original Chinese political elites together and forming a new "pan-establishment camp". In this camp, the opposition's democratic aspirations and visions of shared power were incompatible with the former elites' conservative bent; however, we can at least say that, with British people's coordination and mediation along with their common anti-communist stance, disparate different forces were able to get along with each other even if they didn't cooperate very closely. One of the derived results was a significant increase in the colonial government's investment in social welfare and public services in the transitional period.

Fourth, the British worked to weaken the "executive-led" principle and advocate the concept of "legislature first". Although the Chinese Government insisted on maintaining the "executive-led" regime of the "colonial" period after Hong Kong's return, it was unable to get the active cooperation of the British. The U.K.'s intention was to reduce the strength of the administrative branch and establish a real but nameless "legislature first" regime. That was why China and the U.K. went from cooperating to parting ways on the issue of constitutional development.

The U.K.'s initial plan was to promote a "representative political system" before Hong Kong's return and finally to achieve a legislative branch elected through universal suffrage so that the legislative branch would play a pivotal role in electing the chief executive. This plan could not be practised due to opposition from the Chinese Government. Even so, the last governor Chris Patten still greatly increased the democratically elected components in the 1995 Legislative Council despite strong opposition from mainland China. Although this session of the Legislative Council could not survive the transition to the return, Hong Kong people's demands for democratic election grew greatly because of British people's "generosity", foreshadowing Hong Kong's ensuing political polarisation and conflicts.

More important than the method of electing the legislative branch were the measures which raised the status of the legislative branch deliberately, including: enacting laws that prescribed the legislative branch's rights and privileges; entitling the legislative branch to greater power pertaining to investigating and monitoring the government; separating the Legislative Council from the executive branch so that the legislative branch was no longer under the leadership of the executive branch; terminating the governor of Hong Kong's role as president of the Legislative Council in order to enhance its independence; adopting a highly cooperative if not a de facto subordinate role for the governor and officials vis-à-vis the Legislative Council with the aim of establishing "constitutional conventions" where the executive branch answers to the legislative branch; increasing members' salaries along with operating funds in order to strengthen the political power of members; and encouraging members to put forward "private member's bills", so that they enjoyed more power in making policies. All these were to ensure that the legislative branch and the executive branch were "on an equal footing" and that the Legislative Council was accountable to the government, hoping to bring pressure on the central authority and the SAR Government after Hong Kong's return, thus not giving full play to the aspect of being "executive-led".

Fifth, the British decentralised and subcontracted important undertakings, weakening administrative power. The colonial government handed over power and responsibilities that belonged to the government to a group of independent organisations and statutory bodies, "privatised" some of the government work, and subcontracted some government services to commercial organisations. In order to better supervise the government and protect human rights, a series of organisations aimed at maintaining equality in business opportunities, protecting privacy, preventing the abuse of power and ultra vires actions within the government would be established to form an invisible net encircling the departments of the executive branch. These organisations had the capacity to influence government policies to some extent.

Sixth, the British attempted to give the colonial government's political partners a leadership role in the HKSAR. The British fervently hoped that "Hong Kong people who administer Hong Kong" would be the Chinese political elites nurtured by the colonial government–in order that the systems and procedures established during the "colonial" period could be maintained after British people left Hong Kong, and British and Western interests would be taken good care of after the ending of the "colony", and also to ensure a smooth path for the Chinese political elites.

At first British people hoped the political elites could win public support through election; it would be even better if they were party leaders. However, since the plan for "representative political system" could not be completed with China's opposition, the British had to invent a cultivation program for political talents unprecedented in other colonies; by pushing the "politically neutral" senior civil servants up to the political stage, British people gave them a chance to be the backbone of the "Hong Kong-administering" people from Hong Kong after its return. Due to the shortage of political talents "who loved mainland China and Hong Kong", plus the fact that they had not yet gained the trust of the Hong Kong people and that British people "suddenly" gave up the policy opposing against China, China approved it, agreeing on "smooth transition" of public servants as a whole. Many former "pro-British" people were assigned to important positions. In general, the British did a superb job in making sure the personnel arrangements for "Hong Kong administration" were composed of people from Hong Kong.

Seventh, the British tried to win final adjudication power following the return. Granting Hong Kong's court final adjudication was not included in the negotiation on "the issue of Hong Kong's future". "But after signing the Sino-British Joint Declaration, sophisticated British people were thinking about 'giving up' the crucial power of final adjudication before Hong Kong's transition, the purpose of which was to transit the final appellate body that was established as the Britain wished to the future HKSAR [...] In February 1988, the British side submitted a proposal to set up a court of final appeal, in which major aspects were included such as the structure and composition, adjudicatory power, and judicial proceedings. In the spirit of friendly cooperation, after half a year of careful research, the Chinese side believed that to make early transitional arrangements before 1997 would clarify the arrangements for the future court of final appeal of Hong Kong and was conducive to promoting the regime's smooth transition and enhancing the confidence of the Hong Kong people; therefore, the Chinese side decided to accept British side's proposal, but also requested that the court of final appeal established before 1997 be in full compliance with relevant provisions in the Basic Law and China's requirements."[17]

The jurisdiction of the court of final appeal was one of the focal points of the debate between the two parties. Although the two sides agreed that Hong Kong's court had no right to deal with "state actions", regarding the definition of "state actions", they had different opinions. Chinese official Chen Zuoer recalled: "Whether the expression of 'state actions' refers to 'defence and foreign affairs' exclusively or 'defence and foreign affairs, etc.' is an old issue which has been discussed for years, the result of the discussion was that it could not be denied that apart from defence and foreign affairs, there were truly some affairs that should be classified as "acts of state", including central administrative affairs and affairs concerning the relationship between the central authority and the HKSAR, so the word "etc." should not be omitted.[18] In 1995, the two sides reached an agreement on the issue of the court of final appeal. In the agreement, the British agreed that Article 19 of the Basic Law on "acts of state" be embedded in Hong Kong Court of Final Appeal Bill (Draft), thus the Bill was in compliance with the Basic Law.

In addition, the two sides had different opinions on whether the court of final appeal had the power to inquire into the constitutionality of laws and on how to deal with cases where the court had made erroneous decisions. At last, China's attitude was: "… On the two issues of the power to inquire into the constitutionality of laws and remedial mechanism after sentence, we can adopt a flexible attitude; because

17 Chen Zuoer (2012). *The Negotiation on Hong Kong's Return that I Witnessed* (《我親歷的香港回歸談判》). Hong Kong: Phoenix Publishing House. P222-224. Also in Liang Xinchun. *A Personal Experience of Hong Kong's Return: the Major Events in Post Transition Period* (《親歷香港回歸：後過渡期重大事件始末》). P322-326.

18 Chen Zuoer (2012). *The Negotiation on Hong Kong's Return that I Witnessed* (《我親歷的香港回歸談判》). Hong Kong: Phoenix Publishing House. P233-234.

even if we give up these two requirements in the level of the court of final appeal, the National People's Congress has the highest decision-making authority, which can serve as our guarantee when necessary. "[19] However, experience gained after Hong Kong's return showed that the Hong Kong Court of Final Appeal and the central authority usually had different ideas on major constitutional issues, posing obstacles to the comprehensive and accurate implementation of "one country, two systems".

Eighth, Britain sought to bring in the issue of human rights politics. Britain initiated the "Hong Kong Bill of Rights Ordinance" in 1991; provisions in the United Nations International Convention on Civil and Political Rights that could be applied to Hong Kong were incorporated into the laws of Hong Kong, the key point was that human rights law was positioned as a law "overriding" other laws.[20]One of the most significant consequences was that Hong Kong people's political freedom and space for political activities was greatly enhanced, in particular freedom of the press and the right to protest.[21] Another important consequence was that "human rights" factors had seriously limited the SAR Government's ability to make policies.

On the one hand, human rights considerations were more important than other political, economic, and social factors in the government's administration process. On the other hand, history

---

19 Ibid, P238. Also in Liang Xinchun (2012). *A Personal Experience of Hong Kong's Return: the Major Events in Post Transition Period* (《親歷香港回歸：後過渡期重大事件始末》). P322-327.

20 "Overriding" was reflected in Article 2 Subsection (3) of the Ordinance which was about the explanation and the purpose of the application, Article 3 about "its influence on previous laws", and Article 4 about prescriptions on "interpretation of the future laws". Article 3 Subsection (3) states, "An Ordinance to provide for the incorporation into the law of Hong Kong of provisions of the International Covenant on Civil and Political Rights as applied to Hong Kong; and for ancillary and connected matters." Article 3 is related to its influence on pre-existing legislation, "(1) For all pre-existing legislations, if they can be interpreted in a way that is consistent with the Ordinance, interpretation should be made as it is; for all pre-existing legislations, if they cannot be interpreted in a way that is consistent with the Ordinance, part of the inconsistent legislations should be repealed". Article 4 focuses on the interpretation of future laws, "For legislations enacted on or after the commencement date, if they can be interpreted as consistent with provisions in the International Covenant on Civil and Political Rights that can be applied to Hong Kong, interpretation should be made as it is".

21 Also in Liang Xinchun (2012). *A Personal Experience of Hong Kong's Return: the Major Events in Post Transition Period* (《親歷香港回歸：後過渡期重大事件始末》). P38-165.

showed that the definition of "human rights" tended to broaden, and its meaning tended to be vague.[22] More seriously, the court of Hong Kong could gain the power to overthrow government policies and make public policies on its own through the application and interpretation of human rights law, thus weakening the "executive-led" principle and the SAR Government's authority and governance capacity. Due to the fact that the human rights law's "overriding" nature was in violation of the Basic Law, relevant provisions of the law were abolished by the central authority.[23] However, even so, lawsuits and political struggles caused by human rights issues increased dramatically after Hong Kong's return as compared with the period before its return and Hong Kong people's awareness of human rights was also growing day by day. "Human rights" had become a weapon for defending their interests and challenging the government's policies and decisions. Hong Kong's court claimed to be the defender of human rights, and by doing so, it lowered the threshold for judicial review, thus bringing impetus to promoting the emerging human rights politics and bringing new challenges to the SAR Government's effective administration.

Ninth, the British advocated an alternative interpretation of "one country, two systems" together with the democrats. British people expressed their recognition of the "one country, two systems" policy again and again, and admitted the fact that the Basic Law enacted by China had fully reflected "one country, two systems" and the agreement on the return of Hong Kong reached by the two sides. But in reality, the U.K.'s understanding of "one country, two systems" was quite different from that of China in that the British side interpreted it from the perspective of "maximising Hong Kong's autonomy". During the transitional period, British people had been indoctrinating Hong Kong people with such ideas, in order to better "familiarise" Hong Kong people with

---

22 Hunt Lynn (2008). *Inventing Human Rights: A History*. New York: W.W.Norton.

23 See the *Decision Standing Committee of the National People's Congress on treatment of the laws previously in force in Hong Kong in accordance with Article 160 of the Basic Law of the Hong Kong Special Administrative Region of the People's Republic of China* (passed in the twenty-fourth session of the Eighth National People's Congress Standing Committee in February 23rd, 1997).

their power and rights, allowing them only limited knowledge and even erroneous understandings of the Chinese Government's power and duties. To a certain extent, British people and Hong Kong's opposition forces would advocate these ideas to Hong Kong people if they had some common ground. Because Chinese officials and experts still had some opportunities to express their views and refute skewed theories, Hong Kong people's understanding of "one country, two systems" was not totally biased. However, the U.K. and opposition forces still gained the upper hand since their views were more consistent with Hong Kong people's hopes and desires.

Tenth, the British obstructed and limited the development of "people who love mainland China and Hong Kong", the objective consequence of which was to cause a rift between traditional patriotic forces and mainstream elites and intensify the conflicts between opposition forces and patriotic forces. The reason why British people needed to thwart the forces of "people who love China and Hong Kong" was to lower their chances of acting as administrators of Hong Kong and to protect the opposition activists. This included minimising opportunities by which "people who loved mainland China and Hong Kong" could gain administrative experience, launching campaigns vilifying their image and denying their qualification as Hong Kong people who could "administer Hong Kong". Opposition forces aligned with the "pro-British" elites who remained with the British; they had head-on confrontations with the forces that "love mainland China and Hong Kong", and a great deal of hatred was forged on both sides. The political cleavage caused by these fierce struggles has yet to be healed. Continuous political struggles after Hong Kong's return were closely related to British people's politically divisive strategy in the transitional period.

The political deployments were not limited to the above ten, but these composed the most essential part of the retreat strategy. Other measures also caused some trouble for Hong Kong's smooth transition, such as destroying most of the confidential documents, dismissing the "political department" of The Hong Kong Police under direct leadership of British intelligence agencies that was engaged in intelligence collection and political control, giving 50,000 family members in Hong Kong the "right of abode" in the U.K., changing the land inheritance tradition of the original inhabitants in the New Territories, and trying to "internationalise" the Hong Kong issue to some extent. In general, instead of laying a proper and solid foundation for practising "one country, two systems", British people's political "non-cooperation" created some factors that served as barriers and obstacles in implementing it. Those factors continue to create "negative energy" even today.

The Absence of Decolonisation

After gaining independence from the U.K., almost all of the new leaders of Britain's former colonies, regardless of having received British education or not or whether or not they admired British culture, tried to promote some sort of "decolonisation" plan in terms of culture, thought, institutions, and policies. They strived to strengthen the concept of the state and nationalism, to revitalise certain national and ethnic traditions (original or "transformed", "reconstructed" or "imagined"), to publicise an anti-west awareness[24], to introduce some strategies for socialist development, to set up new political systems and governance models, and to criticise the former governance in the colonies. Even if the former suzerain's laws and judicial system were inherited, they would be "reformed" to safeguard sovereignty and regime security, curb turmoil, and constrain individual rights.[25]

These measures were actually aimed at strengthening national

---

24 See Cemil Aydin (2007). *The Politics of Anti-Westernism in Asia: Visions of World Order in Pan-Islamic and Pan-Asian Thought.* New York: Columbia University Press.

25 Singapore is a successful example, which wins support from the Western countries by maintaining the British common law system. But meanwhile, it skilfully utilises and reforms the laws in British colonisation for political control, highlighting the principle of state superior to individual. See Jothie Rajah (2012). *Authoritarian Rule of Law: Legislation, Discourse and Legitimacy in Singapore.* New York: Cambridge University Press.

confidence, pride and cohesion, preventing ethnic conflicts, propelling national development, repealing former colonists' privileges, enhancing the political reputation and legitimacy of the new regime and forming a diplomatic road map and national defence policies for the new country. Almost every "decolonisation" action was motivated by a certain degree of discontent with the colonial governance, grudges against racial oppression, antipathy toward the policies in the colonial period, and colonial people's desire to govern their own land. The "desire to change" was prominent in many former British colonies.[26]

For Hong Kong, the case was exactly the other way around. Hong Kong has hardly seen real "anti-colonisation" or "independence" movements throughout its history. The general public in Hong Kong accept the history of colonial governance and feel no shame in it. Many of them, particularly the Chinese elites benefiting from colonial gover- nance, have even long opposed the Chinese Government's remarks about or actions toward the resumed exercise of sovereignty over Hong Kong. In accordance with the guidelines for maintaining the status quo and sustaining Hong Kong people's confidence in Hong Kong's future, the "one country, two systems" policy doesn't consist of a "decoloni- sation" plan, and is basically absent, any intention of changing Hong Kong people's post-return outlook, including their remembrance of colonisation, recognition of colonial governance, and resistance against the Communist Party of China and People's Republic of China. Civil servants and political elites that used to participate in governance work continued to undertake important posts, and the "de-Sinicisation"-fea- tured educational curricula and examinations continued to be used in the HKSAR.

However, the "one country, two systems" policy leaves no substantial room for "national education" or "patriotism education". After the Government of HKSAR was established, troubles and crises arisen one after another, strongly affecting the government's authority

---

26　See Margaret Kohn & Keally McBride (2011). *Political Theories of Decolonization: Post colonialism and the Problems of Foundations*. New York: Oxford University Press.

and its capacity to govern. For the government to carry out "national education" is out of the question. Since many people still harbour doubts about the Communist Party of China, "national education" will be ridiculed as "brainwashing education" or "love-the-(Communist)-Party education", and thus would trigger social resistance and add to the political ammunition of the opposition. A lot of civil servants in the Government of the HKSAR hold the anti-communist attitude that prevailed in the colonial period. With no real intention to promote "national education", they overtly agree but covertly oppose it.

In fact, to carry out "national education", it is unavoidable to criticise the former colony's past, including the injustices of the Opium War and imperialism, foreign occupation of China's territory, the U.K.'s incapability to protect Hong Kong from Japan during the Japanese War of Aggression, the absence of the colonial government from Hong Kong, the racism in the colony, the "exploitative" colonial governance, the preference for the suzerain's interests, the role of violent repression in colonial governance,[27] the distortion of colonial people's political personality by reason of colonial governance (xenocentrism, political impotence, political culture of obedient citizens), the lack of imagination and innovation in institutions and policies due to "benevolent" colonial governance, "de-Sinicisation", the weak awareness of state and nation, the subtle national inferiority complex, the contempt for PRC, the CPC, and mainland compatriots, institutions, policies and approaches, the overemphasis on the role of the colonial government's policies and the neglect of the Chinese people's role in Hong Kong's economic achievements, the understatement of the efforts of the new Chinese Government led by the CPC and its contribution to economic and livelihood improvement, the one-sided praise for the contribution of colonial governance, limited and distorted understanding of Hong Kong's post-war "economic miracle" and the shifts of international and domestic situations and so on.

---

27 See Georgina Sinclair (2006). *At the End of the Line: Colonial Policing and the Imperial Endgame.* Manchester: Manchester University Press.

Under the precondition to "maintain the status quo", few people were willing to summarise the past of Hong Kong as a "colony" objectively and critically, actually no easy task, because few people were engaged in research into Hong Kong's history. But more importantly, such research might hinder scholars' individual academic development. During the "colonisation" period, some scholars who had the intention to work on the present or the past of the "colony" from the perspective of "anti-colonisation", "anti-imperialism", "Chinese people" or even from the perspective of Western ideals of freedom, equality, and democracy risked obstruction by the British. But in point of fact, a majority of people in Hong Kong's academia were actually complacent concerning the colonial governance, and formulating a request for research on Hong Kong from a "decolonisation" standpoint would tend to be fruitless approach. What's more, since Hong Kong never underwent any real "independence" movement, the ideas of "independence leaders" would be unavailable in terms of formulating "decolonisation" content.

In the mid-1960s, the pro-communist leftists fomented the Hong Kong 1967 leftist riots, condemning British colonial rule and leveling accusations of injustice, but their resistance was without theoretical support or persuasive argument. In the "communist-phobic", "anti-communist" atmosphere, Hong Kong people excoriated the leftists and the public support for the colonial government was consolidated. Any criticism against the "colony" or its government in such circumstances would be easily considered as stemming from ulterior motives, or even as the CPC's "plot to gain support".

Another practical issue is that "decolonisation" after the return would have definitely led to negative feelings among the U.K.'s "collaborators" (civil servants and police officers included) who used to assist British governance in Hong Kong. "Decolonisation" hence would have been regarded as a sign of the Chinese Government's distrust in the Hong Kong administrators whom it still had to rely on, and "decolonisation" would also have been interpreted by the Hong Kong people as the Chinese Government's intention to break its promise of maintaining the status quo in Hong Kong, stirring up worries and fear about Hong

Kong's future. In view of various misgivings and its limited capacity, the Government of the HKSAR has neither the motivation, nor the political courage and perseverance for "decolonisation".

In the absence of "decolonisation", Hong Kong people still lack a comprehensive grounding on important aspects including: basic Hong Kong history; the deeper and more subtle reasons why Hong Kong became a British "colony"; the complexities of modern China-Hong Kong relations; the interweaving of Hong Kong history and Chinese history; the factors behind the achievements and failures in Hong Kong's development; Hong Kong people's responsibilities and track record as Chinese citizens; Chinese modern and contemporary history; the common fate of Hong Kong and Chinese people; socialist China; the mainland compatriots; Hong Kong people's responsibility for safeguarding national sovereignty, security, and development interests under the "one country, two systems"; the profound traditional Chinese culture and the genes of Chinese society contained in Hong Kong society and so on. The absence of "decolonisation" therefore poses difficulties for the effective settlement of some ideological factors adverse to "one country, two systems".

The Nominal "Executive-led"

The central authority authorises Hong Kong a high degree of autonomy under "one country, two systems", so the mainland's power to directly intervene in Hong Kong's internal affairs is usually constrained, unless in severe crises, in which case the central authority can then execute emergency power in Hong Kong. Thus, the SAR Chief Executive is a crucial position for the central authority's implementation of "one country, two systems" in Hong Kong in ordinary times, and is endowed with a high status and extensive power through the Basic Law. Without "executive-led" put in place, effective governance would be impossible, accurate and comprehensive implementation of "one country, two systems" would be difficult, and the due protection for the security and interests of the country and the central authority would be absent.

Experience after the return indicates that the "executive-led" principle hasn't been fully implemented in the HKSAR and that the Government of the HKSAR has been in a weak position in general. The core reason is that no forceful new regime has been established since the end of the colonial regime in Hong Kong, which is discussed in detail in another book of mine and therefore is only illustrated in brief here.[28]

My basic argument is that the "executive-led" principle cannot merely rely on the chief executive's execution of the power endowed by law. Favourable political conditions are also imperative, including Hong Kong people's respect and trust for the chief executive, the political wisdom and capacity of the chief executive and the leading group, the chief executive's strong leadership of the administrative organisations and senior civil servants, a governing coalition with a solid and broad social base to support the government of the HKSAR, the central authority's strong support for the chief executive and so on. With all those conditions lacking since the return, along with the counterbalance of the Legislative Council, the judiciary's "intervention" in administrative decisions and affairs through interpreting the Basic Law, various social counter forces, and monitoring by the press, governance difficulties in the HKSAR loom even larger.

Equally problematic are the chief executive along with many officials' political feebleness and lack of persuasiveness in political statements. These officials are also under the impression that they have little political legitimacy and, out of the fear of public opposition, dare not to share a stance with the central authority, and thus strike people as timid, indecisive, and jittery. A consistently weak SAR Government in Hong Kong surely lacks the courage, strength and wisdom to carry out the "one country, two systems", not to mention correcting mistakes in

---

28 Lau Siu-kai (2012). *Fifteen-year Governance and New Regime Construction in Hong Kong Special Administrative Region after Return* (《回歸十五年以來香港特區管治及新政權建設》). Hong Kong: The Commercial Press.

its implementation.

To make the situation even worse, a portion of communist-phobic Hong Kong people project upon the chief executive their fears of his being "selected" by the central authority, assigning him a "pro-China" (substitute for "pro-communism" in Hong Kong) "original sin". In a political atmosphere rife with conspiracy theories, the chief executives have to cope with the issues relevant to the interests and security of the central authority, and at the same time they are suspected of partiality to the central authority, and their already low reputation due to the back-lash of the Hong Kong people continues to fall, creating opportunities for the opposition to utilise Hong Kong's discontent and launch political protests. For example, in 2003, the legislation of Article 23 of the Basic Law failed because of mass demonstrations; in 2012, the Government of the HKSAR's introduction of the independent course "national educa-tion" was shelved due to students and teachers' siege of the central authority offices; in 2014 and 2015, the fierce fight over the universal suffrage of the chief executive and the "Occupy Central" movement are also strong examples.

It can be alleged that in the foreseeable future, if other elements remain unchanged, due to the diverse controversial political issues, the downward pressure on the governing authority and capacity of the Government of the HKSAR will continue to mount, and political strug-gles will hover over the HKSAR for a long time; the pressing economic, social, and livelihood issues that haven't received due attention or effec-tive solution will also seriously hamper the SAR administration, adding to the people's decreased public confidence, worries about the future of Hong Kong, and resentment against the government.The experience after the return shows that though the Government of the HKSAR has improved its reputation from some achievements in the economy and people's livelihood, it soon "lost points" in public opinion when it came to addressing political issues. The continuous improvement of gover-nance authority is no easy task.

The Dispersion and Shortage of Talents for the Governance of Hong Kong

The successful implementation of the "one country, two systems" policy requires some supporting conditions, the most vital of which is the adequate supply of political talent for the "governance of Hong Kong". Since the policy was a result of special interactions between the international and domestic circumstances and the situation in Hong Kong, which are ever-changing, the necessary political talents for "Hong Kong people governing Hong Kong" must be equipped with historical and international vision, strategic and long-term insights, rational, pragmatic and flexible ideas, broad mindsets and the capability to gather diverse strengths, as well as the resilience to endure bitterness and carry out important missions, so as to find a strategic position for Hong Kong's existence and development in the changing world. These people can also update Hong Kong's role in China's development in order to ensure Hong Kong's long-lasting value for China. I think of the literati of Qing dynasty and Chen Danran's (陳澹然) precise description of political talents with a clear picture of the situation: "Those who don't consider the long term are incapable of thinking in the short term; those who don't consider the whole situation are incapable of considering a single isolated aspect." This description can be seen as the requirement and expectation for the Hong Kong people to govern Hong Kong.

In addition, since "one country, two systems" is an historic pioneering undertaking carried out in the face of distrust among diverse forces, many people including people outside Hong Kong are looking forward to seeing it fail, which shows the hardships and complexities with regard to implementing "one country, two systems". Thus, the policy places extremely high demands on the Hong Kong people governing Hong Kong in terms of their wisdom, knowledge, experience, ability, and understanding of China and the world. Furthermore, in addition to the adequacy of individuals, the cohesion of talent is also imperative. They must be highly disciplined leaders with one heart and one mind to produce forceful and efficient governance. It is therefore a demanding task to firmly control Hong Kong's political situation and roll out effective governance. Ironically, when Hong Kong was governed by the British, there was no need for local political talent. On the return to China, a thirst for such talent, high-level talent in particular, suddenly burst forth. However, it is almost impossible to produce a large number of political professionals within such a short time under any conditions, without mentioning specific barriers for talent cultivation.

Ideally, qualified Hong Kong administrators should possess both talent and integrity. Integrity refers to not only their general virtues and morals, but to be exact, their willingness to accept the CPC as the ruling party in China and to be loyal to the PRC, their willingness embrace the concept of state and nationalism, their love for Hong Kong, their dedication to protecting Hong Kong's "one system", and their readiness to strive for the well-being of China and Hong Kong. "Talent" refers to their strength to implement "one country, two systems" in Hong Kong and to gain the trust of the central authority, Hong Kong people and the international community. However, such versatile political elites are rather scarce in Hong Kong.

Generally, when a new country or a country that has undergone fundamental changes in its state system or regime, such as revolutions or foreign invasion, it is therefore forced to reform its political system, with the result that the new regime won't be the continuation or the legal successor of the old one, and the new leaders won't be those of the former regime either. Major personnel changes are only natural in such situations. For example, a country independent from the former colonial regime is most often taken over by the "anti-colonisation" or independence movement leaders.[29] After the Second World War, Germany's and Japan's state systems and regimes were changed by the victorious

---

29 Rupert Emerson (1960). *From Empire to Nation: The Rise to Self-Assertion of Asian and African Peoples. Cambridge*, MA: Harvard University Press; Martin Shipway (2008). *Decolonization and its Impact: A Comparative Approach to the End of the Colonial Empires.* Malden: Blackwell.

nations and taken charge of by the anti-Nazis and the anti-milita-rists respectively.[30] After the CPC won the Civil War, it laid off the important officials of the KMT Government and relied heavily on its party members and political alliances to lead the new Chinese Govern-ment.[31] Similar cases are also pervasive in other countries.

The differences between Hong Kong and other colonies prevented it from striving for independence. As the British governance in Hong Kong wouldn't be challenged by "anti-colonisation" or "independent" movements, they didn't keenly cultivate local political elites, but instead feared the rise of local political leaders, particularly those who didn't completely accept colonial governance, including the middle-class elites once edified by Western political values. The colonial government was even more afraid of the Hong Kong leftists who have always supported the CPC, and endeavoured to prevent, curb, suppress, defame and isolate them by any means necessary. Meanwhile, the anti-communist attitude in Hong Kong is always a barrier for promoting the traditional nationalism. The leftists who are the mainstay of traditional nation-alism, however, suffered increased repression in Hong Kong after the Hong Kong 1976 Leftist Riots, which also makes it more difficult for the political force most reliable for the Chinese Government to under-take important governing positions in Hong Kong after its return.

In the transitional period, the British, on the one hand, refused to cooperate with the Chinese Government with regard to cultivation of excellent administrators locally, and by the same token, purpose-fully introduced their chosen Chinese elites to the political stage, the most outstanding of which were the Chinese civil servants. The British intended to establish a state of affairs and force the Chinese Govern-

---

30　Giles Macdonough (2007). *After the Reich: The Brutal History of the Allied Occupation.* New York: Basic Books; Frederick Taylor (2011). *Exorcising Hitler: The Occupation and Denazification of Germany.* New York: Bloomsbury Press; Thomas U. Berger (2012). *War, Guilt, and World Politics after World War II.* New York: Cambridge University Press; Theodore Cohen (1987). *Remaking Japan: The American Occupation as New Deal.* New York: Free Press; Michael Schaller (1987). *The American Occupation of Japan: The Origins of the Cold War in Asia.* New York: Oxford University Press.

31　Suzanne Pepper (1999). *Civil War in China: The Political Struggle 1945-1949.* Lanham: Rowman & Littlefield. P385-422.

ment to "accept" those chosen elites as the leaders to administrate Hong Kong in the future. For the Chinese Government, though the elites already cultivated by the British were surely employable, they were found indifferent or even resistant to the CPC and the PRC. It was also impractical to ask these elites to always care about the interests, security and feelings of the country and the regime. In brief, the Chinese Government could "temporarily" accept them as the administrators in Hong Kong as a practical and expedient method, but they were not at ease with this arrangement. A heart-to-heart relationship of mutual trust was hardly possible. Therefore, the cultivation of political talents who are "people who love China and Hong Kong" remains an arduous strategic task in the long run.

As a result, in spite of the fundamental changes in Hong Kong's political identity, constitutional status, political system and legally constituted authority after the return in 1997 from the theoretical and legal perspectives, there was an obvious continuity in the SAR regime personnel. The handover from the British naturally led to the disintegration of the former regime constituted by the British colonisers and their political elite "collaborators", and of the governance coalition set up by the regime, but the senior Chinese civil servants and many Chinese political elites still remain to govern the SAR Government. Indeed, a minority of the "pro-British" Chinese political elites had already "transformed" into "pro-Chinese" politicians before the return, but the remaining majority didn't undergo such "bleaching" or "transition", particularly the senior Chinese civil servants.

The traditional and veteran compatriots who are "people who love China and Hong Kong", on the contrary, only had a limited share and a limited role in the SAR regime. Evident changes didn't occur until Leung Chun-ying (梁振英) took office, but then more serious division and internal friction among the "people who love China and Hong Kong" ensued. The traditional "people who love China and Hong Kong" mainly refers to the leftists, who began to support and follow the CPC before the PRC was founded, while the "veteran supporters" refers to those who became the pro-Chinese Government soon after the topic of Hong Kong's future arose in the early 1980s, some of whom used to be "collaborators" of the colonial government. The definition of "people who love China and Hong Kong" has been expanded after the return and now incorporates all those willing to accept Hong Kong's return to China and to cooperate with the Chinese Government.

Without a doubt, the traditional "compatriots" are most loyal to the CPC and theoretically the most suitable administrators to exercise the high degree of autonomy under "one country, two systems". Unfortunately they are generally considered as the CPC's satellites and distrusted in Hong Kong. The Hong Kong 1967 leftist riots consolidated Hong Kong's enmity and hostility toward the leftists and the political conflicts and estrangement have yet to be settled. The leftists are subject to constant suppression and exclusion, and prejudice and contempt from the mainstream social elites. As the anti-communist attitude prevails, the camp of "people who love China and Hong Kong" is isolated. It has become a social community separated from the mainstream society and gradually forming its own organisational structure, personal network, economic activities, values, educational inheritance and polity. But constrained by a wide array of unfavourable conditions, the camp has limited space for development and is hard-pressed to attract and cultivate talents, particularly those respected and acknowledged by the ordinary Hong Kong people, let alone the high-level talents capable of governing this international metropolis and winning the confidence of the international community.

After the settlement of the issue of Hong Kong's future, the ideal arrangement would have been that China and the U.K. worked together to cultivate talents for "Hong Kong governance" after the return and assigned them to key positions with a willingness of each to accept the individuals selected by the other. But later practice deviated from the arrangement, with both parties selecting and cultivating people loyal to their own interests. The pro-Chinese elites and the pro-British elites confronted each other over the China-U.K. political conflict, creating a political gap with many personal conflicts and resentment among the pro-British "mainstream" elites (the "old pro-establishment camp"), the opposition (mainly the democrats) and the patriotic force (the "new pro-establishment camp", including the traditional patriots and the newcomers after the return). The estrangement between the "main-stream" elites and the opposition will be difficult to bridge. Prior to the return, both the "mainstream" elites and the opposition supported the colonial government, made up the general "pan-establishment camp" and reluctantly sustained political coordination under British rule. After the return, the formerly pro-British "mainstream" elites and the tradi-tional "pro-China patriotic force" constituted the general new "pan-es-tablishment camp", but the differences and conflicts between them remained difficult to resolve.

Without the coordination and support of the British, the Chinese Government's one-sided cultivation of talents for "Hong Kong governance" ran into a large number of difficulties and made limited achievements. The main reasons were nothing more than the Hong Kong people's distrust of the "pro-China" people, obstructions from the British and many Hong Kong mainstream elites' resistance toward the CPC. If the U.K. insisted on fighting with China until the end, the political talents produced by China would have to undertake the respon-sibility to "govern Hong Kong", which was definitely not the original intention of the national leaders, including Deng Xiaoping.

Since the traditional and veteran "people who love China and Hong Kong" and the political elites cultivated by the British all had their own weaknesses in the eyes of the Chinese Government, it certainly wanted to independently foster a group of Hong Kong people for the future "Hong Kong governance".

Deng Xiaoping was greatly concerned about the shortage of talents in this realm. When discussing the basic policies and guide-lines for settling Hong Kong's future on the enlarged conference of the Central Politburo of the Communist Party of China on April 22nd, 1983, Deng Xiaoping for the first time mentioned "preventing turmoil in Hong Kong" and "ensuring its prosperity, stability and smooth handover" to "create conditions" for the Hong Kong people to "gradually participate in the governance" in the "transitional period", and talked about the talent cultivation for "Hong Kong people governing Hong Kong" after Hong Kong's return in 1997. Deng pointed out that "Hong Kong people should take part in the governance in all sectors step by step, such as law, diplomacy and particularly politics and economy. [...] The Hong Kong patriots should progressively participate in the administration and enter the legislative, judicial and executive sectors. However, candidates should certainly start with the major capitalists, heads of universities, the excellent intellectuals and the ones these people recommend. [...] The current task is to come up with measures to nurture cadres, and to gradually enlarge their presence in governance. Only the Hong Kong people can administrate Hong Kong, not those from the Hong Kong and Macao Work Committee. These people, though mostly Chinese, still operate in the British style to some extent. We should also think about how to train the young by carrying out specific work. The Hong Kong capitalists have also recommended some young talents. I mentioned before that the Hong Kong and Macao Work Committee intends to establish some associations in Hong Kong, which are actually political parties. The British also established some similar associations. In response, we should do the same in order to train a group of politicians. Such work is imperative and should be carried out as soon as possible. [...] It must be noted that these governing personnel shouldn't be sent by Beijing, since they would find no way into the core of Hong Kong governance. People that can find a footing and have profound understanding of Hong Kong must be those recommended by the prominent figures. We must concretely implement this task."[32] Deng also added that "Two years ago (1982), we planned to organise some associations in Hong Kong dedicated to scouting for talents, which is impossible without such platforms. These groups can also cultivate talents. Then why not adopt this method? But we won't establish political parties; instead we establish associations such as work committees, colloquia, and academies, or hold relevant activities like dinner parties."[33]

---

32 Qi Pengfei (2004). *Deng Xiaoping and the Return of Hong Kong* (《鄧小平與香港回歸》). Beijing: Huaxia Publishing House. P189.

33 Ibid, P191.

In 1984, Deng Xiaoping reiterated the importance of political talent cultivation in the transitional period: "It is important for two purposes: to maintain stability and prosperity for 13 years, and to pave the way for Hong Kong people to govern Hong Kong for the next 13 years, so as to achieve a smooth return. Otherwise, an abrupt change on July 1st, 1997 will absolutely trigger chaos. The key to the smooth return is Hong Kong people's participation. [...] I heard that the British are planning for a representative system and looking for representatives for political and legal affairs. But who are these representatives? Who are their successors? Do they adhere to and safeguard China's sovereignty? Can they really maintain a stable and prosperous Hong Kong? So we cannot listen to the British. Hong Kong should have its own plan, for which the central authority should offer its help. [...] Its help is also seen as its participation in Hong Kong's planning. The central authority must grasp the overall picture, and decide how to select the participants during a state of disunity. Who is to be selected? To what extent should the participation be? And in what aspects? These are the questions to be answered from now on, and to be discussed with the British. The focus is not the sovereignty, but a better arrangement for the next 13 years with regard to maintaining stability and to realizing a smooth handover."[34] Deng even demanded, "Four years from now, suitable people should be selected for these preparatory tasks. People in their forties and fifties should be chosen for direct governance, as well as some relatively young people for better succession. Good participation goes hand-in-hand with smooth succession."[35]

For Deng, the British were apparently untrustworthy in the cultivation of governance professionals. When he met with the then British Foreign Secretary, Geoffrey Howe, on July 31st, 1984, Deng issued a clear-cut warning to the U.K.: "I hope the British Hong Kong Government won't set up a government on its own and force it on the Hong Kong Special Administrative Region in the future."[36]

---

34 Ibid, P191.
35 Ibid, P194.
36 Deng Xiaoping (1993). *How Deng Xiaoping Views Hong Kong Issue* (《鄧小平論香港問題》).

Deng's wisdom and strategies on key issues can be seen from his statements about the talent cultivation for Hong Kong governance, but his good intentions were hindered by the existing conditions. The British insisted on independently establishing a government and importuned the Chinese Government to accept the political elites they chose and trained. They also refused to include or train the politicians selected by China in the colonial governance framework. An evident example is how the British ignored China's demand to appoint a Lieutenant Governor of Hong Kong.[37] Another example is that the British refused to cooperate with the Preparatory Committee for the HKSAR and prohibited the colonial government officials' contact with the Committee. Moreover, the British basically held a negative view towards those whom China was going to nurture and attempted to constrain the enhancement of their reputation. Some of them might have even fallen prey to political "plots". The U.K.'s refusal to cooperate is understandable. Besides a lack of confidence in and antipathy toward the pro-Chinese people, the British were reluctant to allow those talents to learn the deployments of the U.K., hamper the U.K.'s profit-making activities, and complicate the governance in the transition period, or let China know the U.K.'s intentions.

The Tiananmen Square Protests in 1989 worsened China-U.K. relations, and the U.K.'s unilateral transformation of Hong Kong's political system made the China-U.K. cooperation on political reform impossible. The negotiated political transition was derailed, and the Chinese Government began to deal with the return issue in an independent manner. With regard to training political talents, no room for cooperation remained on either side.

Hong Kong: Joint Publishing. P10.

37 Lu Ping recalls, "China went to great pains for Hong Kong's smooth transition and once planned to groom an executive leader. As the governor of Hong Kong was certainly a British, China hoped there was a lieutenant governor undertaken by a Chinese citizen of Hong Kong recognised and selected by both sides. And the lieutenant governor would naturally become the first chief executive of the HKSAR after July 1st, 1997. Unfortunately, before we were able to go into detailed discussion with the U.K., the plan was totally rejected by Chris Patten." See Lu Ping (2007) *Lu Ping's Verbal Recollections of the Reunification of Hong Kong* (《魯平口述香港回歸》). Hong Kong: Joint Publishing. P67-68.

Given the hostile relationship between China and the U.K., the British accelerated its plan to cultivate talents, with a focus on turning senior Chinese civil servants into political talents, thus altering, or destroying, to be exact, the "political neutrality of civil servants" it had long touted. As many Hong Kong citizens lacked trust and respect for politicians but had confidence in civil servants' abilities and integrity, British indoctrination of civil servants, although never implemented during the U.K.'s retreats from other colonies, was undertaken with success in the special environment of Hong Kong.

In contrast, Hong Kong people's confidence in the Chinese Government and Hong Kong's future was severely undermined by the Tiananmen Square Protests in 1989, and more unluckily, politicians close to the Chinese Government also lost their favour and support. The situation undermined all achievements in China's talent cultivation and held back relevant work into the future. Despite all of the hindrances, however, the political talents China nurtured on its own when it "started all over again", especially the Hong Kong affairs advisors, members of the Preliminary Working Committee and the Preparatory Committee for the HKSAR, and the district affairs advisors assigned by China, would have probably needed to step forth to undertake the important mission of governing Hong Kong if the U.K. had insisted on confronting China. But these people were undeniably unqualified in experience and capability. If they had to shoulder that task, they wouldn't have been able to deliver exemplary performance in the face of Hong Kong people's distrust. In that case, the last situation Deng Xiaoping wanted to see would probably have emerged after Hong Kong's return, which would unavoidably threaten Hong Kong's prosperity and stability.

However, as a formidable political veteran, the U.K. took a U-turn in its China policy, as it realised that the CPC wouldn't collapse along with the Communists in the Soviet Union and the Eastern Europe, but instead keep to the national policy of Reform and Opening up in the increasingly strengthening regime. Except in the field of political reform, the U.K. was cooperative in all aspects, and China-U.K. relations improved significantly. As for the more specific issue of "Hong

Kong people governing Hong Kong", the Chinese Government, short of alternatives, agreed that the British-nurtured political elites would become the main body of Hong Kong people's self-governance. All the major officials, except for the Secretary for Justice Elsie Leung Oi-sie and the first chief executive Tung Chee-hwa, were formerly senior Chinese civil servants during the period of colonisation, and the rest of the political framework after the return (the Legislative Council, the Executive Council, judicial organs, legal organs and consultative organisations) was also led by the elites of the former colonial regime. The traditional and veteran "compatriots for both China and Hong Kong" were allocated only a limited share of power in the new regime, and they harboured deep bitterness and disappointment. The two Hong Kong forces governing Hong Kong (elites of the colonial regime, and the traditional and veteran "compatriots for both China and Hong Kong") fought fiercely, with resulting hostility not only between the two camps, but also among individuals.

Besides the internal division between political elites, some problems arose with the less important labour and grassroots representatives of the Hong Kong people's governance of Hong Kong after its return. In spite of their acknowledgement of the "one country, two systems" policy and the CPC's leadership, in the face of an increasingly acute class rift, they encountered a gap too wide to bridge, with the interests and the stances of representatives and mainstream elites widely separated, particularly along the lines of Hong Kong society, economy and people's livelihood. The frequent conflicts and disputes among the representatives from different social classes of the political forces that governed Hong Kong had so far seriously crippled the unity of the camp of "compatriots for both China and Hong Kong", adding to the difficulties in the governance in the SAR Government.

Additionally, due to the absence of a political hierarchy among the Hong Kong administrators and the scarcity of renowned authoritative Hong Kong politicians, leaders of all parties "shared the same status" and were "unwilling to be subordinated to others". As a result, it was hardly possible for the Hong Kong people governing Hong Kong headed by a "widely supported" leader to gather together, carry out coordinated responses and form into a forceful power. It was difficult to change the loose and fragmented organisation of the Hong Kong people governing Hong Kong.

It can be concluded that the major reason for the post-return challenges in Hong Kong's governance and development was not the rampant opposition faction, but the general force "loving both China and Hong Kong" possessing the power of governance and profound economic resources but without cohesion and a unified leadership. In the early days after the return, the central authority pursued a "non-intervention" policy and constrained its role in cultivating, unifying, and leading the force of "compatriots for both China and Hong Kong". After the large-scale Hong Kong demonstrations against Article 23 of the Basic Law in 2003, the central authority has strengthened its participation in Hong Kong affairs, while still lacking the resolve to establish a "governing coalition" with "compatriots for both China and Hong Kong" at its core. It has preferred the "reliable" political conservatives for the united front, refusing to make room for more talents. In terms of maintaining a self-disciplined camp of "compatriots for both China and Hong Kong", the central authority has been inefficient and unable to fully utilise the existing resources for rewards and punishments, thus failing to harness the scattered force of the "compatriots for both China and Hong Kong".

In general, over the 20 years since the return, the cultivation and organisation of political talents who constitute "people who love China and Hong Kong" has been stumbling forward without breakthroughs. There is still a long way to go before Hong Kong is governed by people possessing both "ability and integrity".

Challenges from the Opposition to "One Country, Two Systems"

The "one country, two systems" policy encountered numerous difficulties and challenges after Hong Kong's return. The most evident reason has been the obstruction, resistance, defamation and interruption from various opposing forces. Prior to the return, the opposing forces had grown substantially with the help of the colonial government. They became, to some extent, "allies" in the transitional period, glossing over the previous mutual suspicion, constituting an "anti-communist unholy alliance", and jointly resisting the Chinese Government's "intervention" in Hong Kong affairs. After the return, the opposition has continued to summon political capital with "anti-communist" and "communist-fearing" sentiments obstinately pervading Hong Kong society.

The opposition in Hong Kong is a fusion of anti-communists, democrats, pro-Western forces, civil societies and pressure groups striving for social reforms, religious organisations (Catholic and Christian in particular), as well as people from the fields of law, education, social welfare, media, the college and middle school students and so on. Their major supporters are the relatively well-educated middle-class professionals and administrators.

There are three common points for their opposition. Firstly, they refuse to accept the CPC as the ruling party in China and desire to change it by any means necessary. Secondly, they oppose Hong Kong's return to China and only show half-hearted belief in and support for "one country, two systems" as a solution for "Hong Kong's future". Thirdly, they deny the political legality and legitimacy of the SAR Chief Executive, whom they think is not authorised by the Hong Kong people since the executive is not elected through universal suffrage.

These different groups of people meanwhile have different priorities in their goals and modes of action. The most prominent goals include ending the CPC's "single-party regime", leading China to a pro-Western "peaceful evolution" that imitates the Western development model, propelling Hong Kong to become an "independent political entity" through democratic reform, gaining the maximum degree of autonomy from "one country, two systems", seizing the HKSAR regime, reforming social and economic systems and conditions, enhancing equality, justice, human rights and freedom in China, promoting the guidelines and principles of "nativism", and advocating "Hong Kong independence", to name but a few.

The opposition also lacks definition, with political activities ranging from District Council elections and fights on the platform of the Councils to campaigns utilising the linkage between the inside and outside forces of the Councils, from peaceful movements of collective resistance (demonstrations, protests, sit-ins, press conferences, petitions), to fierce ones (public occupations, violent confrontations with law enforcement personnel, violence against political enemies, traffic blockage); they have also extended their activities to online hacking; collusion with foreign political forces; endeavouring to win support from Western media; and verbal or behavioural violence in the Councils. A portion of the opposing forces carry out actions obstructive for the administration including: utilising the rules of procedure of the Legislative Council for filibusters or taking advantage of judicial review or legal actions to restrict the government; adopting Western values as Hong Kong's core values while disregarding the equally important traditional Chinese values; initiating public opinion warfare; taking advantage of opportunities to incite disputes between Hong Kong people and the central authority as well as the Government of the HKSAR; pushing for "alternative interpretations" of "one country, two systems" and the Basic Law to enhance their own control over them; using the issues relevant to human rights, rule of law, and freedom in the mainland to stir up Hong Kong people's dissatisfaction with the central authority; intensifying the populist and welfarist demands in Hong Kong; politicising economic, social and livelihood issues; spreading the message of "Hong Kong people first" and promoting the exclusion of mainland compatriots by opposing the fusion of mainland and Hong Kong; forging a "separatist" identity of "new" Hong Kong people; advocating, rewriting and rediscovering Hong Kong history in a Hong Kong-centred fashion; promoting the idea that Hong Kong people are a unique "nation" or "ethnic group" distinguished from the Chinese nation; rejecting all kinds of national education; opposing exchanges of young Hong Kong people to the mainland and the learning and speaking of Mandarin; carrying out promotions in schools; motivating the young to be active in political campaigns; involving associations and other groups in political disputes and so on.

In terms of social support, the opposition has the upper hand in terms of public opinion, media, schools and the general public. Compared to the camp of "people who love China and Hong Kong", the opposition is better acknowledged by the middle class, intellectuals, and the young, so they are better organised and more forceful campaign fighters. They also have stronger voice in interpreting "one country, two systems", which has led to many public misunderstandings about the policy and the Basic Law, and particularly the central authority's power and responsibility in the HKSAR. The opposition's efforts to ignite conflicts between Hong Kong and the central authority, on the contrary, have not gone as well as anticipated. Although the opposing forces have successfully aroused Hong Kong people's antipathy against the central authority, the general public have less and less desire for confrontation, which they believe will do harm to themselves, and thus in turn resist the extreme confrontations advocated by the opposition.

Nevertheless, the opposition's aggressive speech and actions have unavoidably brought about a series of negative effects upon Hong Kong including political divergence and confrontation, a chaotic and unstable political landscape, ineffective governance of the SAR Government, opposition between the executive and the legislature and difficulty in making or implementing major policies. The negative effects also include: waste of public money caused by administrative delay, poor morale among civil servants, public anxieties, worsened business and living environments, declining competitiveness in the international economy, failure to recognise opportunities brought along by national development, constant social conflicts, inefficient administration and operations, worries about Hong Kong's prospects, a declining impression of Hong Kong within the international community, and complaints from mainland compatriots about Hong Kong people. More seriously, due to the impact from the opposition, the "one country, two systems" policy and the Basic Law haven't been implemented as planned, most revealingly evidenced by a lack of acknowledgement and respect for the powers of the central authorities on the part of Hong Kong people, judicial trespassing, and the central authority's losing its grip on the power of interpretation of "one country, two systems" and the Basic Law. Even though the opposing forces cannot justifiably assume power, they are potent in influencing or obstructing the administration of the Government of the HKSAR.

Seen from another perspective, the continued unbridled confrontations of the opposition and their supporters with the central authority indirectly reflect their confidence that the central authority won't respond with severity. The apparent "communist-phobic" attitude among the older generation in Hong Kong makes this older generation afraid of being dragged into political whirlpools or suffering as a consequence of offending the CPC; as a result, they avoid any activities that challenge the CPC, dare not make public anti-communist speeches, and constantly warn the younger generation to "steer clear" of the CPC. In contrast, the opposing forces and the young in particular obviously don't share the same apprehensions and continue to assail the CPC's authority while fomenting fear among the Hong Kong people. But it is exactly their confrontational speech and behaviours that indicate that they feel sure that the CPC is now much more tolerant and open and will put up with them. They also think the CPC won't "take harsh measures" and crack down in response to public utterances because of the opinion of the international community and because Hong Kong is too valuable to the central authority. More importantly, the opposition and their Hong Kong supporters believe to a certain extent that the central authority is quite sincere in implementing "one country, two systems", "Hong Kong people governing Hong Kong", and a high degree of autonomy, and thus will never suppress them for their rebellion, lest various sectors should lose confidence in the central authority's policy.

Therefore, the opposition's confident and fearless attitude is an indirect reflection of the successful implementation of "one country, two systems" and the mitigation of "communist-phobic" sentiments after the return. There are even Hong Kong people who feel that the central authority has brought Hong Kong into its favour, constantly boosted Hong Kong's development, and offered immediate help when difficulties have arisen.

Before the return, some of my political acquaintances thought the opposing forces would restrain themselves after the return due to the "communist-phobic" atmosphere and the disappearance of the British buffer, and thus would not be much of a threat. Seeing the unexpected rise of the opposition's political dynamism and momentum after Hong Kong's return, the central officials admitted to having underestimated the post-return political landscape in Hong Kong, because of which there was no efficient precaution to cope with the opposing forces.

Constant Conflicts over Major Political Issues

The constant political conflicts in Hong Kong after its return have been triggered by some major political issues.[38] The reason why various opposing forces have been able to mobilise the masses, to whip up conflicts between Hong Kong people and the central authority as well as the Government of the HKSAR, and to accumulate political capital is that long-standing political issues have been harnessed in order to expand the opposition's political ammunition.

As I have mentioned above, "one country, two systems" and the Basic Law are both concessionary by nature. One evident concessionary behaviour is to shelve a dispute unsolvable in the short term, allowing for the coexistence of diverse interpretations of the issue while post-

---

38 See Hao Tiechuan (2013). *Review on Disputes about the Basic Law of Hong Kong* (《香港基本法爭議問題評述》). Hong Kong: Chung Hwa Book Company.

poning a settlement until the proper moment. One of the three aspects of compromise put forward by mainland legal academic Qiang Shigong (強世功) is related to "residual power". Should "residual power" be vested in the central authority or the HKSAR? "Due to disputes on the subject, the Basic Law doesn't clearly stipulate the location of the 'residual power'".[39] The second type of compromise rests with the ties between the Constitution and the Basic Law and involves the effects of the Constitution in Hong Kong. "Due to the relevant divides, the Basic Law doesn't specify the issue. But its annex clearly enumerates the mainland laws applicable to Hong Kong, such as the nationality law and the national flag law, but certainly not the Constitution, which is thus considered invalid in Hong Kong by some Hong Kong people."[40] The third concerns the interpretative power of the Basic Law. "The Hong Kong members of the Basic Law Drafting Committee contended that the Basic Law should follow Hong Kong's traditional common law system and that the legal interpretative power should be vested in Hong Kong courts, while the mainland members contended the Law should adhere to the mainland traditions and that as a law legislated by the NPC, the Basic Law should conform to the Constitution and its interpretative power should be vested in the NPC Standing Committee. Both sides compromised in the end by establishing a complicated interpretative mechanism for the Basic Law. Within this mechanism, the NPC Standing Committee has the power to interpret the Basic Law but authorises the courts of the SAR to interpret on their own the provisions within the sphere of Hong Kong's autonomy. Further, the provisions outside its autonomy should be submitted to the NPC Standing Committee for interpretation. The original aim was a sound interaction between the 'Two Systems' with regard to legislation, judiciary, interpretation and judgment. After the return, however, the Hong Kong legal circle tried to deny the NPC Standing Committee's power of interpretation, instigating the long-lasting dispute over the NPC Standing

---

39 Qiang Shigong (2014). *Hong Kong, China: the Political and Cultural Visions* (《中國香港：政治與文化的視野》). Beijing: Joint Publishing Company. P250.

40 Ibid, P250-251.

Committee's interpretations of the Basic Law."[41]

The major political issues since Hong Kong's return involve the division of power between the central authority and the SAR; the allocation of power between the executive and the Legislative Council; the power relations between the NPC, the NPC Standing Committee, and the Hong Kong Court of Final Appeal; the methods for elections, particularly for selecting the chief executive through universal suffrage; the right of abode, and so on. The opposition, along with a portion of the legal circle, makes every effort to define and settle the political issues above by using "Hong Kong as an independent political entity" as a starting point, while outwardly acknowledging China's sovereignty in Hong Kong. Their overall goal is to deny, weaken, and break away from the central authority's governance, and to enhance the depth and breadth of Hong Kong's "high-degree of autonomy". Their understanding and interpretations, as compared with the central authority's, are inconsistent and sometimes even poles apart. However, in accordance with the "non-intervention" policy after the return, the central authority has seldom expressed its stance on those issues or established authoritative decisions or pronouncements as the final say. As a result, the "alternative interpretations" by the opposition and some people from the legal circle have gained influence over time, leading to misunderstandings and misinterpretations of the "one country, two systems" policy and the Basic Law among many Hong Kong people, which prevents the "one country, two systems" from being implemented in full accordance with the Chinese Government's blueprint.

Below is my brief commentary on several relatively important political issues:

---

41 Ibid, P250-253; the third point (about the power of interpretation of the Basic Law) refers to Li Hou (1997). *The Return of Hong Kong* (《回歸的歷程》). Hong Kong: Joint Publishing. P155.

1. Political Reform[42]

"Political reform" refers to gradually transforming the method of electing the chief executive and Legislative Council into one of universal suffrage. The opposition and many other Hong Kong people demand a relatively rapid reform, while the central authority and the camp of "compatriots for both China and Hong Kong" prefer a steady one. This is not a difference in principle and compromises are possible.

However, behind the many proposals for the specific elective methods are two entirely different ideas. One starts with the general principles or beliefs and firstly sets requirements for the universal suffrage, based on which, the method for the universal suffrage is to be designed. Although these requirements originate in Western countries, or Western political ideas and theories, they are regarded or touted as "international criteria" or "universal values". This idea can be best represented by the slogans among the general public such as "one man, one vote, and one value" and "universal and equal right to nominate, to elect and to be elected". The electoral system and procedure are paramount for this idea and the result doesn't matter much since all the election results are "certainly" fair and reasonable. Those who support this idea claim that elected officials don't have to be among the opposition figures, but they stress there must be adequate and meaningful choices for Hong Kong people.

The other idea is built on the "one country, two systems" policy's strategic goals set and elaborated on by Deng Xiaoping. It focuses on how universal suffrage as a goal should be compatible with other goals of the "one country, two systems" policy, and how universal suffrage as the tool should serve the implementation of the policy. Supporters of this idea emphasise that the main goals of the policy include maintaining Hong Kong's prosperity and stability, its original capitalism, investors' interests, and Hong Kong's economic value for China; enhancing the

---

42 This sector mainly refers to Lau Siu-kai. (2015). Disputes on Political Reforms and the "Battle" between Two Interpretations of "One Country, Two Systems" ("政改爭議及兩種 '一國兩制' 理解的 '對決' "). *Hong Kong & Macao Studies* (《港澳研究》). 7(2). P19-28.

sound relationship between Hong Kong and the central authority as well as the mainland; and preventing Hong Kong from becoming a base for "subversive activities". From this stance and starting point, the democratisation of the Hong Kong political system must be approached not in dissociation or isolation but in a prudent and proper way based on Hong Kong's historical background, the central authority's policy towards Hong Kong and the current state of Hong Kong. According to this idea, the method for selecting the chief executive and the Legislative Council must be in the interests of, or at least should not obstruct, achieving the major goals of the "one country, two systems" policy.

The touch-and-go conflicts between both sides regarding "political reforms" are mainly caused by the fundamental differences in their basic political stands, with the result that their interpretations of the "one country, two systems" policy is divergent and irreconcilable. The opposition's interpretation, regarding Hong Kong as an "independent political entity", has always prevailed in Hong Kong's public opinions over the other interpretation of the policy which was formulated by Deng Xiaoping and long contended by the central authority, the core of which is "the high degree of autonomy authorised by the central authority". It's not that the opposition objects absolutely to everything about the policy, but their stance on some core issues in understanding the policy (particularly the policy goals, Hong Kong's political system, the power relations between the central authority and the SAR, and national security) are a far cry from the central authority's.

The core of the "one country, two systems" policy of the central authority is actually "a high degree of autonomy authorised by the central authority" with several crucial political principles: China has sovereignty over Hong Kong; the central authority has the right to all-round governance in Hong Kong; Hong Kong's high degree of autonomy is authorised by the central authority; the "residual power" doesn't belong to Hong Kong; a high degree of autonomy doesn't mean absolute autonomy; The powers to direct and decide on Hong Kong's political system belong to the central authority; the political legitimacy of the Government of the HKSAR comes from the Basic Law and the central authority's appointments; the power of interpretation of the Basic Law is vested in the NPC Standing Committee; the central authority bears the power and responsibility to ensure the comprehensive and correct implementation of the "one country, two systems" policy in accordance with the policy; Hong Kong should not become a base for "subversive activities" and so on.

Before the return, the central leaders, officials, and mainland scholars tirelessly reiterated those principles, and many people in Hong Kong also had a general understanding of them. However, under the guidelines of "no intervention" and "no governance is good governance" (不管便是管好), there were few chances for Hong Kong people to hear the viewpoint of the central authority and the mainland. Faced with the opposition's constant denials and distortion of those principles, the central authority and the mainland refuted them from time to time, albeit rather feebly and ineffectively. The central authority gradually lost influence on public understanding of the "one country, two systems" policy over time, resulting in serious consequences. It has led to deviation from implementing the policy, undermined the HKSAR Government's reputation and governance capability, and now and then sparked political conflicts between the central authority and Hong Kong people. The more evident negative result is that the younger generation is only exposed to the opposition's interpretation of the policy without an understanding of its history.

Although the opposition outwardly accepts the policy and "respects" the central authority's power, its members actually deliberately misinterpret the Basic Law and deny the central authority's authority, power and functions. They believe that the central authority, though powerful, will execute its power in a prudent and restrained way, and no execution would be more preferable. They maintain a constant effort to restrict the central authority's power and functions and stir up conflicts between Hong Kong people and the central authority, with the aim of achieving full or absolute autonomy in Hong Kong and gradually transforming Hong Kong into a real "independent political entity" though without the nominal title.

The opposition holds that the major goal of the policy is to maintain Hong Kong people's confidence in Hong Kong and to serve the interests of Hong Kong. Thus, they don't take Hong Kong's responsibility toward and duties to China seriously. In line with this idea, Hong Kong only undertakes limited responsibility for the protection of national security, and should place a higher priority to the effective and concrete safeguarding of Hong Kong's human rights and freedom when fulfilling such responsibility and duties. The opposition perceives that the power of the Government of HKSAR comes from the Hong Kong people; that the political system, the electoral system in particular, should be determined by Hong Kong people; that the NPC Standing Committee's interpretations of the Basic Law are inappropriate as they are harmful to the rule of law and judicial independence of Hong Kong; and that as an "independent political entity" with a Western democratic political system and values, Hong Kong may serve as a model for the mainland in order to lead China to peaceful evolution and to become a satellite of the Western Bloc. If Hong Kong is to become an "independent political entity", it is natural to elect the chief executive through universal suffrage in accordance with the Western "universal values" and "democratic principles". When striving for Western-style universal suffrage, some opposing forces are even open to external funds, support, and cooperation.

The opposition's understandings of the "one country, two systems" policy certainly collides with that of the central authority and the people "loving both China and Hong Kong". The opposition has a stronger voice in the "universal suffrage for selecting chief executive" and "one country, two systems", and thus is able to repeatedly incite social protests against the central authority and the Government of the HKSAR. They have, to some extent, damaged Hong Kong people's confidence in "one country, two systems", stirred up conflicts between the central authority and Hong Kong people, harmed the Government of the HKSAR's administrative reputation, and negatively affected Hong Kong's prosperity, stability, and rule of law and order. Nevertheless, with the regime of the HKSAR still remaining in the hands of the "compatriots for both China and Hong Kong", full rein has never been given for constructing an "independent political entity", and thus the opposition must try their utmost to mobilise all the people for "true universal suffrage", so as to realise their long-cherished wish to take over the SAR regime.

2. NPC Standing Committee's Interpretations of the Basic Law

There are debates across Hong Kong about the extent of the power of the NPC Standing Committee in "one country, two systems". Since Hong Kong's return, the NPC Standing Committee's interpretative power and validity, in Hong Kong or in the mainland, when the central authority formulates, revises, or interprets the Basic Law, has always been controversial. The issues involved so far are as follows:

(1) Are there any restrictions on the NPC Standing Committee's power of interpretation?

(2) Does the NPC Standing Committee have the power of interpretation for the provisions within the SAR autonomy when the courts of the SAR are already authorised to interpret them on their own? Already in the consultation stage of the drafting of the Basic Law, some people held that 'the NPC Standing Committee should constrain itself from interpreting the provisions that only involve the internal affairs of the SAR', which is still supported by some scholars.

(3) Who has the right to decide whether a provision should be submitted to the NPC Standing Committee for interpretation? Although the NPC Standing Committee has shown through interpretation that it maintains the right, in the case of Ng Ka-ling, the Hong Kong Court of Final Appeal demonstrated that it considered itself the only party with the right to decide.

(4) Is the submission by the SAR courts the only procedure the NPC Standing Committee may employ to exercise the power of interpretation of the Basic Law? The NPC Standing Committee has shown with the existing interpretations that it can exercise the power

of interpretation through submission by the State Council, the Standing Committee, or the Hong Kong Court of Final Appeal. However, in the case of Lau Kong-yung, the Hong Kong Court of Final Appeal regarded itself as the single rightful party to make the submission.

(5) To what degree are the NPC Standing Committee's interpretations valid? In the case of Chong Fung-yuen, the Court of Final Appeal held that the NPC Standing Committee's interpretations should be taken as similar to the judicial verdicts of the Court of Final Appeal and that both contain legally binding...and non-binding contents... (And the latter are not legally binding for Hong Kong courts, which is apparently not in line with the mainland's view)."[43]

The issues above are yet to be settled. Though they are legal issues in nature, they contain, to a great extent, political elements. With a lack of mutual trust, a "common consensus" is hardly attainable.

According to the Basic Law, the power to interpret the Basic Law is vested in the NPC Standing Committee, who authorises part of the power to Hong Kong courts. But that doesn't mean that the NPC Standing Committee has given up or lost the power to interpret all the provisions of the Basic Law, as clearly illustrated in Article 158 of the Basic Law.[44] Undoubtedly, the NPC Standing Committee's interpreta-

---

43 Dong Likun (2014). *Relationship between the Governance Power of the Central Authorities and the High Degree of Autonomy of Hong Kong Special Administrative Region* (《中央管治權與香港特區高度自治的關係》). Beijing: Law Press. P42-44.

44 Article 158: The power of interpretation of this Law shall be vested in the Standing Committee of the National People's Congress. The Standing Committee of the National People's Congress shall authorise the courts of the Hong Kong Special Administrative Region to interpret on their own, in adjudicating cases, the provisions of this Law which are within the limits of the autonomy of the Region. The courts of the Hong Kong Special Administrative Region may also interpret other provisions of this Law in adjudicating cases. However, if the courts of the Region, in adjudicating cases, need to interpret the provisions of this Law concerning affairs which are the responsibility of the Central People's Government, or concerning the relationship between the Central Authorities and the Region, and if such interpretation will affect the judgments on the cases, the courts of the Region shall, before making their final judgments which are not appealable, seek an interpretation of the relevant provisions from the Standing Committee of the National People's Congress through the Court of Final Appeal of the Region. When the Standing Committee makes an interpretation of the provisions concerned, the courts of the Region, in applying those provisions, shall follow the interpretation of the Standing Committee. However, judgments previously rendered shall not be affected. The Standing Committee of the National People's Congress shall consult its Committee for the Basic Law of the Hong Kong Special Administrative Region before

tion of the Basic Law is an important component of Hong Kong's legal system under the "one country, two systems" policy, and an indispensable means to ensure that Hong Kong implements the policy as planned by the central authority after the return. With the highest authority for interpreting the Basic Law, the NPC Standing Committee is able to define the intent and correct understanding of the provisions in the Basic Law with active or requested interpretations. Through interpretations, it can also settle disputes in Hong Kong about the Basic Law, and rectify Hong Kong courts' misinterpretations of the Basic Law. However, the NPC Standing Committee's interpretations cannot alter final verdicts made by Hong Kong courts and thus won't change Hong Kong Court of Final Appeal's status or power of "final appeal".

However, after the return, Hong Kong has been generally unwilling to ask for the NPC Standing Committee's interpretations. As I see it, the Court of Final Appeal strongly resists the NPC Standing Committee's interpretations, and suspects, disdains and distrusts the NPC Standing Committee. It thinks that the Standing Committee is not qualified in terms of legal knowledge as a supreme authority, and always puts political considerations above the law, leading to unfair and unjust judicial interventions. Judges of the Court of Final Appeal also worry that submission to the NPC Standing Committee for interpretation will damage the confidence of Hong Kong, the international community and particularly foreign legal circles in Hong Kong's judicial system, hindering the reputation-building process of the newly established Hong Kong Court of Final Appeal.

Similarly, many people from Hong Kong's legal circle and the opposition feel unaccustomed to the "statutory interpretation" system in the mainland and regard the NPC Standing Committee as a political institution that serves the CPC regime and suspect and assail its judicial capability and intent. They refuse to acknowledge the NPC Standing Committee's interpretations as a complete component of the Hong Kong legal system, and perceive that the NPC Standing Committee

giving an interpretation of this Law.

has delegated all the power of interpretation of the Basic Law provisions concerning "the high degree of autonomy" to Hong Kong courts and thus loses the power to interpret those provisions. There are even those who make the demand of no interpretation by the NPC Standing Committee on any occasion, in order to avoid harming the rule of law, judicial independence and the confidence in "the high degree of autonomy" of Hong Kong. Quite a few Hong Kong people, even in the Government of the HKSAR, agree with this opinion. Due to these opinions and the fear of public resistance, the Government of the HKSAR is extremely reluctant to ask the central authority for the NPC Standing Committee's interpretations even in the face of cases violating the Basic Law in Hong Kong.[45]

The power of interpretation of the NPC Standing Committee and Hong Kong courts is actually linked to their right of jurisdiction. More and more relevant studies have been made by mainland legal scholars in recent years. They disagree with the negative opinion of the Hong Kong Court of Final Appeal about the NPC Standing Committee's interpretations and criticise the "egoist" Court for not only deliberately derogating the power and status of the NPC Standing Committee, but also hampering the correct implementation of "one country, two systems".

The mainland legal expert Dong Likun makes the criticism that "Since Hong Kong's return, there have been constant forces challenging the NPC Standing Committee's power of interpretation, disparaging the validity of its interpretation, and trying to supplant or dominate the NPC Standing Committee's interpretations of the Basic Law with Hong Kong courts' judicial interpretations, so as to damage the reputation of the Central governance."[46] He further illustrates the criticism by saying, "In the case of Chong Fung-yuen the Hong Kong Court of

---

45 Since the mainstream viewpoints of the opposition and the legal circle in Hong Kong, as well as relevant academic and commentary materials are pervasive, they won't be elaborated here. On the contrary, only recently have the clear and systematic explanations of the mainland opinions been brought up, which will thus be quoted more often in this book.

46 Dong Likun (2014). *Relationship between the Governance Power of the Central Authorities and the High Degree of Autonomy of Hong Kong Special Administrative Region* (《中央管治權與香港特區高度自治權》). Beijing: Law Press. P78.

Final Appeal divided the NPC Standing Committee's interpretations into legally binding ones and non-binding ones, in order to constrain the validity of the NPC Standing Committee's interpretation. Such conduct is absolutely incorrect."[47]Dong further expounds upon the "subordinating" relationship between NPC Standing Committee and the Hong Kong Court of Final Appeal: "When the Court of Final Appeal cites a provision interpreted by the NPC Standing Committee, and the Court must adhere to the NPC Standing Committee's interpretation. The Basic Law doesn't stipulate that Hong Kong courts have the right to supervise the NPC Standing Committee's interpretation....The interpretation of the NPC Standing Committee and that of Hong Kong courts are not of equal nature and the latter is subject to the former's supervision."[48]

Dong advises Hong Kong courts, "There is something I want to clearly point out. If Hong Kong courts seek loopholes in the NPC Standing Committee's interpretations in order to claim new power in making verdicts as some have advocated, the NPC Standing Committee will make new interpretations to limit the courts' power. These are by no means sound interactions, but local revolts against the central authority, and local judicial challenges against state sovereignty. If an interactive relationship of this nature is built between the NPC Standing Committee and Hong Kong courts, how can NPC Standing Committee and the Basic Law safeguard their authority? In this case, it will be even more difficult for the NPC Standing Committee's interpretations to find a footing."[49] Dong also contends, "Hong Kong courts shall not go beyond the Basic Law to exercise the power they are not granted, or to undermine state sovereignty. This basic principle has become more and more explicit under constant conflicts. For example, it is further clarified in the disputes over the Congo (Kinshasa) case, in which the NPC Standing Committee interpreted relevant provisions at the request of the Hong Kong Court of Final Appeal. New problems will certainly emerge in practice and I believe the jurisdiction of Hong Kong courts will be a

---

47 Ibid, P97.
48 Ibid, P98.
49 Ibid, P98.

topic of debate for a long time to come."[50]

Bai Sheng (白晟), another mainland legal scholar, carried out a special study on the interpretation of the Basic Law, criticising the Hong Kong Court of Final Appeal for disrespect towards the NPC Standing Committee. He carefully and critically studied the reason why the Court refuses, when it is in fact necessary, to submit cases to the NPC Standing Committee for interpretation of relevant provisions in the Basic Law. He concludes that "In spite of the jurisprudential evidence put forward by the Court, the major reasons are perhaps political ones. It's because of the Court's distrust in the NPC Standing Committee's interpretation and its endeavour to ensure the complete judicial sovereignty of Hong Kong courts."[51]

He claims that, "The reason why the Hong Kong legal circle opposes the NPC Standing Committee's interpretation is not so much about the traditional common law system as their unwillingness to acknowledge the judicial sovereignty of the central authority. [...] They actually advocate the Hong Kong Court of Final Appeal having its own judicial sovereignty. Although the Basic Law clearly stipulates that the NPC Standing Committee has the power to interpret the Law, Hong Kong legal specialists refuse to acknowledge the provision and hope to abolish it through judicial practices, making it a nominal stipulation."[52] In fact, the Basic Law possesses the features of both the continental law and the common law. While recognising the central authority's judicial sovereignty, it also authorises a considerable but less than full degree of judicial sovereignty, with the central authority retaining a minimum degree of judicial sovereignty. This deployment requires in-depth dialogues about jurisprudence between the two traditions of legal interpretation and between the central sovereignty and the authorisation of a high degree of local autonomy. It is actually a challenge from the 'one

---

50 Ibid, P137.

51 Bai Sheng (2015). *A Study on the Interpretation of the Basic Law: A Jurisprudential Perspective* (《基本法釋法問題探究：從法理學角度剖析》). Hong Kong: The Commercial Press. P98.

52 Ibid, P272-273.

country, two systems' policy to the jurisprudence in the mainland and Hong Kong."[53]

"[...] the acceptance of the NPC Standing Committee's interpretation is an issue not so much about judicial sovereignty or a high degree of autonomy as about the acknowledgement of the judicial sovereignty of the central authority. [The legal circle in Hong Kong] recognises complete judicial sovereignty of the U.K. but not the minimum judicial sovereignty of the central authority, the intervention of the Judicial Committee of the Privy Council (JCPC) of the U.K. in daily affairs but not the occasional supervision of the NPC Standing Committee in exceptional cases. And it is apparently not because the legal interpretation of the JCPC is better than that of the NPC Standing Committee in terms of jurisprudence or natural justice. Therefore, in the early days after the return, the distrust the Hong Kong legal specialists cultivated in a common law system in NPC Standing Committee's interpretation actually amounts more to a lack of confidence towards China's legal system than towards the continental legal tradition. Their rejection of the NPC Standing Committee's interpretation doesn't mean their acceptance of China's Supreme Court's interpretation of the Basic Law. The key is to cultivate 'political identity', which is the overarching political task in Hong Kong under 'one country, two systems'."[54]

Unfortunately, the Court of Final Appeal has neglected its jurisprudential duty to request or abide by the NPC Standing Committee's interpretation, resulting in the NPC Standing Committee objecting to its verdicts. Evident examples are the case of Ng Ka-ling in 1999 which was submitted to the NPC Standing Committee for interpretation of the Basic Law, and the case of Chong Fung-yuen in 2001 to which the NPC Standing Committee objected. Ironically, while the NPC was blamed by some Hong Kong legal specialists, it won support from the general public in Hong Kong, who thought that the central authority understood the situation in Hong Kong better than Hong Kong courts and was more

---

53 Ibid, P274.
54 Ibid, P275.

willing to protect Hong Kong's prosperity and stability.

The Congo (Kinshasa) case in 2011 which the Court of Final Appeal actively submitted to the NPC Standing Committee for interpretation didn't mitigate the Hong Kong courts' rejection of the NPC Standing Committee's interpretation. The Court of Final Appeal only did so because the case was relevant to national interests. Additionally, the central authority then publicly took a stand on the case. A verdict out of accord with national interests would definitely be rectified through the NPC Standing Committee's interpretation, which would inevitably weaken the authority of the Court of Final Appeal. Even under these considerations, the submission only narrowly passed, with the judges of the Court voting 3 to 2.

3. Hong Kong Courts' Power of Constitutional Review

Power of "constitutional review" refers to the right of the court to decide that some laws violate the constitution or constitutional documents, and thus declare the laws null and void and repeal them. The Basic Law allows Hong Kong courts to interpret those provisions relevant to "the high degree of autonomy". If the court considers that some provisions violate the Basic Law, it can simply avoid using them in verdicts, but it won't declare them invalid or repeal them. It is yet to be decided whether Hong Kong courts have the power of constitutional review. If they do, will it challenge the inherent power of constitutional review of the NPC Standing Committee, the most authoritative organ for interpreting the Basic Law? And if the courts have that power, does it mean they are also granted the power to formulate and override the government's policies? If yes, will it mitigate the effectiveness of "executive-led"? Therefore, whether Hong Kong courts have the power of constitutional review is an issue of vital importance.

The main focus of this book is on whether Hong Kong courts have the power to make constitutional review of local laws. Before we deal with this topic, let's first look back on the political storm triggered by the remarks of the Hong Kong Court of Final Appeal in 1999. In a case about the right of abode in Hong Kong (the case of Ng Ka-ling), the Hong Kong Court of Final Appeal claimed that it "has the right to review according to the Basic Law whether the legislative behaviours of the NPC and the NPC Standing Committee contradict the Law, and the duty to declare them null and void when they do." According to Bai Sheng, "It is already something more than a large-scale expansion of power within the constitutional framework in Hong Kong. It may also lead to excessive intervention by the SAR courts' jurisdiction in the legislative institutions nationwide, or even to severe damage in national unity."[55] The Court was immediately blistered by the mainland legal circle and forced to withdraw its arrogant statement. This incident established the unchallengeable status of the NPC Standing Committee's decisions for Hong Kong courts.

Do Hong Kong courts have "the power of constitutional review" regarding Hong Kong laws? Many legal specialists believe they do since it conforms to the common law traditions. And only when they do can they protect the judicial independence and dignity, the high degree of autonomy and the existing system in Hong Kong.

Legal experts on the mainland have also diverse opinions about the issue. But I feel that the majority of them believe that Hong Kong has no power of constitutional review, as it is exclusively vested in the NPC Standing Committee. Wang Zhenmin (王振民), a legal expert on the mainland, as an instance of one who thinks that Hong Kong has the power of constitutional review, contends that "prior to the return, in the common law system, Hong Kong had formed a system in which the general judicial institutions, i.e. courts, are responsible for reviewing constitutionality, i.e. the judicial review system. In this system, Hong Kong courts have the power of constitutional review. After the return, according to (relevant) stipulations in the Basic Law, this judicial review system is naturally preserved. [...] Seen from all aspects, courts of the HKSAR should continue to possess the power of constitutional

---

55   Ibid, P80.

review, a certain basis relevant to which can be found in the Basic Law."[56] "Hong Kong's Court of Final Appeal used to be the JCPC of the U.K. [...] which was substituted by the Court of Final Appeal of the HKSAR after the return. Thus the Court of Final Appeal inherited the JCPC's power of constitutional review."[57] "The key is that the mainland doesn't specifically classify the NPC or the NPC Standing Committee as an institution for constitutional review. Although the constitution clearly stipulates the NPC Standing Committee's power to interpret law, we have to acknowledge that the NPC Standing Committee has never interpreted any law, whether in the mainland or at the international level. In this case, how can we demand the Court of Final Appeal of the HKSAR to willingly and reasonably request a legislative institution to 'interpret laws', even though it is an institution of supreme authority?"[58]

Dong Likun, another mainland legal expert, sees it differently. He stresses that "there is no stipulation in the Basic Law that allows Hong Kong courts to review the laws made by Hong Kong legislative organs, to declare those laws null and void for violating the Basic Law, or to review Hong Kong's original laws. Its original common law also doesn't mention that Hong Kong courts have the power of constitutional review. However, Hong Kong courts have arbitrarily established in various cases their power to review the laws formulated by the legislative organs of the HKSAR as well as Hong Kong's original laws, and to invalidate any reviewed laws that contradict the Basic Law. That Hong Kong courts establish their own constitutional power through cases is totally inconsistent with Hong Kong's legal status as China's special administrative region."[59] "From the Basic Law, Hong Kong's original common law traditions or its rule to refer to legal precedents, no basis

56 Wang Zhenmin (2002). *Relations between the Central Authorities and the Special Administrative Region: An Analysis of the Legal Structure* (《中央與特別行政區關係：一種法治結構的解析》). Beijing: Tsinghua University Press. P362.

57 Ibid, P362-363.

58 Ibid, P375.

59 Dong Likun (2014). *Relationship between the Governance Power of the Central Authorities and the High Degree of Autonomy of Hong Kong Special Administrative Region* (《中央管治權與香港特區高度自治權》). Beijing: Law Press. P77.

can be found for Hong Kong courts' power of constitutional review for the violation of the Basic Law. According to the Basic Law, it must be stipulated by law if Hong Kong courts have the power to review violation of the Basic Law."[60] "The power of final appeal and the power of constitutional review are different constitutional powers in nature, authority, function, and subject. [...] Whether the courts are allowed to exercise the power of final appeal and the power of constitutional review depends on the specific authorisation from the constitution. The Basic Law can only authorise Hong Kong the power of final appeal, while the NPC Standing Committee retains the power to review violations of the Basic Law. That Hong Kong Court of Final Appeal derives its power to review violation of the Basic Law from the power of final appeal confuses the two powers, contradicting the NPC Standing Committee's power to review violations of the Basic Law, and exceeds the scope of power of the final power of appeal granted by the Basic Law.[61]

However, Chen Xinxin (陳欣新), another mainland legal scholar, considers that the central authority and the HKSAR both have "constitutional power" in general terms with regard to the Basic Law, but Hong Kong courts are not authorised to "repeal" the local laws that they deem inconsistent with the Basic Law. "Because the power to decide to abolish, revise or repeal legislation belongs to legislative power, which is not part of the power of judicial institutions according to the strict principle of separation of power in the common law. The conclusion that the Court of Final Appeal's declaration (that certain legislation is without validity in trials) amounts to repealing them is in accordance with the case law system in Hong Kong, which stipulates that the verdicts of the Court of Final Appeal are legally binding for all courts. It is not the "repeal" specified in the provision of the Constitution of the People's Republic of China that the organ of supreme power is empowered to repeal laws that contradict the constitution. The two "repeal" actions are different in essence. The latter means the repealed law is no

---

60 Ibid, P167.
61 Ibid, P186-187.

longer considered law, binding for any institution, group or individual, or serving as the basis for lawful action. The former has no influence on the "nominal legal status" of the law declared null and void; the amendment should conform to the legislative procedure; institutions, groups and individuals can take action based on the law; without lawsuits in the statutory period, the legal effect of relevant action is no longer the subject to question."[62]

The divergent opinions among the legal scholars with regard to whether Hong Kong courts have the power of constitutional review demonstrate that it is no easy task to coordinate the continental law system and the common law system.

4. Article 74 of the Basic Law

In order to ensure the chief executive's firm grip on the power to dominate, propose, and decide, the Basic Law only endows the Legislative Council the power to veto the bills put forward by the chief executive, but not the power to formulate policies or allocate fiscal resources, with the overarching aim of maintaining the original prudent fiscal policies in Hong Kong and to prevent welfarism and populism.

Article 74 of the Basic Law stipulates that "Members of the Legislative Council of the Hong Kong Special Administrative Region may introduce bills in accordance with the provisions of this Law and legal procedures. Bills which do not relate to public expenditure or political structure or the operation of the government may be introduced individually or jointly by members of the Council. The written consent of the Chief Executive shall be required before bills relating to government policies are introduced." This article's legislative intent is to strictly limit the Legislative Council members' efficiency in formulating

---

62 Chen Xinxin (2015). "Coordination between Hong Kong and the Central Authorities in 'Constitutional Review'" ("香港與中央的 '違憲審查 '協調"). *Perspectives on Hong Kong Basic Law* (《香港基本法面面觀》). (Chen Hung-yee & Zhou Xueping (eds.). Hong Kong: Joint Publishing. P106-122; P115-116.

public policies (including fiscal policies) through the so-called "private member's bill", so as to ensure that the chief executive won't lose the power to formulate policies. Though there is a social consensus that the "private member's bill" proposed by Legislative Council members is subject to the constraint of Article 74, as for whether it is also true of their amendments to government bills, no authoritative decision has been made yet.

In fact, after the return, the opposition among the Legislative Council members put forward various amendments to government bills from time to time, in attempt to change, overthrow or introduce policies, or to hinder government administration through filibusters. Most amendments apparently contradict Article 74. The opposition certainly regards Article 74 as inapplicable to the government bill amendments brought forward by Legislative Council members, but the Government of the HKSAR holds the opposite viewpoint. However, according to the legislative intent of the Basic Law, if its Article 74 is targeted to secure the chief executive's power to command, propose and decide in terms of formulating policies, fiscal policy in particular, it is hard to imagine that the Basic Law would open a "back door" for the legislative council members to acquire certain policy-formulating powers through proposing amendments to government bills. The Government of the HKSAR has always confronted the "illegal" behaviour of Legislative Council members through political means. One of the methods is that when the government is unwilling to accept amendments proposed by the legislative council members, it simply repeals the bills to nip their plots in the bud, which, however, poses difficulties when it comes to implementing the government's policies. Another approach is to change the proposed amendments into the ones "actively" proposed by the government itself so that they can be passed in the Legislative Council. The government's attempt to "avoid the controversial issues" in order to quell conflicts indicates that the government is also willing to compromise with the Legislative Council in order to pass the policy. Although the amendments may make the policy less comprehensive and complicate its implementation, the policy can still be carried out. Meanwhile, the government can sometimes be praised for its "openness to different opinions".[63]

In recent years, however, some radical Legislative Council members have proposed a high volume of meaningless, trivial, and unreasonable amendments to government bills, for which they squander considerable time discussing and voting, trying to delay the passing of bills, or even shelving some bills when the time for review has run out. These "filibusters" are the core activities of the opposition's "non-co-operation" movement against the government administration in the HKSAR, severely hampering governance and "executive-led". The issue that ensues is whether Article 74 is applicable to Legislative Council members' proposals of amendments to government bills. If not, many "filibusters" are illegal and should be curbed. But so far this issue hasn't been settled once and for all through legal means.

Another important issue Article 74 involves is: who has the power to decide whether the Legislative Council's proposals of "private member's bills" and amendments to government bills are in line with the Basic Law? According to Article 74, such power is vested in the chief executive. But theoretically, the Rules of Procedure of the Legislative Council formulated based on Article 74 also endow the Legislative Council president and all members with the power. Mainland legal scholar Dong Likun perceives that the Rules of Procedure of the Legislative Council violates Article 74 in this aspect and actually deprives the chief executive of an important power.

Dong said, "[A]s this item of the Rules of Procedure involves the key stipulations *vis-a-vis* the power of the chief executive and the Legislative Council, the Rules of Procedure of the Legislative Council made by the Legislative Council is revised. Article 31 of the Rules of Procedure stipulates '[a] motion or amendment, the object or effect of which may, in the opinion of the President or Chairman, be to dispose of or charge any part of the revenue or other public moneys of Hong

---

63 Lau Siu-kai (2000). "Political System of Executive-Led: Blueprint and Reality" ("行政主導的政治體制：設想與現實"). *Hong Kong Blueprint in 21st Century* (《香港二十一世紀藍圖》). Hong Kong: the Chinese University Press. P1-36.

Kong shall be proposed only by (a) the chief executive; or (b) a designated public officer; or (c) a member, if the chief executive consents in writing to the proposal.' Article 51(3) stipulates that '[m]embers may not either individually or jointly introduce a bill which, in the opinion of the President, relates to public expenditure or political structure or the operation of the government.' Article 51(4) stipulates that '[i]n the case of a bill which, in the opinion of the president, relates to government policies, the notice shall be accompanied by the written consent of the chief executive in respect of the bill.' The above content in the Rules of Procedure undoubtedly violates Article 74 of the Basic Law. In other words, the clause stating that a bill involves government policies are to be decided by 'the chief executive' has been changed into 'by the Legislative Council president and chairman' of the whole Council. Such a shift seriously cripples the chief executive's power and expands the power of the Legislative Council president and chairman of the whole Council, which should not have happened. It has entailed conflicts between administration and legislation, affecting the performance of the political system in Hong Kong."[64]

Actually, the Rules of Procedure of the Legislative Council was drafted by the provisional Legislative Council before the return and continues to be applied after the return. As early as on the eve of the return, Professor Xu Chongde (許崇德), a mainland legal authority, showed me how the draft violates Article 74. However, it remains a suspended legal dispute as to whether the Legislative Council can determine on its own its members' proposals of "private member's bills" and amendments to government policies that conform to Article 74 of the Basic Law.

As long as the three issues (namely, whether the LegCo members' proposals of amendments to government policies are subject to the stipulation of Article 74 of the Basic Law; whether the Rules of Procedure of the Legislative Council conforms to the Basic Law; and who

---

64 Dong Likun (2014). *Relationship between the Governance Power of the Central Authorities and the High Degree of Autonomy of Hong Kong Special Administrative Region* (《中央管治權與香港特區高度自治權》). Beijing: Law Press. P73.

has the power to decide) remain unsettled, and it remains impossible to effectively solve "filibusters" or to fully implement "executive-led" and effective governance.

5. Right of Abode

Since the return of Hong Kong, the most immediate and bothersome issue for Hong Kong people has been the right of abode. Before the return, they worried that the intensive influx of mainland compatriots would exceed Hong Kong's capacity. To reassure Hong Kong people, the central authority made a commitment that following the principle of "one country, two systems", the immigration policy of Hong Kong would remain unchanged and would continue its strict administration of mainland compatriots who came to Hong Kong for settlement. In the Sino-British Joint Declaration, the specification on the Chinese Government's basic policy towards Hong Kong stipulated in particular what constitutes permanent Hong Kong residency. [65]Article 24 of the Basic Law cited the exact content of this stipulation and legalised the provisions concerning the right of abode in the Sino-British Joint Declaration. Provisions one, two and three are particularly "controversial". These three provisions defined the permanent residents of the HKSAR as follows: (1) Chinese citizens born in Hong Kong before or after the

---

65 The content is as follows: "The following categories of persons shall have the right of abode in the Hong Kong Special Administrative Region, and, in accordance with the law of the Hong Kong Special Administrative Region, be qualified to obtain permanent identity cards issued by the Hong Kong Special Administrative Region Government, which state their right of abode: all Chinese nationals who were born or who have ordinarily resided in Hong Kong before or after the establishment of the Hong Kong Special Administrative Region for a continuous period of 7 years or more, and persons of Chinese nationality born outside Hong Kong of such Chinese nationals; all other persons who have ordinarily resided in Hong Kong before or after the establishment of the Hong Kong Special Administrative Region for a continuous period of 7 years or more and who have taken Hong Kong as their place of permanent residence before or after the establishment of the Hong Kong Special Administrative Region, and persons under 21 years of age who were born of such persons in Hong Kong before or after the establishment of the Hong Kong Special Administrative Region; any other persons who had the right of abode only in Hong Kong before the establishment of the Hong Kong Special Administrative Region."

establishment of the HKSAR; (2) Chinese citizens who have ordinarily resided in Hong Kong for a accumulated period of no fewer than seven years before or after the establishment of the HKSAR; (3) persons of Chinese nationality born outside Hong Kong to residents listed in categories (1) and (2).

Before the return, I learned that the provisions above were obscure and might bring about misunderstandings. Taking illegal immigration as an example, if both parents are Chinese citizens without a permanent ID and bear a child in Hong Kong, should their son or daughter gain the right of abode or not in accordance with provision (1)? If someone was born in mainland China and his or her parents were not permanent residents of the HKSAR when he or she was born, but gained the right of abode after his or her birth, should he or she gain the right to abode in accordance with provision (3)? In accordance with the immigration policy before the return of Hong Kong and the policy made by the Chinese Government that placed strict limitations on mainland compatriots settling in Hong Kong, illegal immigrants and their children born in Hong Kong as well as those whose parents were not permanent residents of Hong Kong when they were born are not granted the right of abode.

However, the Hong Kong Basic Law Drafting Committee was aware that the different specifications of the right of abode between the Basic Law and the Joint Declaration would bring about misunderstandings and anxieties in Hong Kong, so the Committee decided to cite the exact words from the Joint Declaration. [66]After the establishment of the Preparatory Committee for the HKSAR, the members of the committee worried that the articles about permanent residency would lead to misinterpretation, so on 10 August 1996, *The Preparatory Committee's*

---

66 The Basic Law of The Macao Special Administrative Region of The People's Republic of China made a clear specification on the permanent residency which obviously drew from the experience of the draft of Hong Kong Basic Law. The second paragraph of Article 24 stipulated Macao permanent residents as follows: "Chinese citizens who have habitually resided in Macao for a continuous period of not less than seven years before or after the establishment of the Macao Special Administrative Region and their children of Chinese nationality born outside Macao after they have become permanent residents."

Opinions on Implementation of the Article 24 paragraph 2 of the Basic Law of the Hong Kong Special Administrative Region of the People's Republic of China was approved, which clarified the legislative intent of that article. Two things are essential according to the opinions of the committee. First, "The Chinese citizens born in Hong Kong in accordance with Article 24 Provision 2 Category 1 refer to children both of whose parents or one of whose parents legally settled in Hong Kong; children whose parents are illegal immigrants, over stayers and temporary stayers are excluded." Second, Article 24 Provision 2 Category 3 specifically stipulates that persons born outside Hong Kong of Chinese nationality whose parents or one of whose parents gain the right of abode in accordance with Article 24 paragraph 1 or 2."

The opinions of the Preparatory Committee for the HKSAR was approved by the Standing Committee of the National People's Congress, which means that China's highest organ of state power confirmed the legislative intention of the articles regarding right of abode of the Basic Law. This was what happened shortly after the return of Hong Kong.[67]

However, in the "Ng Ka Ling Case" of 1999, the interpretation of the Hong Kong Court of Final Appeal on Hong Kong permanent residency was very different from that of the central authority's. On 29 January 1999, the Hong Kong Court of Final Appeal made a final judgment on the right of abode for Hong Kong residents' children born on the mainland. The Court of Final Appeal ruled that the children of Hong Kong permanent residents have the right of abode in Hong Kong whether his parents become permanent residents before or after his birth. This judgment evoked a strong reaction on the mainland and in Hong Kong because it was completely different from most people's interpretation and, what was worse, it would attract many mainland Chinese to settle in Hong Kong, which would bring a heavy burden to the education, housing, welfare and medical care system of Hong Kong. In fact, this judgment denied the solemn commitment made by

---

67 See the *NPC Standing Committee's Decision on the Approval of the Proposal to Dissolve the Preparatory Committee* (approved by the Twenty-sixth Session of the Eighth NPC Standing Committee on 3 July 1997).

the central authority following "one country, two systems" that after the return there would not be a large flow of mainland Chinese to Hong Kong. The judgment of the Final Court laid too much emphasis on the literal interpretation of the articles from the Basic Law, standard of human rights and equality but underestimated the legislative intention of the Basic Law and the spirit of the central authority's policy towards Hong Kong without considering Hong Kong's capacity.

It is interesting that the Hong Kong Court of Final Appeal was reluctant to adopt the opinion from the Preparatory Committee for the HKSAR on the right of abode because they thought that this opinion was issued after the Basic Law was promulgated so it was unclear whether the opinion reflected the legislative intent of the Basic Law. To solve the crisis spurred by the judgment of the Court of Final Appeal, the HKSAR Government was forced to ask the central authority for interpretations of the Basic Law. On 6 June, 1999, the Standing Committee of the National People's Congress interpreted the Basic Law and ruled that if persons of Chinese nationality born outside Hong Kong wish to gain the right of abode, his or her parents or one of his or her parents must be (1) Chinese citizens born in Hong Kong before or after the establishment of the HKSAR; (2) Chinese citizens who have usually resided in Hong Kong for an accumulated period of not fewer than seven years before or after the establishment of the HKSAR. The interpretation of the Standing Committee of the National People's Congress denied the interpretation of Hong Kong Final Court and confirmed that the opinion of the Preparatory Committee for the HKSAR represented the legislative intent of the Basic Law. After that, the Court of Final Appeal was subjected to the interpretation of the Standing Committee of the National People's Congress. But the previous judgments were free from this interpretation so the final adjudication was still complete and the dispute brought the Hong Kong Final Court's judgment to an end.

In another case concerning the right of abode in 2001, the Hong Kong Court of Final Appeal ignored the opinion of the Preparatory Committee for the HKSAR again and made a controversial judgment. On 7 July, 2001, the Hong Kong Court of Final Appeal reached an unanimous verdict on the "Chong Fung-yuen case" that Chinese citizens born in Hong Kong shall have the right of abode in accordance with the Basic Law, be their parents Hong Kong permanent residents or not. This judgment was fully interpreted in accordance with the Basic Law and the judges of the Final Court believed that the relevant expressions were clear. The Hong Kong Government also estimated that the judgment would not cause serious social consequence. However, this judgment was entirely different from the Hong Kong people's understanding of the Basic Law. They always believed that children born in Hong Kong by mainland Chinese should not have the right of abode if their parents do not have the right of abode. Strangely enough, the Standing Committee of the National People's Congress did not correct this "mistake" by the interpretation of the law or other methods, but merely expressed disapproval.

On 22 July, 2001, the Legislative Affairs Committee of the NPC made a statement on the final judgment of the Court of Final Appeal, expressing that the judgments of Hong Kong Court of Final Appeal did not fully conform to the *Interpretation by the Standing Committee of the National People's Congress Regarding Provision 4 in Article 22 and Category (3) of Provision 2 in Article 24 of the Basic Law of the HKSAR of the PRC* and expressed concern about this. This interpretation indicated that "the legislative intent of other articles" in provision 24 had already demonstrated in the opinion of the Standing Committee of the National People's Congress and the Preparatory Committee for the HKSAR's opinion on the implementation of article 24 provision 1 of the Basic Law. This viewpoint argued that provision 24 in article 2 and category (1) ruled that the Chinese citizens born in Hong Kong refer to "children born in Hong Kong whose both parents or one of the parents are legal Hong Kong residents excluding children of illegal immigrants, over stayers and temporary stayers."

However, the development of this situation was beyond the expectations of the central authority, the HKSAR Government, the Court of Final Appeal and the Hong Kong people. In order to gain the right of abode for their children, a large number of mainland pregnant women without right of abode came to Hong Kong for childbirth through various means. Therefore, the medical care system of Hong Kong came under enormous pressure and many Hong Kong pregnant women complained about the lack of access to appropriate care. Under pressure from the general public, the HKSAR Government adopted administrative measures to constrain Hong Kong's hospitals from taking pregnant women from the mainland and forbade them to enter Hong Kong. Although the number of mainland pregnant women coming to Hong Kong was put under control, the judgment of the Court of Final Appeal that caused these problems was still in effect in Hong Kong. Therefore, the right of abode is still a serious problem that distresses Hong Kong.

Limitations of the Mechanism of "Clarifying Doubts and Resolving Disputes"

The reason a large number of disputes on some major political issues remained after the return is that the mechanism for "clarifying doubts and resolving disputes" did not perform its due role at that time. Some major political issues concerning the relationship between the central authority and the HKSAR Government, the relationship between administration and legislation as well as the right of abode mentioned above remained unresolved. This situation hampered the operation of the "one country, two systems" policy and made it possible for the opposition activists to use these disputes in rallying political power and setting the masses in opposition to the central authority and the HKSAR Government. The question is: why do those major political issues still remain unsettled, with compounding consequences, eighteen years after the return? As far as I am concerned, it is not because of the lack of the mechanism of "dispelling doubts and resolving disputes", but the lack of a strong political will and determination to solve the problem.

The central authority, in particular, as the biggest "stakeholder" of the implementation of "one country, two systems" and the Basic Law, carries the final responsibility.

Some foreign countries have some common methods for solving constitutional problems, but they are not applicable to Hong Kong at present. For one thing, there are no "constitutional courts" in China to solve lawsuits and disputes regarding the Basic Law. For another, since Hong Kong is part of China, the Chinese Government would not allow foreign or arbitrary international authority to interfere with the political issues concerning "one country, two systems" and the Basic Law, thus raising the question of Hong Kong becoming "internationalised".

In theory, the Basic Law itself is the legal measure for settling political disputes. But since the disputes usually involve diverse under-standings of the articles of the Basic Law, the Basic Law is useless for "dispelling doubts and resolving disputes". Although the chief executive is responsible for the comprehensive and accurate implementation of "one country, two systems" and the Basic Law, the lack of confidence in the seat and long-standing weakened position of the chief executive have caused him to shrink from such major constitutional disputes. The exception occurred in 1999 when the judgment of the "Ng Ka-ling case" threatened Hong Kong with an unsustainable population burden, and the chief executive was thereby "forced" to refer to the central authority for interpretation; otherwise, he is normally unwilling to refer such major political issues to the court. There are three reasons for this. First, the chief executive is afraid that losing in a lawsuit would harm the authority of the HKSAR Government. Second, the chief executive is afraid that if the judgments were unfavourable, the NPC Standing Committee's interpretation would have to be sought, which would trigger a backlash from Hong Kong people and legal professionals and make the HKSAR Government guilty of destroying the independence of the Hong Kong judiciary. Third, if the chief executive directly asks for the NPC Standing Committee's interpretation, he would face a severe political impact. Even if the NPC Standing Committee is willing to interpret the Basic Law, the HKSAR Government fears becoming implicated.

Moreover, even when the HKSAR Government was aware that some actions of the members of the Legislative Council and the Rules of Procedure did not conform to the Basic Law, or firmly believed that there was a discrepancy between the judgment of the court and the Basic Law, the chief executive and senior officers would not express their standpoints publicly. The chief executive usually responds "passively" and "mechanically" by pledging to "respect the decision of the court" or "accepting the judgment of the court". From this point of view, the HKSAR Government not only passively tolerated issues that did not conform to "one country, two systems" and the Basic Law, but gave up the great responsibility of promoting and explaining "one country, two systems" and the Basic Law. The reason why the opposition forces had more say in the interpretation of "one country, two systems" and the Basic Law is largely due to the fact that the chief executive and other officials were afraid to express their standpoints and argue with opposition forces.

Established in accordance with the regulations of the Basic Law, the Hong Kong Basic Law Drafting Committee should have performed certain functions of "clarifying doubts and resolving disputes". However, after the return, the Hong Kong Basic Law Drafting Committee has adopted a low profile and consulted the Standing Committee of the National People's Congress only when necessary. The Hong Kong Basic Law Committee barely took the initiative publicly to express opinions about the disputes on the Basic Law in the name of a collective or institution. Although some members of the committee expressed their opinions in public about the Basic Law on occasion, the people and media in Hong Kong regarded their words as personal opinions, which received little attention. To make matters worse, members of the Hong Kong Basic Law Committee occasionally hold diverse opinions, so the committee could hardly become an authorised institution of arbitration.

Therefore, the remaining mechanism that can serve as the determining factor of major political disputes is the Standing Committee of the National People's Congress. In mainland China, interpretation of law is rare. However, after the return, there have been five interpretations of the Basic Law of Hong Kong. Although the Hong Kong opposition forces and some legal professionals regarded the interpretations by the Standing Committee as "dreadful monsters", they were well-received by Hong Kong people, who thought that these interpretations were helpful for the settlement of the problems. Nevertheless, the central authority still hesitated in issuing an interpretation as it might cause the backlash of Hong Kong people and the international community, which might grant the opposition certain opportunities. The notion that failure to conform to "one country, two systems" and the Basic Law will harm the interest of the nation and Hong Kong must be seriously considered from all angles.

It must be admitted that even if the central authority gives the final word on disputes pertaining to "one country, two systems" and the Basic Law, the Hong Kong opposition and a portion of the Hong Kong people will not accept it readily but would only try to oppose it. Although some Hong Kong people are not fully behind them, the courts of Hong Kong must meet the central authority's bottom line. When the courts refer to the central authority for interpretation, the opposition would have difficulty in using political disputes to wage struggles unless the interpretation by the Standing Committee of the National People's Congress sparks an outrage among Hong Kong people.

So it can be said that following "one country, two systems" policy and the Basic Law, the mechanism of "clarifying doubts and resolving disputes" surely exists, but it cannot perform its due role because the central authority and the HKSAR Government still have doubts concerning the launching of the mechanisms. In the final analysis, the problem is all about political will, political determination and political courage.

The Unsound Mechanism of Preventing
"Well Water" from Interfering with "River Water"

During the "colonial period", the British Government and the colonial government actually played a role in preventing Hong Kong from becoming the "bridgehead" threatening the politics and security of mainland China. The reason why the British were willing to act in this way is that—should China gain the ability to take over Hong Kong at any time, and if Hong Kong had become a "base for subversion" against the Communist Party of China—it was necessary to keep Hong Kong as a "colony".

For example, even though the U.K. and the U.S. were allies (or more exactly, the U.K. was the "client state" of the U.S. in diplomacy and military)[68] during the Cold War, the U.S. took advantage of the Korean War and allied with other countries to conduct "political containment" and "economic blockade" on China in the name of United Nations. The U.S. also used Hong Kong as a base for gathering political and military intelligence and for spreading political propaganda against mainland China.[69] While cooperating with the U.S. in deployment and action, the U.K. also exercised constraints on the U.S.'s actions of "subversion" and "anti-communism" and constructed a "barrier" between Hong Kong and the mainland.

Chi-Kwan Mark (麥志坤), a Hong Kong historian who currently resides in the U.K., has read many declassified documents in his research on the relationship between the U.K. and the U.S. regarding Hong Kong.[70] As an ally that shared a special relationship with the

---

68 Guy Arnold (2014). *America and Britain: Was There Ever a Special Relationship*. London: Hurst & Co. The author believed that although many British leaders and officials often talked about the "special relationship" of the U.K. and America, the Americans actually regard the U.K. as the "Trojan Horse" and the "outpost" for maintaining its global interest.

69 After 1945, with the Cold War, Hong Kong became the most important spy centre in the world. In 1950s and 1960s, both America congress and the Pentagon (United States Department of Defence) regarded Hong Kong as the most essential part of British overseas territories. See in Calder Walton, *Empire of Secrets: British Intelligence, the Cold War, and the Twilight of Empire* (New York: The Overlook Press, 2013), P331-333.

70 Mark Chi-Kwan (2004). *Hong Kong and the Cold War: Anglo-American Relations 1949-*

U.S., the U.K. worked closely with the U.S. in many affairs especially in the embargo of China led by the U.S. and supported by the United Nations in order to keep a sound relationship with the U.S.. But as a free port that had long relied on import and export, the economy and the society as well as people's livelihood of Hong Kong would suffer if the embargo continued. In this way, the U.K. did not fully appreciate or cooperate with the U.S.'s policies and action in some cases, for the interest of Hong Kong and for keeping Hong Kong out of danger in the event of offending China.

For example, in the cases of Korea, Indo-China Peninsula and Taiwan, the U.K. and the U.S. held different opinions because the U.K. did not want the policy of the U.S. to have an adverse effect on Hong Kong, and the U.K. disapproved of the U.S. for expanding the Korean War and did not offer military support to the U.S. in the cases of Taiwan and the Indo-China Peninsula. There were also many debates between the two on whether China should join the United Nations. Sometimes the positions of the U.K. upset the U.S. For example, the U.S. sent a great number of people to Hong Kong to conduct intelligence gathering, espionage, and instigation, spread anti-communist propaganda, and "take care of refugees" who fled to Hong Kong, in addition to personnel engaging in "China Research". America also utilised Hong Kong for sabotaging and subverting mainland China. Although it was impossible for the U.K. to be fully aware of those activities and therefore to forbid them, the U.K. tried to constrain those activities, especially all secret and illegal activities, such as intelligence gathering, anti-government guerrilla activity on the mainland and the operations of the "Third Force" (第三勢力) seeking refuge in Hong Kong that opposed the KMT and the CPC. In 1951, regardless of America's opposition, the U.K. banned the broadcasting of Voice of America in Hong Kong through the colonial government's radio station and prevented the U.S. from handing out materials for political propaganda in Hong Kong schools.

What worried the U.K. most was the possibility that the Chinese

1957. Oxford: Clarendon Press.

Government would regard Hong Kong as "a base for subversion" and launch an attack or take over Hong Kong. The U.K. was not willing to let the CPC gain opportunities to carry out destructive activities in Hong Kong. "In short, except for a brief period during the Korean War, the United States utilised Hong Kong mainly for intelligence gathering, rather than covert action against the PRC. The British would not allow the 'American Cold Warriors' to turn Hong Kong into a base for subversion."[71] Of course, in order to maintain the support of the U.K. during the Cold War, the U.S. tolerated the U.K.'s unwillingness to irritate or upset China.

With the exception of the U.S., the British would not allow other countries to utilise Hong Kong to conduct unfavourable activities against China. The Soviet Union was not approved to establish a consulate in Hong Kong because of its terrible relationship with China and the background of the Cold War between the East and the West.

After the founding of new China, the remnant force of KMT in Hong Kong (usually called the Rightists) was still significant, and they often stirred up trouble against the leftists of Hong Kong. Taking Hong Kong as its base of operations, the Taiwan authority also worked with the U.S. to disturb and sabotage the mainland. In order not to upset China, the U.K. tried to restrict the actions of Taiwan in Hong Kong. The colonial government also maintained a neutral position between the KMT and the CPC but never allowed them to harm the security and order of Hong Kong. For example, both parties were not allowed to conduct political propaganda in Hong Kong's schools (except for the schools run by the leftists).

The colonial government worried more that the conflict between the KMT and the CPC would lead to the Chinese Government's interference in Hong Kong affairs. The U.K. Cabinet Papers on 3 April 1950 pointed out that: "The survival of Hong Kong depends on our not being dragged into the political disputes of China and conducting impartial

---

71 Ibid, P193-194.

administration and adhering to the rule of law."[72] In the 1950s, the conflicts between the KMT and the CPC constantly appeared. "Caught between the two rival regimes, the British tried to maintain a neutral posture, while demonstrating non-provocative firmness from time to time...While not firmly taking sides, the British Government tended to be more accommodating to the Chinese Communists, or at least less sympathetic towards the Chinese Nationalists."[73]

Sun Yang (孫揚), a mainland historian, offers a similar description on how carefully the U.K. dealt with the rivalry between KMT and the CPC in Hong Kong.[74] He believes that in order to maintain the authority of its administration, the colonial government prevented the KMT and the CPC from meddling with Hong Kong affairs and instigating nationalism and anti-colonial consciousness. The colonial government was also afraid to see the KMT use Hong Kong against the mainland government, and was reluctant to see the internal contradiction evolve into a Sino-British conflict and cause anti-British sentiments.[75] It tried to remain neutral, between MT and the CPC and constrain their political activities in Hong Kong. If the two parties "transgressed" politically, the colonial government would respond harshly and it would certainly have anticipated the possible reactions of the Chinese Government.

The reason why I elaborated on the role played by the U.K. that "separated" Hong Kong and the mainland in the "colonial" period is that after the return of Hong Kong, a replacement for the U.K. has not yet been found. The colonial government spared no efforts to prevent Hong Kong from becoming a "base for subversion" in order not to render any excuse to China to interfere with Hong Kong affairs and even to take over Hong Kong. Besides not letting any foreign countries or external

---

72 Ibid, P86.

73 Ibid, P113-114.

74 Sun Yang (2014). *Sino-British Negotiation on the Hong Kong Issue 1945-1949* (《無果而終：戰後中英香港問題交涉》). Beijing: Social Sciences Academic Press.

75 One of the most serious incidents happened in the beginning of 1948. The colonial government tore down the Kowloon City without Beijing's permission which led to a diplomatic crisis and the Shamian event, shocking the world. In the Shamian event the parade of Guangzhou people was beyond control, and people destroyed the British consulate in Shamian.

forces to use Hong Kong to deal with China, the colonial government would also not let things that would upset China transpire in Hong Kong, for example, a few anti-communist movies were banned in Hong Kong.

However, considering the large number of anti-communists in Hong Kong and the freedom of news and speech advocated by the colonial government, the British had to allow many anti-communist media outlets to operate in Hong Kong, including some newspapers and magazines with an American and Taiwan background. In general, during the "colonial" period, anti-communist speeches were allowed, but activities posing a threat to China were restricted or stifled.

After settling the issue of Hong Kong's future, Hong Kong people were increasingly involved in political activities, and the anti-communists and opposition activists were also frequently engaged in actions that provoked China. But because of the declining capability of the administrative authority of the colonial government, it was difficult for them to constrain the activities unfavourable to the Sino-U.K. relationship. Moreover, the colonial government actually allowed or even incited some anti-communist behaviour for a number of reasons; they hoped to remain independent through the remaining governance; they wished to work for the biggest share of interest for the U.K.; they wished to prevent "intervention" by the Chinese Government; and finally, they were motivated to counterbalance China. The Hong Kong radio station of the colonial government, as a result, turned into an official voice of anti-communists.

After the return of Hong Kong, the free and liberal political and legal environment led to an increasing number of anti-communist activities that had once been common in the "colonial" period. Topics that gave some political and civil groups excuses to set out protests against the CPC included "Tiananmen Square Protests", the status of human rights and legal system in the mainland, the situation of mainland dissidents, the operations of Hong Kong media in the mainland, the central authority's policy towards Hong Kong, the NPC Standing Committee's interpretations of the Basic Law, and political reforms.[76]

---

76 See Ho-fung Hung and Lam-chong Ip, (2012) "Hong Kong's Democratic Movement and the

After the British left Hong Kong and returned its sovereignty back to China, the only way to prevent Hong Kong from becoming a "base for subversion" was relying on the mechanisms of Hong Kong and the mainland, the most powerful one of which is the law, Article 23 of the Basic Law, in particular. From today's perspective, Article 23 could only deal with a narrow range of national crises, which was far from meeting needs pertaining to guarding national security with the increasingly growing conventional and non-conventional threats. The differences between the National Security Law of the People's Republic of China approved by the NPC Standing Committee in 2015 and Article 23 of the Basic Law can be found if they are compared with each other.[77]

In 2003, the HKSAR Government's attempt to legislate locally according to Article 23 came to nothing because of the rejection of Hong Kong residents for fear that their human rights and freedom would be damaged and due to the ensuing large-scale protests by the opposition activists.[78] After the failure to legislate locally according to Article 23, the number of speeches and activities challenging China and fulminating against the CPC was on the rise, and the Liaison Office of the Central People's Government in the HKSAR became the target of many protests. In the second half of 2014, the break out of the illegal 79-day "Occupy Central" movement symbolised the peak of the activities against the central authority. Although those activities did not actually pose substantial threats to national security and were not even banned by law, they indeed impaired the Hong Kong residents' respect

---

Making of China's Offshore Civil Society," *Asian Survey*, Vol. 52, No. 3, P504-527.

77 On 1 July 2015, the Fifteenth Session of the Twelfth NPC Standing Committee approved the new *National Security Law*. Article 20 goes like this: "The state shall adhere to developing an advanced socialist culture, enhance the education and promotion of core values of socialism, maintain its ideological domination, inherit and carry forward the fine traditional culture of the Chinese nation, prevent and resist the impact of harmful culture, enhance the overall cultural strength and competitiveness.

78 Reference: Ngok Ma (2005) "Civil Society in Self-Defence: the Struggle against National Security Legislation in Hong Kong." *Journal of Contemporary China*, Vol. 14, No. 44, P465-482; Francis L.F. Lee and Joseph M. Chan (2011). *Media, Social Mobilisation, and Mass Protests in Post-Colonial Hong Kong: The Power of a Critical Event*. London: Routledge; and Elaine Chan and Joseph Chan (2014), "Liberal Patriotism in Hong Kong," *Journal of Contemporary China*, Vol. 23, No. 89, P952-970.

for national sovereignty and the central authority. As for some mainland compatriots, they saw Hong Kong people as shirking their responsibility to safeguard national security and even felt a strong dislike for such activities that challenged the central authority, thus reducing their political sympathy towards "one country, two systems".

It is true that even if Article 23 was legislated successfully, it is still hard to tell whether it could effectively prohibit movements against China such as are commonly seen in Hong Kong, with such taglines as the "rehabilitation of June 4", "ending the one-party regime", or whether it could effectively punish those who provide material or moral support for anti-government groups in the mainland. The attitude of the Hong Kong court is even more difficult to predict.

The other legal means that people have paid less attention to is the surrender of fugitive offenders between Hong Kong and the mainland. Eighteen years after the return, no agreement has been reached on the exchange of fugitive offenders between Hong Kong and the mainland. One of the reasons is that the death penalty that still exists in the mainland has already been abolished in Hong Kong. "In accordance with *The Criminal Law of the People's Republic of China*, the death penalty was included in the principal penalties while in Hong Kong death penalty has already been abolished. Convenient transportation allowed those who had committed crimes flee to jurisdictions where the sentences are lighter or can be exempt. Without agreements on the surrender of fugitive offenders between Hong Kong and mainland China, Hong Kong may become the paradise for criminals escaping responsibilities, which is not in the interest of both sides. But after the "Cheung Tze-keung" case and "Li Yuhui" case, the death penalty drew great attention to the issue of exchanging fugitive offenders. "[79] "However, the *Fugitive Offenders Ordinance* and *Mutual Legal Assistance in Criminal Matters Ordinance* are still not applicable to Hong Kong and mainland China, so there is still room for Hong Kong and

---

79 Lao Qichang (2000). "On the Analysis of the Death Penalty in Agreement on the Surrender of Fugitive Offenders between Mainland and Hong Kong" ("死刑在中國內地及香港兩地達成移交逃犯協定上的問題研究"). Xiao Weiyun (2000) *The Successful Implementation of the Basic Law* (《香港基本法的成功實踐》). Beijing: Peking University Press. P172-181.

mainland China to discuss the surrender of fugitive offenders or the existence of the death penalty in *Mutual Legal Assistance in Criminal Matters Ordinance.*"[80] However, Hong Kong opposition activists, human rights activists, and anti-communist activists worried that agreements on the surrender of fugitive offenders would open the door for those "dissidents", "pro-democracy activists" and "human rights activists" to be repatriated by the HKSAR Government even if they had foreign nationalities or permanent residency in Hong Kong, so they treated agreements on the surrender of fugitive offenders with reservations or even stood against them. But they were not concerned about whether mainland China would become a refuge for Hong Kong criminals, because they believed that even though there were no agreements, the mainland government would surrender the criminals who fled to the mainland to Hong Kong for the interest of Hong Kong. In this case, the lack of agreements on the surrender of fugitive offenders rendered Hong Kong a loophole in national security to some extent.[81]

The other legal means is the operation of national emergency power in rarely seen emergency situations. Article 18 ruled that: "In the event that the Standing Committee of the National People's Congress decides to declare a state of war or, by reason of turmoil within the HKSAR which endangers national unity or security and is beyond the control of the Government of the Region, decides that the Region is in a state of emergency, the Central People's Government may issue an order applying the relevant national laws in the Region." However, the operation of national emergency power would have an adverse effect on the "one country, two systems" policy, and it would lower the confidence of Hong Kong people and the international community in Hong Kong and even damage the international image of China. The actual utilisation of emergency power is limited in that it cannot prevent Hong

---

80 Ibid, P173.
81 On the other side, the Independent Commission against Corruption of Hong Kong enhances the cooperation with mainland anti-corruption agencies to prevent Hong Kong from becoming the refuge for mainland corrupt officials. The cooperation is beneficial for the anti-corruption of mainland and it encourages mainland enterprises in Hong Kong to improve their administration to meet the international standard.

Kong from becoming "a base for subversion" in normal cases but more likely create fright among the populace.

The central authority can also protect national security through legislation for the HKSAR or introduce mainland laws to Hong Kong since it is responsible for issues including national defence and diplomacy under "one country, two systems", but those methods can only be accepted by Hong Kong people in very extreme cases or emergencies, and their effectiveness is also limited in normal scenarios.

Article 48 provision 8 of the Basic Law requires the chief executive "to implement the directives issued by the Central People's Government with respect to the relevant matters provided in this Law", including issues related to national security, but the implementation of this power is rare even today; accordingly, it has yet fallen short of being able and competent enough to shoulder the grave task of responsibility for national security.

Besides legal means, there are also political, administrative and military means available to the central authority and the HKSAR Government for reducing the possibility of Hong Kong becoming a "base for subversion". The significance of the Chinese People's Liberation Army stationed in Hong Kong is to demonstrate national sovereignty over Hong Kong and to eliminate the threat to national security posed by Hong Kong. Article 14 of the Basic Law ruled that: "Military forces stationed by the Central People's Government in the HKSAR for defence shall not interfere in the local affairs of the Region. The Government of the HKSAR may, when necessary, ask the Central People's Government for assistance from the garrison in the maintenance of public order and in disaster relief." But the People's Liberation Army Hong Kong Garrison only operates when necessary, so it cannot assume the daily responsibility for safeguarding national security.

In fact, there are many laws in Hong Kong that can be applied for maintaining public security and public order and restraining the activities in Hong Kong which pose a threat to the central authority and mainland governance. In order not to meet challenges in courts or face accusations from human rights groups, the HKSAR Government applied those laws with caution. Experience after the return has showed that those officials of the Department of Justice or among the Hong Kong judges who regard maintaining universal human rights as their prime responsibility showed a tendency not to approve enforcement measures in the HKSAR.

Struggle to Construct New Identity

Following colonial independence, most new leaders devoted themselves to reshaping national identity. The main purpose is to construct national identity and loyalty, a feeling that the whole nation is a solitary political community with a shared destiny, a feeling of pride towards national tradition and culture, and a sense of confidence in mastering the destiny of the nation and the state. Meanwhile, the new leaders have also wanted to reduce the sense of inferiority along with both the blind worship and fear of the West through criticism of colonial governance and Western values. Since such reconstruction of national identity was related to national unity, security and development, it received much attention from the new independent states. Some newly independent countries were left with lines of demarcation, drawn during the colonial period, which separated ethnic groups and created geographical divisions, separating people into groups that remained alien or even came into conflict; in such cases, it is politically important to construct a new identity that unites people of all ethnicities.

Before independence, some previous colonies had already laid down the groundwork of uniting people in the colony and constructing a new identity beyond that of individuals, ethnicities and regions through ideological construction, political propaganda, armed struggle and anti-colonial and independence movements. In order to build a new identity, the leaders of independent movements, such as Gandhi of India and Ho Chi Minh of Vietnam, selected and glorified the tradi-tional culture while selectively taking advantage of religious beliefs, and also, adopting some Western anti-colonialism ideas, appropriating Western radicalism (socialism in particular) or fostering anti-colonialist discussions. In other colonies, such as Singapore, the construction of new identity was delayed because independence was achieved without a protracted struggle, so time and opportunity were insufficient for constructing a new identity before independence. Therefore, they promoted construction more vigorously after independence.[82]

Hong Kong is a special case. Since there were no independence movements, anti-colonialism movements, or charismatic leaders in Hong Kong, the construction of a new identity could not be achieved through movements or by those leaders. Although the British tried to consciously enhance Hong Kong people's sense of belonging to Hong Kong after 1970, it made little progress due to vague policies, weak implementation and the political concerns of the colonial rulers. More importantly, the colonial government deliberately reduced Hong Kong people's sense of belonging to China and separated "Hong Kong people" from "the Chinese people" on purpose. As a result, it is true that after 1970, most Hong Kong people came to regard Hong Kong as their home, their own place for themselves and their next generations, and their sense of belonging to Hong Kong grew accordingly.

However, their identity as people of Hong Kong was not strong or clear even at the eve of Hong Kong's return. To some extent, the so-called "qualities of Hong Kong people" was the product of the contrast between Hong Kong people and mainland compatriots. For example, Hong Kong people are hardworking, flexible, deeply influenced by Western culture. They have an international perspective, a love for freedom; they value human rights and the legal system, advocate clean governance and independence from government. They favour fair competition, adhering to the law, valuing systems and procedures, respecting individuals and have a sense of public morality, etc. Some Hong Kong people even feel proud of being "colonial" people, thinking

---

82 Reference: Margaret Kohn and Keally McBride (2011). *Political Theories of Decolonization: Post colonialism and the Problem of Foundations*. New York: Oxford University Press.

they are therefore superior to their mainland patriots. Yet most people still regard themselves as Chinese from the perspective of ethnicity, bloodline, history, culture and geography. They politically disapprove the People's Republic of China and the CPC mainly because many Hong Kong people came to Hong Kong to escape the governance of the CPC or they dislike the CPC because they themselves or their relatives suffered from political persecution on the mainland.

In general, before the return of Hong Kong, the so-called identity awareness of Hong Kong people was neither strong nor clear. Hong Kong people could not be regarded as a group of people sharing a "common fate" or a "national destiny" different from the Chinese nation. There were just many commonalities between Hong Kong people. Before the reunification, a large number of Hong Kong people migrated to foreign countries, and many of those who chose to stay in Hong Kong merely did so because they did not have opportunities to leave. Waves of mass migration from Hong Kong are the best portrayal of the weak sense of common identity of Hong Kong and their lack of a sense of belonging.

Objectively, the return of Hong Kong to China means that the political identity of Hong Kong people would change significantly from "colonial subjects" or "Hong Kong residents" to "citizens of the People's Republic of China", which, theoretically speaking, would cause Hong Kong people to reflect on their personal identity. The question of "Who are We?" should have been brought up. Hong Kong people as "Chinese citizens", their relationship with China as well as mainland compatriots, and Hong Kong people's responsibilities and obligations towards China, should have become hot topics immediately. The emergence of and the discussion about these topics would have had an actual effect on the construction of the new identity of Hong Kong people. However, except for some intellectuals, only a small number of Hong Kong people were interested in those topics and they were not bothered about their identity, nor did they comprehend their oncoming identity crisis. They even found it difficult to understand or resonate with the "propositions" that deliberately separated Hong Kong from mainland compatriots, such as "Hong Kong Nation Theory", "City-State Theory", "Hong Kong Reform Theory" or "Hong Kong Agency Theory".[83] Many people did not see the need for changing their identity, and thought it acceptable just to continue with their ideas and behaviours from the "colonial" period, which they understood to be the original intent and main idea of "remaining unchanged for fifty years."

In fact, the "one country, two systems" policy of the Chinese Government did not explicitly set the target of changing Hong Kong people's identity because, for one thing, it would not help to reassure Hong Kong people, and for another, it was not necessary. To be specific, the "one country, two systems" policy does not require Hong Kong people to be "patriotic" or "love the CPC". It only requires people who "administer" Hong Kong to be "patriotic", but the definition of "patriotic" is also rather broad. The policy only requires Hong Kong people not to harm the country or the regime, not to engage in activities that would harm the "one country, two systems" policy or the implementation of the Basic Law in Hong Kong.

From the perspective of implementing "one country, two systems" policy, after the return, the Chinese Government should have adopted various policies and measures to promote the identity change of Hong Kong people in order to defend state sovereignty, security and development and to achieve the goal of "Hong Kong people governing Hong Kong". The Chinese Government should also have taken this opportunity and cultivated a large number of "patriotic" political talents in order to ensure the long-term implementation of "Hong Kong people governing Hong Kong". So the national education and patriotism education should have been regarded as the strategic focus of the "one country, two systems" policy. It was equally important for the country to inculcate the concept of the state and instil a national consciousness in Hong Kong people and to undergo a process of "decolonisation" in their ideas and mentalities. Considering that "patriotic" political talents

---

83 See Chow Wing-sun (2015). *Hong Kong People's Identity and Values* (《香港人的身份認同和價值觀》). Hong Kong: Chung Hwa Book Co.

were rare before the return, the training of those talents should also have been the top priority.

However, the Chinese Government did not put changing Hong Kong people's identity at the top of its agenda, unintentionally missing the opportunity to reconstruct the Hong Kong people's identity and allowing the opposition forces to gain a "monopoly" position. There are several important reasons for this.

First, as mentioned above, any intentions or activities smacking of "ideological transformation" from the Chinese Government would definitely have caused Hong Kong people to panic, which would be detrimental to instilling confidence in "one country, two systems" and Hong Kong's future. In light of the deep anti-communist sentiment in the Hong Kong people, transforming their political identity would be no easy task.

Second, given the sizeable gap between the "one system" in Hong Kong and the "one system" on the mainland, whether Hong Kong people were patriots or not did not much matter so long as they did not damage the "one system" on the mainland.

Third, the rights and obligations of Hong Kong people as "Chinese citizens", differs greatly from those of their mainland compatriots but remain hard to define. In view of the intentional contrast between the identity of Hong Kong people and that of mainland people, it is difficult to convince Hong Kong people that they share a "common destiny" with mainland compatriots. In such an unfavourable environment, no positive outcomes would result in carrying out national or patriotic education.

Fourth, Hong Kong enjoys a high degree of autonomy and governs itself following the "one country, two systems" policy. Although the Chinese Government is responsible for national defence and foreign affairs, in a relatively peaceful international environment, along with the increasingly growing Chinese national power, Hong Kong people are generally unaware of the crisis posed by "internal and external threats," so they lack the patriotic feeling that usually arises when a country faces serious difficulty. Under the normal operation of "one country, two systems" Hong Kong people would not realise the direct and immediate influence of the nation and the central authority, especially in the event that mainland policies and laws are not generally applied in Hong Kong.

Fifth, the growing prosperity, rising international status and influence of China and the greater respect Chinese people receive from around the world make Hong Kong people feel prouder about China and the Chinese nation, but these factors haven't really caused Hong Kong people to feel a greater identification with the People's Republic of China.

Sixth, the values and political ideals cherished by Hong Kong people have been incompatible with those held by the mainland, which hampers their identification with the mainland and the CPC. This obstacle to the construction of a new identity is hard to overcome within a short period of time.

Seventh, as the economic cooperation between the mainland and Hong Kong is becoming increasingly close following the "one country, two systems" policy, Hong Kong people also realise the importance of national prosperity for Hong Kong, thus feeling a sense of "community of shared destiny" between Hong Kong and the mainland. Despite this, the frequent economic interaction has only had a limited positive impact on changing Hong Kong people's attitude towards the CPC. Some Hong Kong people even regard themselves as victims of the economic integration of the mainland and Hong Kong.

Eighth, in the past the mainland over-emphasised idealism and romanticism. Socialism, which emphasised egalitarianism and socialism did attract quite a following among the younger generation. But the reform and opening-up policy which brought prosperity and improved people's well-being also bred a socialist market economy that advocated material returns and phenomena like venality, selfishness, extreme disparity between the rich and the poor, corruption, and declining morality, which made it hard for Hong Kong people, especially the younger generation, to embrace the Chinese nation. Lacking emotional appeal, ideals, ambitions and a sense of mission that attracted Hong Kong's younger people, they felt alienated from or even disliked the national spirit represented by mainland China.

Ninth, after the return, channels for Hong Kong people to participate in international affairs have been extremely limited, which has had an adverse impact on enhancing the relationship between Hong Kong people and China. The mainland scholar Qiang Shigong made a comment on this topic. He stated: "However, Hong Kong residents cannot join the army to defend the motherland, cannot take the mainland college entrance examination for national education, cannot take the civil service examination to serve the people, and they are regarded as overseas students on the mainland; an average resident cannot seek justice if he is treated unfairly; a poor person in a desperate situation cannot qualify for relief from the nation.

In a word, as long as the compulsory juristic identity disparity exists, how could it be possible to nurture the loyalty of Hong Kong people to the nation? The idea of resident identity is not an abstract legal concept but a natural feeling that is intertwined with people's life and destiny. "A nation is not an abstract symbol, but an entity essential to daily life. Nowadays, the 'non-national treatment received as Chinese nationals' Hong Kong people are facing may be the exact reason for the slow progress of winning their recognition."[84] Of course until today, the number of Hong Kong people looking to engage in national and international affairs is not very large, but if more Hong Kong people take vital roles in the national and international affairs as well as national defence, Hong Kong people would notice the attention they receive from the mainland and thereby develop a favourable impression of the PRC and the CPC.

In all fairness, the Chinese Government and the HKSAR Government did not completely ignore the importance of the construction of Hong Kong people's identity. Some forms of "national education", such as facilitating exchanges among the younger generations between Hong Kong and the mainland, promoting the Basic Law, sending

---

84 Qiang Shigong (2014) *Hong Kong China: Visions on Politics and Culture* (《中國香港：政治與文化的視野》). Beijing: SDX Joint Publishing Company. P198.

Olympics medalists to visit Hong Kong, and sending space flight elites to Hong Kong to meet with Hong Kong people have had great influence. However, national education efforts haven't maintained great momentum due to various obstacles. To sum up, the limited investment of resources and the shortage of determination and perseverance constrained the effect.

In fact, the more severe trouble brought by the implementation of "one country, two systems" was the identity "crisis" experienced by a certain number of Hong Kong people. To be specific, Hong Kong people started to question the identity which had supported their pride and confidence in the past and started to look for a new identity.

As I have mentioned above, in many aspects the identity of Hong Kong people stems from the exaggeration of the differences between Hong Kong people and mainland compatriots, smearing and belittling others in order to elevate their own status. Hong Kong people have emphasised their concept of anti-communism and identification with capitalism, believing in and practising so-called "universal values" (Western values actually) and Chinese traditional culture and values, and thought that the regulations, laws and ways of doing things in Hong Kong were more advanced, open, and fairer than in the mainland. All of these made Hong Kong people proud and complacent and led them to assume that the disparity of the development and living standards between Hong Kong and the mainland was natural and would never change. These ideas support Hong Kong people's sense of superiority and arrogance toward their mainland compatriots. This kind of identity gradually developed after the war especially among the younger generation. In this case, Hong Kong people had a stronger sense of belonging to Hong Kong and felt more aliened to their mainland compatriots. According to this identity, the colonial governance is not an evil, but a blessing Westerners gave to Hong Kong, and the British are the "torchbearers" for Hong Kong people.

In past decades, following the reform and opening-up policy, mainland China has made considerable advances in all respects. Its development and substantial prosperity have provided Hong Kong's economy with continuous support and opportunities. Many countries and regions in the world eagerly pursue a relationship with China in order to share in China's development. The beneficial location and the special care from the central authority should have enabled Hong Kong people to gain a stronger sense of connection to and confidence in the country, the central authority and the mainland compatriots; however, the situation turned out to be more complicated. There was no doubt that most Hong Kong people had more confidence in the country, the central authority and the "one country, two systems" policy due to the achievements and growing international status of China and were pleased to see the accomplishments of their mainland compatriots in many respects. However, there were still many Hong Kong people, young people in particular, who got lost in an "identity crisis" because of the rise of China. They lost their confidence and sense of pride in Hong Kong and themselves. Feeling depressed, helpless, resentful and unfairly regarded, they developed as sense of resistance towards the country, the central authority and the mainland compatriots. Some people even held a sense of hostility; they treated the "new immigrants" from the mainland and the mainland compatriots who came to Hong Kong for tourism or shopping with abusive words, cursing them as "locusts". A very small number of people even denied their identity as Chinese people and proposed "Hong Kong independence".

The rise of China, the growing competitiveness of the mainland economy, the flow of capital and talent from the mainland to Hong Kong, mainland compatriots' habit of binge shopping among the luxury shops of Hong Kong, Hong Kong's growing dependence on the mainland have all affected the core part of Hong Kong people's identity. Obviously, the continuously rapid development of the mainland economy forms a sharp contrast with the slow development of Hong Kong and Hong Kong's monotonous industrial structure, and strongly affects Hong Kong people's pride in their economic achievements. There is a huge disparity between the bright prosperity of the mainland economy and Hong Kong's depressed economy. And the significance of Hong Kong in the national economy continued to decrease as it simultaneously became more and more dependent on the support of national favourable policies.[85] People were depressed by media coverage of the ostentatious shopping habits of wealthy mainlanders in Hong Kong. The high living standards of mainland compatriots called attention to the plight of middle and lower class Hong Kong people who suffered due to the widening income gap. All of these made Hong Kong people question and become dissatisfied with themselves, Hong Kong's capitalism, the once-effective "positive non-interventionism" policy, and the fairness and the rationality of Hong Kong society.

The glory of Hong Kong values, Hong Kong systems and the lifestyle Hong Kong people had been long proud of and enjoyed showing off to mainland compatriots faded with the rise of China, and a sense of inferiority and insecurity emerged. There were even those worried that elements from the mainland would gradually replace or destroy those of Hong Kong, but what they worried most are that the adverse events such as corruption, rule of people and privileges would corrode the integrity, the legal system and justice principles that they had always cherished.

Since some Hong Kong people, especially a small number of scholars, being no longer maintain their sense of superiority over their mainland compatriots, have tried their best to dig around for new advantages and features of Hong Kong that can be used to reconstruct Hong Kong people's identity and recall the confidence and pride of Hong Kong people. Their efforts included idealizing the "colonial" period of Hong Kong. The colonial administration was approved which made some people, young people who never experienced the colonial governance in particular, believe that the "merciful" colonial governance was a "blessing" to Hong Kong people and it was the reason why Hong Kong surpassed the mainland in many areas. Others turned to "non-material" aspects to find the advantages in Hong Kong, claiming that the legal system, human rights, liberty, and pluralism were the "core values" that Hong Kong had apparently always exemplified, and

85 Although the "preferential policy" towards Hong Kong was beneficial to the development of the country and it suited the overall development strategy, many Hong Kong people and mainland compatriots still regard this as "great gift" or "blood transfusion".

those people proudly believed that the mainland lacked these values and advantages.[86]

While some Hong Kong people were reconstructing their "identity" in order to regain confidence, their efforts also intensified the contradiction between Hong Kong people and the mainland compatriots and internally divided Hong Kong people. Aiming to highlight the differences between Hong Kong people and mainland compatriots and praising Hong Kong people while criticising mainland compatriots, the reconstruction of Hong Kong people's identity would definitely tend to cause the mainland to dislike Hong Kong people and cause contradictions and conflicts to arise. In Hong Kong, the new Hong Kong "identity," characterised by the exclusion of mainland compatriots and the central authority, slandering the mainland, could easily evolve into various forms of "localism" and even the spread of ideas about "Hong Kong's independence from China". However, the majority of Hong Kong people did not accept these "new things" as they had strong emotional ties to China. They were impressed by the achievements made in the mainland over the previous thirty years and thought that some attributes of the mainland, such as decision-making mechanisms, executive capacity, economic planning and the central authority's care for the underprivileged were worth learning from. In this way, some people's "localism" and "separationism" which were not accepted by the majority, would lead to discussion about conflicting definitions of "Hong Kong people". More seriously, to express their anger, draw others' attention and attack dissidents, some people would resort to abusive language or even violence. Young college students and high school students in particular set a bad example and the "Occupy Central

---

86 The rise and rapid development of China has posed influence not just in Hong Kong. China used to be impoverished and backward, and had always been looked down upon by the developed countries. As a result, people in the developed countries "felt unbalanced" when they saw China's continuous growth of economy and enhancement of competitiveness. About the situation of Japan and America, please refer to Brad Glosserman and Scott A. Snyder (2015). *The Japan-South Korea Identity Clash: East Asian Security and the United States.* New York: Columbia University Press, P47-48 and Benjamin I. Page (2010). *Living with the Dragon: How the American Public Views the Rise of China.* New York: Columbia University Press respectively.

Movement" lasting 79 days in the second part of 2014, was another. The barbarian behaviour and illegal actions of some Hong Kong people during the movement displayed their denigration and trampled on the "core values" characterised by"inclusiveness", "adherence to the law", "treating people with courtesy" and "peace". Their actions also aggravated the greater part of Hong Kong people's anxiety towards Hong Kong's current situation and prospects, and caused them to fear, in particular, that the "core values" they were once proud of would disintegrate. The conflict between Hong Kong people and "localism" as well as the growing aversion towards "loyalists" are exacerbating the social rifts in Hong Kong society.

In the process of constructing and reshaping Hong Kong people's "identity", the opposition forces have always acted as leaders. In the period of 1997–2003 when the central authority and the HKSAR Government ignored the identity issue, the opposition forces were experiencing the "loneliness of being at the top" of the battlefield of constructing and reshaping Hong Kong people's "identity". They successfully guided Hong Kong people's identity towards the "separation" from mainland compatriots and set Hong Kong people against Chinese people and the Chinese nation. In fact, before the return, the difference in identification between Hong Kong people and Chinese people was insignificant since most Hong Kong people still considered themselves Chinese people.[87] But with the unremitting efforts of the opposition forces, some people who named themselves "Hong Kongers" became increasingly convinced that they were not Chinese and even discriminated against or rejected the notion of being "Chinese" both conceptually and in real action. To achieve their aims, the opposition activists rejected all forms of "national education" and smeared it as "brainwashing education". With the emergence of the "crisis" of Hong Kong people's "identity", some opposition activists absorbed some discontent and demands from the "localism" into their political stance

---

87   Lau Siu-kai (1997). "'Hong Kong People' or 'Chinese People'": Hong Kong Chinese's identity 1985-1995 ("'香港人' 或 '中國人' : 香港華人的身份認同"). *The 21ˢᵗ Century* (《二十一世紀》), Vol. 41, P43-58.

and further developed the "localism" and Hong Kong people's "sense of agency". Although the majority of the opposition activists did not approve of "Hong Kong's independence from China", it is still obvious that they led Hong Kong people, the younger generation in particular, to the opposition side and away from the central authority and mainland compatriots.

What worried people more is that the emergence of some Hong Kong people's new "identity awareness" came at a time when mainland compatriots' patriotism and sense of national pride were increasing and when they felt sceptical about the Western values and development models. The rise of China and the notable success made mainland compatriots more respectful of and confident about Chinese culture and Chinese values.

Beginning in 1999, the "patriotic education" whose main content was "never forget national humiliation" was vigorously promoted in the mainland. The education achieved instant results and greatly changed young people's ideas, reducing their blind worship of the West and boosting their respect for Chinese civilisation.[88] In recent years, the mainland advocated traditional Chinese culture and listed "cultural security" as part of national security. By linking "cultural security" with patriotic education, the mainland improved people's awareness and recognition of Chinese culture. Meanwhile, many mainland compatriots felt that the Western world was caught in an irreversible process of

---

88 According to the research of the Chinese scholar Wang Zheng who stayed in the U.S., "In the past thirty years, China's opening and the close interaction with the international community witnessed the emergence of a new generation of anti-western patriots." "After the Tiananmen Square Protests and the breakdown of Communism in eastern Europe, the patriotism and the theory of national humiliation became a vital part of China. "By emphasising that the CPC was the backbone of fighting for the national independence, the patriotism education consolidated the authority of the CPC." "The democratic campaigns in 1980s which were characterised by introversion, anti-corruption and anti-dictatorship principles turned into extroverted, anti-western nationalism in the 1990s". See Zheng Wang (2012). *Never Forget National Humiliation: Historical Memory in Chinese Politics and Foreign Relations* New York: Columbia University Press. P2, P96, P100 and P116. Another Chinese overseas student Zhao Shuisheng in the U.S. had the same opinion. He thought that the CPC regarded itself as the protector of the interest and honour of the nation, the nationalism was the moral support of China. See Zhao Suisheng (2004). *A Nation-State by Construction: Dynamics of Modern Chinese Nationalism*. Stanford: Stanford University Press.

decline, and Western systems and the values did not receive extensive praise.

A growing number of people agreed that the long history of China, the rich culture and the unique conditions of China made it vital that China should explore a development path that suited itself instead of rigidly copying Western models. Due to the growing national confidence and pride, mainland compatriots were not willing to follow the Western path, which once prosperous, showed a lot of weaknesses at present. The "Socialism with Chinese characteristics", the "Chinese model", [89] the "Beijing consensus" and the "Chinese dream of rejuvenation of the Chinese nation"[90] revealed that to achieve national prosperity, well-being of people and open politics, Chinese people must and can only resort to their own specific development path. It is true that there were no authorised descriptions of the path of China's development or its significance for other countries, and yet some elements with Chinese characteristics have been perceived in many areas. In economy, it has become an unquestioned fact that the market competition, state

---

89 Pan Wei (2009). *A New Development Model from the Sixty Years of the People's Republic* (《中國模式：解讀人民共和國的 60 年》). Beijing: Central Compilation & Translation Press.

90 See Xi Jinping (2013) *The Chinese Dream of the Great Rejuvenation of the Chinese Nation* (《習近平關於實現中華民族偉大復興的中國夢論述摘編》). Beijing: Central Party Literature Press. In different parts of this book, Xi illustrated his ideas about "Chinese dream". "To build a well-off society in an all-round way, to build a prosperous, strong, democratic, culturally advanced and harmonious modern socialist country, to achieve the Chinese dream of the great rejuvenation of the Chinese nation, is to achieve the prosperity, national revival, people's well-being. These reflect Chinese dreams and the honourable tradition that our ancestors constantly pursuit." (P4-5) "To achieve the great rejuvenation of Chinese nation is the greatest dream since modern times which we called 'Chinese dream', the basic content of which is to achieve national prosperity, national rejuvenation and people's well-being." (P5) "In the new historical period, the essence of the Chinese dream is national prosperity, national rejuvenation and people's well-being. Our goal is by 2020, the GDP and the income of urban and rural residents per capita doubles compared with that of 2010 and build a moderately prosperous society. By the middle of the 21st century, a prosperous, strong, democratic, culturally advanced and harmonious modern socialist country will be in place and the Chinese dream of the great rejuvenation of the Chinese nation will be realised." (P7) "We are a great people. In our long history, with their diligence, courage, wisdom and intelligence, the Chinese people have created a beautiful homeland with all groups of people living harmoniously together and nurturing a great culture. Our people enjoy life, hope for better education, long for more stable jobs, more satisfying incomes, more reliable social security, higher standard of medical care, more comfortable living conditions, more pleasant environment, hope for better prospects, better jobs, and better living standards for their children. To fulfill the people's longing for a beautiful life is our goal." (P13)

enterprises and the "visible hand" of the government should co-exist. "Collective leadership" and "reasonable decision" is another bright spot.[91] The Chinese political system is described as a system of "meritocracy" which is the counterpart of the Western democratic political regime and shares distinguished features, and which suits developing countries in particular.[92]

While mainland compatriots vigorously advocated Chinese culture and doubted Western culture, some Hong Kong people flaunted the Western values on purpose and even praised colonialism, so the "cultural conflict" and the "moral criticism" between mainland compatriots and some Hong Kong people was hard to contain. The estrangement, misunderstanding and friction between people were definitely bad for the economic cooperation and emotional communication between Hong Kong and the mainland.

In short, the "one country, two systems" policy does not require Hong Kong people to love the country or the party. All it expects is that Hong Kong people refrain from confronting the central authority and the people on the mainland, and keep from turning Hong Kong into a "base for subversion". However, the central authority and mainland compatriots still expect that Hong Kong people show more concern for their own country and nation. Such expectations are reasonable, especially as the central authority and the country have shown so much kindness, caring and respect for Hong Kong people. However, instead of injecting more Chinese elements into the construction of the identity of Hong Kong people, the reunification fosters sentiments of separation

---

91 See Hu Angang (2014). *Research on the Collective Leadership of China* (《中國集體領導體制探究》). Hong Kong: Chung Hwa Book Co.; Wang Shaoguang, Fan Peng (2013) *The Chinese Model of Consensus Decision Making: a Case Study of Healthcare Reform* (《中國式共識型決策: "關門" 與 "磨合"》). Beijing: China Renmin University Press; David M. Lampton (2014) *Following the Leader: Ruling China, from Deng Xiaoping to Xi Jinping.* Berkeley and LA: University of California Press.

92 See Daniel A. Bell (2015). *The China Model: Political Meritocracy and the Limits of Democracy.* Princeton: Princeton University Press. According to Bell's analysis, the meritocracy of China can be categorised at three levels: democracy at the grassroots level, in-between the localities and the central government are incessant experimentation and innovation, and meritocracy at the top. The Chinese model guaranteed that the top leaders are equipped with talents and morality as well as rich administration experiences.

from the country and the nation, posing potential threats to the security and the interests of China. Such sentiments are intolerable to the mainland compatriots. It will certainly arouse their antipathy and undermine their support for the "one country, two systems" policy. Therefore, an intense battle between the central authority and the SAR Government over the construction and reshaping of the identity of the Hong Kong people is inevitable.

CHAPTER FOUR

Changes in the International Situation in Mainland China and in Hong Kong during the Transitional Period and after the Handover

After two years of painstaking negotiation, China and Britain signed a joint declaration in 1984, under which Britain agreed to return Hong Kong to China, while China was committed to implementing the "one country, two systems" policy in Hong Kong. China also made a solemn promise to keep the social and economic systems as well as the lifestyle of the people in Hong Kong unchanged for 50 years, meaning, as they were found in the mid-1980s. However, the political system was an exception since both countries agreed that the political system in Hong Kong should democratize progressively. The Chinese Government also pledged to follow through with the policies and guidelines stated in the Joint Declaration by stipulating them in the Basic Law and thus lent them legality. For the Chinese Government, the status which would remain unchanged for 50 years was basically the "status quo" in 1984, when the Joint Declaration was signed. One important part of that status quo is that Hong Kong was not a "base for subversion". Of course, when the Basic Law was promulgated in 1990, it selectively adopted some "changes" in Hong Kong's political system that had been introduced by the departing colonial government between 1984 and 1990. The Chinese Government hoped and demanded the British Government not to change the "status quo" in Hong Kong during the transitional period between 1984 and 1997. If the change had to be made for any inevitable reasons, it should not be implemented until negotiated and approved by both sides.

However, as was mentioned before in Chapter 3, the British Government had decided unilaterally to introduce a multitude of political reforms in Hong Kong even before the signing of the Joint Declaration. Britain's overall objective was to preemptively set up the layout of the political system and personnel arrangement in Hong Kong in post-1997 Hong Kong, so that the HKSAR would have no alternative but to implement the "British version" of "one country, two systems". The political intention and actions of Britain became more blatant after the "June 4 event". Confident that the Chinese communist regime would collapse in no time, Britain decided to take proactive actions and if necessary even confront the Chinese Government. Although Britain continued to cooperate with China in many areas during the remaining years of the transitional period, its non-cooperative or even confrontational gestures in the political arena did create some adverse factors that hindered or even distorted the practice of the "one country, two systems" policy after the handover. The after-effects of Sino-British non-cooperation are still visible even today.

What affected the practice of the "one county, two systems" even more was the dramatic change in the international environment after China and Britain signed the Joint Declaration in 1984. With the collapse of communism in Eastern Europe, the disintegration of the Soviet Union, the isolation of China by the West, the rise of the U.S. hegemony and the worldwide dominance of the Western ideas of liberal democracy and the free market, the influence of socialism plummeted, and only a few countries still continued to uphold socialism. In this unipolar world, the U.S. attempted to reshape the global system and world order in its self-image, deploying the enormous political, military and economic power at its disposal. However, drastic changes soon reappeared to reshape the international landscape, clearly proving that human history had not yet reached the end and that there was no "final destination" in social, political, economic and ideological development. Human history would continue to evolve and progress with different countries and civilisations competing with, learning from and promoting each other.

In the past 18 years since the return of Hong Kong, the most important changes affecting the practice of the "one country, two systems" policy are other than China's rapid rise in all dimensions, the geographical shift of global economic power from the West to the East (particularly East Asia), the financial and economic crises plaguing Western countries, the decline of the appeal of Western values and the increasingly fierce competition between the East and the West, particularly between China and the U.S.. The drastic changes in the surrounding environment of Hong Kong will inevitably bring about changes in the territory and give rise to new and unanticipated challenges and difficulties. Hong Kong's relationship with mainland China and with the West have also witnessed some subtle changes, posing new issues and challenges to the practice of the "one country, two systems" policy as well.

The Rise of China

After the problem of Hong Kong's future was resolved, the reform and opening up processes in China also accelerated, and remarkable achievements were reached in the fields of economy, politics, military and diplomacy. The rise of China is based on the construction of "socialism with Chinese characteristics". China's development path is quite unique with its ingenious combination of a strong collective political leadership, market-driven economy, economic and social planning, state-owned enterprises, a robust private sector, emphasis on both export and domestic demand, and the positive role of the government in delivering public benefits and services. China's development strategy continues to be highly and widely praised for achieving continuous and steady economic growth, having even been hailed as the "Beijing

Consensus"[1] or the "Beijing Model"[2]. The focus of the strategy keeps flexible in the face of needs and goals of national development, among which the transition from "bringing in" to "going out" is the most striking. The rise of China has brought about significant changes in the international configuration, order and rules. For Hong Kong, the rise of China changes the relationship between mainland China and Hong Kong, the role and importance of Hong Kong in terms of national development, as well as the international environment for Hong Kong along with the nature of the relationship between Hong Kong and the Western countries.

With regard to the economy, China's overall economy as measured by GDP exceeded that of Japan in 2011, making China the second largest economy following the U.S. Hu Angang (胡鞍鋼), a mainland scholar, described the process of Chinese economic development since "Reform and Opening up" in the following words: "In April 1987, Deng Xiaoping elaborated thoroughly on the strategy to achieve the national target in "three steps". In the first step, the GDP per capita will reach 500 dollars, doubling the 250 dollars of 1980. In the second step, China's GDP per capita will further double to 1000 dollars by the end of 20th century. In the third step, the most important, the GDP per capital will achieve a further fourfold increase, reaching about 4000 dollars." "In the 2002 *Report of the 18th CPC Congress,* President Jiang Zemin clearly set out the ambitious goal of building a more advanced, better-off society that would benefit over one billion people by 2020. It was proposed that by 2020, the GDP should quadruple as compared with 2000 and fundamentally achieve industrialization. The specific idea and main target at that time was to ensure that the GDP per capita reached over 3000 dollars, reaching the approximate level of middle-in-

1   See Joshua Cooper Ramo (2004). *The Beijing Consensus.* London: Foreign Policy Centre; Stephen Halper (2012). *The Beijing Consensus: Legitimising Authoritarianism in Our Time.* New York: Basic Books.

2   Not everyone agrees that the development model of China is unique. Some people think that China is only copying the route of "The Four Asian Tigers" and she can only be regarded as a huge dragon at most. See Huang Yasheng (2014). *How Unique is the Chinese Model* (《中國模式到底有多獨特》). Hong Kong: Chung Hwa Book Company.

come countries, to raise urbanisation to over 50% while lowering the rate of agricultural employment to about 30%. In 2007, President Hu Jintao suggested new requirements for building a well-off society, and the grand blueprint drafted for 2020 became clearer and more complete. The target was further revised, requiring a quadrupled per capita GDP reaching 40000 RMB (5000 U.S. dollars) by 2020. This target was higher than that set by the 16th CPC National Congress."[3]

The mainland economist Lin Yifu (林毅夫) described China's economic achievement as follows: "China was one of the low-income countries in 1979. In accordance with the unwavering prices in 2000, the income per capita was only 175 dollars, lower than that of one-third of the poorest sub-Saharan African countries. In 2012, the income per capita surged to 6000 dollars, five times that of Saharan African countries. The excellent performance of China's economy had lifted more than 600 million people above the poverty line of 1.25 dollars per day per person as defined by the Word Bank."[4] "By acknowledging its comparative advantages in different periods of development and capitalizing on the advantages of the latecomer in technological innovation and structural transformation, any developing country can speed up its economic growth. It is very likely that China would also achieve the grand goal introduced in the 18th CPC Congress: doubling the GDP and income per capita by 2020 with 2010 as the base for comparison, thus achieving a well-off society by the 100th anniversary of the CPC, and developing China into a prosperous, strong, democratic, culturally advanced and harmonious modern socialist country."[5]

Li Yining (厲以寧), a mainland economist, offered a brief overview of China's economic transformation: "From 1979 on, China entered into a dual transitional phase, which means a combination or

---

3 Hu Angang, Yan Yilong & Wei Xing (2011). *2030 China: Towards Common Prosperity* (《2030 中國：邁向共同富裕》). Beijing: China Renmin University Press, P8-9.

4 Lin Yifu (2013). "China's Rejuvenation: Experiences, Challenges and Outlook on the Future" ("中國的復興之路：經驗、挑戰與未來的展望"). In Cheng Siwei, Li Yining, Wu Jinglian & Lin Yifu (2013). *Reform is the Biggest Bonus* (《改革是中國最大的紅利》). Beijing: People's Publishing House. P51-72, 53-54.

5 Ibid, P68-69.

overlapping of both system transition and development transition. What is system transition? It is the transition from a planned economic system to a market-driven economic system. What is development transition? It is the transition from a traditional agricultural society to an industrial society. The combination and overlapping of the two transitions are neither precedented nor discussed in traditional development economics. [...] In the reform and transition of the system, property rights reform is the key to breakthrough, as well as the central theme. In the development transition, delimitation and clear definition of property rights are the dynamic mainspring. [...] In a dual transition, people's lives must be improved whilst the economy also grows, which is an important means to narrow regional income differences and the income gap between urban and rural areas. [...] Employment should never be neglected at any time. [...] Expansion of domestic demand is closely related to the improvement of people's lives and is the only way to lead the Chinese economy into a virtuous cycle. Urbanisation will present the greatest opportunity for investment and domestic expansion in the next several years, which will allow the Chinese economy to maintain a relatively fast and continuous rate of growth. [...] The strategic goals of developing the private economy are not only aimed at relieving employment pressure, but also, more importantly, at mobilising the private sectors, including realising the potential of private capital. [...] The strategic goals we are seeking are specific: from the perspective of system transition, to establish a mature socialist market economy; in terms of development transition, to achieve industrialisation, establish a modernised society, promote common prosperity, and build a harmonious society. Reform needs to be deepened and development needs to be strengthened with the utmost resolve, or all previous efforts will be wasted."[6] He also warned that "inheritance of jobs" and the "dual urban-rural structure" must be prevented from hindering national economic development.

The mainstream view of the international community is that,

---

6   Li Yining (2014). *The Road to Dual Transition of the Chinese Economy* (《中國經濟雙重轉型之路》). Hong Kong: Chung Hwa Book Company. P2-6.

despite the large number of difficulties China is facing, including the great difficulty of transforming the mode of economic development, economic upheaval, financial instability, excessive debt burden, as well as resistance from people who have benefited from past economic practices, a majority of observers still are confident that China's economic growth will still proceed at a relatively high rate in the foreseeable future.[7]

In the field of politics, what disappoints and frustrates Western countries, especially their anti-communist and anti-China elements, is that despite the existence of favourable factors such as rapid economic development, increasing market competition and the emergence of a middle class in China, China has so far failed to follow the anticipated path of "peaceful evolution". Instead, under the influence of its historical and political tradition and in order to meet the practical needs of present circumstances, China has gradually and successfully charted a political development path with Chinese characteristics and has established a political system with a bundle of preeminent features that are founded on the premise of the CPC's long-term rule. The main characteristics of the Chinese political system include: a powerful government with solid political authority; a government widely supported and trusted by the Chinese people all over the country; a powerful leadership practising collective consultations and decisions; effective and forceful crisis management; a combination of elite leadership and public participation; and a leadership group that can chart the nation's development path and strategies from macro, historical, rational, pragmatic and long-term perspectives.[8] Daniel Bell, an American scholar teaching

---

7   See the World Bank & the Development Research Centre of the State Council, the People's Republic of China (2013). *China 2030: Building a Modern, Harmonious, and Creative Society*. New York: The World Bank; and Hu Angang (2015). *The Grand Strategy of 13ᵗʰ Five-Year Plan* (《十三五大戰略》) Hangzhou: Zhejiang People's Publishing House.

8   See Hu Angang (2014) *A Research on China's Collective Leadership System* (《中國集體 領導體制探究》). Hong Kong: Chung Hwa Book Company; Wang Shaoguang & Fan Peng (2013). *The Chinese Model of Consensus Decision-Make: a Case Study of Healthcare Reform* (《中國式共識型決策：" 開門 " 與 " 磨合 "》). Beijing: China Renmin University Press; and David M. Lampton (2014). *Following the Leader: Ruling China, from Deng Xiaoping to Xi Jinping*. Berkeley and LA: University of California Press.

in Beijing, even depicts the Chinese political system as "political meritocracy", which, according to his analysis, comprises three tiers: grassroots democracy, continuous experimentation and innovation in the region between the central government and the local government, and meritocratic leadership at the top. A political system like this ensures both the political integrity and professional competence of the top leaders, as well as their rich experience in running the country.[9]

In the eyes of Daniel Bell, despite the inherent defects of "political meritocracy", the Chinese political system, while not necessarily superior to Western democracy, is still a vigorous political system that facilitates China's development, is compatible with China's tradition and national conditions, and possesses its own advantages as compared with Western democracy. Although many people in the West still slander the Chinese Government as a dictatorship or a totalitarian regime that threatens the whole world, even the results of the opinion polls conducted in China by Western pollsters show unmistakably that the Chinese Government is highly supported and trusted by the Chinese people. Such feats as achieved by the CPC Government are definitely rare in today's world.[10] In any case, as long as China is rising and its values and interests are different from those in the West, and notwithstanding the incessant proclamations by China that she is a force for peace and stability in the world, the CPC regime is considered a long-term political rival and threat by the Western governments.

Recently, Chinese leaders have come up with the goal of "modernisation of the national governance system and governance ability" also known as "the fifth modernisation", which is about devising methods and models of governance in China that befit China as a large modernising country and that can improve and upgrade China's bureaucratic efficiency and effectiveness.[11] "The Chinese national governance

---

9   Daniel A. Bell (2015). *The China Model: Political Meritocracy and the Limits of Democracy*. Princeton: Princeton University Press; and John Micklethwait and Adrian Wooldridge (2014). *The Fourth Revolution: The Global Race to Reinvent the State*. New York: Penguin.

10 See the Pew Research Centre's series of opinion polls on public trust in their governments in various countries.

11 See Yu Keping (2014). *Essays on the Modernisation of State Governance* (《論國家治理現代

system and capability in terms of governance is a concentrated reflection of the country's system and its executive capacity. The national system is precisely governance under the leadership of the Chinese Communist Party. This system is integrated, closely inter-connected and coordinated, covering the institutions, mechanisms, laws and regulations in different fields, including economy, politics, culture, society, the environment and party building. The capability in terms of national governance refers to China's ability to utilize national institutions to guide a wide variety of social processes, including reform, development, stabilisation, internal affairs, diplomacy, national defence and governance of the party, the country and the army. "[12] "To modernise its national governance system and governance capability, China has to adapt to the changing times by reforming its institutions, mechanisms, laws and regulations that fail to keep pace with the needs of development, while at the same time establishing new ones, ensuring institutions that are rational and robust and that the governance of the Party, the nation, and social affairs is increasingly institutionalised, standardised and procedure-based. China needs to work harder to improve its governance, be more mindful of the need to act on the basis of institutions and in accordance with law, and make better use of institutions and laws in governing the country. It needs to use all the strength of its system to make its governance more effective and enhance the Party's capacity to govern scientifically, democratically, and in accordance with the law."[13]

With its rapid economic development, China has accumulated a huge foreign exchange reserve and massive individual savings, along with excess production capacity in numerous enterprises, a large labouring population and a few industries with international competitiveness (for example, railway, energy, construction and telecommunications). China's domestic market cannot fully utilise its production

化》). Beijing: Social Sciences Academic Press; Hu, Angang et al. (2014). *The Modernisation of the China's Governance* (《中國國家治理現代化》). Beijing: China Renmin University Press.

12　CCCPC Party Literature Research Office (eds.) (2014). *Xi Jinping's Comments on Comprehensive and Profound Reforms* (《習近平關於全面深化改革論述摘編》). Beijing: Central Party Literature Press.

13　Ibid, P25.

capacity. In recent years, it has become increasingly clear that China can no longer rely on export and foreign investment because of the global financial crisis, the declining demand for China's export commodities and services in developed countries, the faltering progress of another wave of the trade globalisation and rising protectionism. It is imperative that China boosts domestic demand to maintain high levels of growth and employment.

All these years, China has used most of its trade surplus and the proceeds of investment abroad to buy U.S. Treasury bonds and other foreign assets. In this way, China can keep its currency of the RMB (Renminbi) at a relatively lower rate and boost exports. However, it also suffers from the devaluation of its foreign assets because the U.S. and Europe have adopted some "irresponsible" fiscal policies. In the meantime, the economic growth pattern characterised by a massive economy and serious damage to the environment is outdated and requires immediate change. China has become increasingly reliant on a foreign supply of energy, resources and food. All these factors call for China to broaden and deepen its "going out" strategy, which has aroused concern among many Western countries, especially the U.S., because of China's increasing global influence.[14]

Currently, China's "going out" strategy consists of five important components. First, given that China lacks the resources to support its rapid development, China is actively seeking to acquire oil, natural gas, oil and gas fields, mineral products, agricultural products and arable land in other countries.[15] Naturally, China's strenuous efforts to procure resources around the world have brought up the prices of commodities and agricultural products and incurred some complaints in the international community.[16] Due to its acquisition of land and resources in some

---

14 David Shambaugh (2013). *China Goes Global: The Partial Power*. New York: Oxford University Press. This book offers a thorough description of China's "going global" strategy.

15 See Elizabeth C. Economy and Michael Levi (2014). *By All Means Necessary: How China's Resource Quest is Changing the World*. New York: Oxford University Press; Michael T. Klare (2012). *The Race for What's Left: The Global Scramble for the World's Last Resources*. New York: Metropolitan Books; and Deborah Brautigam (2015). *Will Africa Feed China?* New York: Oxford University Press.

16 On the other way round, the slower growth of China's economy will lead to lower demands

African and Latin American countries, China is even unfairly accused of practising "new colonialism" in developing countries.[17]

Second, as China's foreign exchange reserve is rising and the operational capability of its state-owned enterprises and private companies continues to grow, its domestic market is becoming unable to satisfy the investment needs of the country and its enterprises. Besides, in order to upgrade its economic structure, China also needs to acquire new knowledge and technologies, new markets and industries, as well as advanced managerial and marketing methods from abroad. In recent years, China has acquired, merged with or bought shares in many foreign enterprises. Some countries, Western countries in particular, have a "love-hate" attitude towards China's actions, but they nevertheless still need China's investment to boost their economies.

Third, more and more Chinese people are going abroad. Both elites and ordinary citizens are taking advantage of opportunities to do business, work, study, travel or migrate abroad. [18] The Chinese people are indeed quite hard-working. However, while this industriousness does bring benefits to the development of the local economies outside China, it also to some degree alters its social and physical environments and ethnic relationships, causing some dissatisfaction among local people.

Fourth, some mainland enterprises with international competitiveness have achieved results through developing foreign markets, particularly those in the high-speed railway, construction, infrastructure, power generation and information communication sectors.

Fifth, the RMB is increasingly internationalized. The post-WWII international financial order dominated by the U.S. dollar was forcibly imposed by the U.S., disregarding the concern and dissatisfac-

---

for the commodities and their prices in the world will fall, and that will be detrimental to global economic growth.

17 Deborah Brautigam (2009). *The Dragon's Gift: The Real Story of China in Africa.* New York: Oxford University Press.

18 See Howard W. French (2014). *China's Second Continent: How a Million Migrants are Building a New Empire in Africa.* New York: Alfred A. Knopf; Juan Pablo Cardenal and Heriberto Araújo (2013). *China's Silent Army: the Pioneers, Traders, Fixers and Workers Who are Remaking the World in Beijing's Image.* New York: Crown.

tion of Britain and other allies, though perhaps bringing some benefit to developing countries.[19] Although this new financial order primarily served the interests of the U.S., it also facilitated the economic development of Western countries and some developing countries after the war. However, as the fiscal conditions in the U.S. worsens with long-lasting deficits in its current accounts and continuous depreciation of the dollar, many countries suffer severe losses. The 1997 Asian financial crisis and the global financial crisis in 2008 further exposed the shortcomings and unfairness of an international financial order dominated by the U.S. dollar and attested to the "collapse" of neoliberalism as an orthodox doctrine.[20] Meanwhile, as China has become a major producing and trading nation, China believes that the RMB should have a corresponding international status. The internationalisation of the RMB will definitely improve China's international influence and soft power. Since the 2008 global financial crisis, China has accelerated the internationalisation of the RMB. Though China still lacks the proper conditions to fully open the capital account and needs to guard against financial risks, it has taken various effective measures to promote the internationalisation of the RMB, including deepening the reform of the

---

19 In the Bretton Woods conference, the U.S.'s determination to reduce the importance of the pound and to make the U.S. dollar the primary international currency for trade, investment and reserves was obvious. The U.S. aimed to equate the status of the dollar to gold, establish a fixed exchange rate regime, strictly limit other countries' ability to promote export through currency depreciation, maintain the stability of the international financial order, facilitate the recovery of the American economy and control other countries' fiscal and financial policies. Another hidden aim was to allow the U.S. to arm itself when it needed to wage financial wars. See Benn Steil (2013). *The Battle of Bretton Woods: John Maynard Keynes, Harry Dexter White, and the Making of a New World Order*. Princeton: Princeton University Press; Benn Steil and Robert E. Litan (2006). *Financial Statecraft: The Role of Financial Markets in American Foreign Policy*. New Haven: Yale University Press; James Rickards (2011). *Currency Wars: The Making of the Next Global Crisis*. New York: Portfolio/Penguin; and Eric Helleiner (2014). *Forgotten Foundations of Bretton Woods: International Development and the Making of the Postwar Order*. Ithaca: Cornell University Press.

20 Gérard Duménil and Dominique Lévy (2011). *The Crisis of Neoliberalism*. Cambridge, MA: Harvard University Press; Carmen M. Reinhart and Kenneth S. Rogoff (2009). *This Time is Different: Eight Centuries of Financial Folly*. Princeton: Princeton University Press; Raghuram G. Rajan (2010). *Fault Lines: How Hidden Fractures Still Threaten the World Economy*. Princeton: Princeton University Press; Charles W. Calomiris and Stephen H. Haber (2014). *Fragile by Design: The Political Origins of Banking Crisis and Scarce Credit*. Princeton: Princeton University Press; Alan S. Binder (2009). *After the Music Stopped: The Financial Crisis, the Response, and the Work Ahead*. New York: The Penguin Press.

domestic financial system, establishing the Shanghai Free Trade Zone, promoting Hong Kong as an offshore RMB business hub, and allowing other international financial centres such as Singapore, London, New York and Frankfurt to do more RMB business. Furthermore, China has also expanded the usage of the RMB in global trading, investment and foreign reserves around the world, signed currency swap agreements with more and more countries and used RMB for more out-bound investment and financial aid. China encourages Chinese people to use RMB for spending and investment and allows the inflow of overseas RMB for domestic investment. It has also changed the way it sets the value of the RMB, allowing market forces to play a bigger role, and obtains an IMF Board decision to include the RMB in the Special Drawing Rights (SDR, also called "paper gold") basket.[21] It is particularly important that China strengthens the status and usage of the RMB in the regional cooperative organisations in which China participates actively. In this way, the internalisation of the RMB has started in Southeast Asia, and is gradually expanding to other countries and regions around the world.

Complementary with the "going out" economic strategy is China's international strategy based on the experience of its failures and successes in the past. Wang Gonglong (王公龍), a mainland scholar, made a comprehensive introduction of China's international strategy as follows: "Entering the new stage of 'Reform and Opening up', economic development and modernisation have become the most important strategic goals of the country, thus creating a favourable international environment which has been a subject of strategic importance in China. Chinese leaders believe that a favourable international environment refers to a peaceful and stable international environment, a friendly neighbouring environment, a cooperative environment

---

21 Cheng Siwei (2014). *The Road of Internationalisation of the RMB* (《人民幣國際化之路》). Beijing: CITIC Press Group; Robert Minikin and Kelvin Lau (2013). *The Offshore Renminbi: The Rise of the Chinese Currency and Its Global Future*. Singapore: Wiley & Sons; and Barry Eichengreen and Masahiro Kawai (eds.) (2015). *Renminbi Internationalisation: Achievements, Prospects, and Challenges*. Tokyo: Asian Development Bank Institute and Washington, D.C. Brookings Institution Press.

featuring equality and mutual benefit, and a public opinion environment that is objective and friendly. These four aspects from an organically integrated system for China's development, of which the peaceful and stable international environment is the foundation. Without the peace and stability of the greater international environment, it is hard for China to create a favourable international environment for its development. Even if it can maintain a good relationship with the external world for a short period of time, it won't last long. A friendly neighbouring environment is the most important part of the system, which is determined by the unique geopolitical environment surrounding China. The key to creating a favourable international environment is to achieve stability and prosperity in the neighbouring environment.

A cooperative environment characterized by equality and mutual benefit is an indispensable component of the system as well as an important means of achieving and pursuing national interests. An objective and friendly public opinion environment plays an increasingly crucial role in China's development. As a rising power, China need not only to maintain and pursue its national interests, but also to establish its international image. In over 30 years of "Reform and Opening up", the international landscape, especially in terms of China's relationship with the outside world, has undergone many changes. During these years, the Chinese Communist Party has adjusted the goals of its international strategy and diplomatic policies constantly. It has turned its focus from itself to the world as a whole, from pursuing its own interests to promoting win-win cooperation with others. However, what remains unchanged is that China has always considered that diplomacy should serve economic development and that creating a favourable international environment for its modernisation project is at the core of its goals. Therefore, striving for a favourable international environment is the central theme of China's development and its international strategy in the new period."[22]

22　Wang Gonglong (2012). *Research on the International Strategic Thought System with Chinese Characteristics* (《中國特色國際戰略思想體系研究》). Beijing: People's Publishing House. P110; Gibert Rozman (2010). *Chinese Strategic Thought toward Asia*. New York:

The specific content of China's international strategy is rich and being constantly modified in line with the changes in China's external environment, in order to ensure that national sovereignty, security and development interests are maintained at an optimal level.[23] In the following section, I shall discuss those components, particularly those pertinent to the practice of "one country, two systems" in Hong Kong.

First, China is committed to contributing to a fairer and more reasonable international order. Since World War II, the international order has been crafted and overseen by the U.S. based on its interests, values and strategies and those of its allies.[24] Under this international order, the dominant powers are the U.S. and the Western countries, which is reflected in the choice of the leaders of important and weighty international organizations. For example, the head of the International Monetary Fund must be someone from Europe and that of the World Bank must be an American. The distribution of power and the rules of the game under this order work to the disadvantage of developing countries in many aspects. China not only actively participates in international affairs, fulfils international responsibilities and obligations, but also strives for reform so as to build a fairer and more reasonable international order for the whole world.[25]

Second, China strives to establish exemplars in certain countries and regions to serve as models for a renewed international order. China hopes to work with other countries to establish some international cooperative organisations at the regional level and some international bodies

---

Palgrave Macmillan; and Bates Gill (2007). *Rising Star: China's New Security Diplomacy*. Washington, DC: Brookings Institution Press.

23 See State Council Information Office of the PRC (ed.) (2014). *New Philosophy of Chinese Diplomacy* (《解讀中國外交新理念》). Beijing: China Intercontinental Press; Yan Xuetong (2013). *Inertia of History: China and the World in the Next Ten Years* (《歷史的慣性：未來十年的中國與世界》). Beijing: China CITIC Press; Mark Leonard (2008). *What Does China Think?* London: Fourth Estate; and Daniel C. Lynch (2015) *China's Futures: PRC Elites Debate Economics, Politics, and Foreign Policy*. Stanford: Stanford University Press.

24 Anne-Marie Slaughter (2004). *A New World Order*. Princeton: Princeton University Press; and David Ekbladh (2010). *The Great American Mission: Modernisation and the Construction of an American World Order*. Princeton: Princeton University Press.

25 See Alastair Iain Johnston (2008). *Social States: China in International Institutions, 1980-2000*. Princeton: Princeton University Press.

concentrated on particular areas, so as to demonstrate to the world the form and operating modes for a renewed international order. Two typical examples of the former are the Shanghai Cooperation Organisation and the "free trade zone" comprising China and the ten ASEAN countries. A good example of the latter is "BRICS", a group of countries comprising Brazil, Russia, India, China and South Africa cooperating on economic and security matters. As a member of such cross-national cooperative organisations, China makes great efforts to embody in them the principles of fairness, mutual benefits, win-win outcomes, inclusiveness, openness and equality.[26]

Third, China opposes any unreasonable behaviour that could have a harmful effect upon a fair and peaceful post-war international order. Insisting on the inviolability of national sovereignty, China opposes placing human rights and humanitarian concerns above national sovereignty, and using these as the excuse to change the regimes of other countries by military force or by other means.[27] China also is against exporting Western democracy by the Western nations to other countries,[28] and violations by any country (particularly the U.S. and Japan) of the international agreements (especially the Potsdam Declaration) worked out by victorious nations after World War II,[29] and any country's violation of the sovereignty of another country by taking unilateral actions that ignores or sidelines the United Nations.[30]

26 See Shi Ze (eds.) (2014). *The Countries and Cooperative Organisation in China's Periphery* (《中國周邊國家與合作組織》). Beijing: People's Publishing House.

27 Rein Müllerson (2013). *Regime Change: From Democratic Peace Theories to Forcible Regime Change*. Leiden: Martinus Nijhoff; Michael MacDonald (2014). *Overreach: Delusions of Regime Change in Iraq*. Cambridge, MA: Harvard University Press.

28 William Blum (2013). *America's Deadliest Export: Democracy—The Truth about U.S. Foreign Policy and Everything Else*. London: Zed Books; F. William Engdahl (2009). *Full Spectrum Dominance: Totalitarian Democracy in the New World Order*. Joshua Tree: Progressive Press.

29 The Chinese people regard the handover of Diaoyu Islands to Japan by the U.S., Japan's military expansion and its engagement in overseas military operations as violations of the Potsdam Declaration. Regarding the disputes over Diaoyu Islands, see Han Jiegen (2014). *The Historic Truth of the Diaoyu Islands* (《釣魚島歷史真相》). Shanghai: Fudan University Press.

30 Robert O. Hilderbrand (1990). *Dumbarton Oaks: The Origins of the United Nations and the Search for Postwar Security*. Chapel Hill: The University of North Carolina Press and Paul Kennedy (2006). *The Parliament of Man: The Past, Present & Future of the United Nations*.

Fourth, China wishes to alter the global financial order, which is currently dominated by the U.S. dollar. The short- and medium-term objective of RMB internationalisation is to promote the diversification of the major currencies around the world and enhance the role of the RMB in international finance, while for the time being recognising the hegemony of the U.S. dollar. With respect to the prospects of the U.S. dollar, different scholars hold quite different opinions. The American scholar of finance Prasad asserted that after the financial tsunami, "The U.S. dollar's position as the foremost reserve currency was more secure. Financial assets measured by U.S. dollars, especially the U.S. Government securities, were still the top choice for investors who intended to retain the value of their assets." He went on to say, "The post-World War II Bretton Woods system was in fact an attempt to bring order to international trade and finance by limiting national governments' use of competitive devaluations as a tool to promote [the export of goods and services and] domestic growth." Although the Bretton Woods system collapsed in August, 1971 because the U.S. dollar was unpegged to gold, the U.S. dollar retained its strong international status and didn't collapse despite repeated depreciations and financial crises caused by America and the U.S. dollar. The reason is that there is no better alternative to the dollar. The European Union lacks a unified fiscal policy, which is not conducive to the stability of the exchange rate of the Euro. RMB has a potential for long-term appreciation and "is a prime example of a currency that is increasingly being used in international transactions, even though China exercises restrictions on capital flows."[31]

However, quite a few scholars hold a contrary opinion. They think that the international status of the U.S. dollar would inevitably be challenged by other currencies, particularly the RMB, and neither

---

New York: Random House.

31 Eswar S. Prasad (2014). *The Dollar Trap: How the U.S. Dollar Tightened Its Grip on Global Finance*. Princeton: Princeton University Press, quoted respectively from Pxi, 135 and 231. Another American scholar believed that the Euro and the RMB couldn't replace the status of the U.S. dollar in the world because the U.S. dollar is the only currency that has the political and economic advantages that are needed to support a strong currency. See Benjamin J. Cohen (2015). *Currency Power: Understanding Monetary Rivalry*. Princeton: Princeton University Press.

the lack of free convertibility of the RMB nor the absence of complete openness of China's capital accounts can stop the RMB internationalisation. The American scholar Barry Eichengreen pointed out candidly that the international status of the U.S. dollar grants it exorbitant privileges and hence tremendous benefits for the U.S., but that often brings sufferings to other countries.[32] As a result, many countries hope to have other currencies to choose from, so as to depend less on the U.S. dollar.

History has shown that a country's monetary policy is closely related to its domestic situation and the international environment. In other words, the monetary policy and exchange rate policy of a country are closely tied to its internal affairs as well as diplomatic policy.[33] The American scholar James Rickards criticised America for devaluing the U.S. dollar in order to launch a worldwide currency war: "The effects of printing [a high number of] dollars are global; by engaging in quantitative easing, the Fed has effectively declared a currency war in the world." "[American senior] military and intelligence officials have now come to the realisation that America's unique military predominance can be maintained only with an equally unique and predominant role for the dollar. If the dollar falls, America's national security falls with it." "A currency war, fought by one country through competitive devaluations of its currency against others, is one of the most destructive and feared outcomes in international economics. [...] Whether prolonged or acute, these and other currency crises are associated with stagnation, inflation, austerity, financial panic and other painful economic outcomes. Nothing positive ever comes from a currency war."[34] He mentioned three currency wars in the history, namely Currency War I (1921–1936), Currency War II (1967–1987) and Currency War III (2010 till now). The recent "CWIII [Currency War III] [...] will be fought over the relative value of the Euro, the U.S. dollar and the Yuan,

32 Barry Eichengreen (2011). *Exorbitant Privilege: The Rise and Fall of the Dollar and the Future of International Monetary System*. New York: Oxford University Press.

33 Jeffry A. Frieden (2015). *Currency Politics: The Political Economy of Exchange Rate Policy*. Princeton: Princeton University Press.

34 James Rickards (2011). *Currency Wars: The Making of the Next Global Crisis*. New York: Portfolio/Penguin, quoted from Pxiv, 11, 37 and 100.

and this will affect the destinies of the countries that issue them as well as their trading partners. [...] Today the risk is not just of devaluating one currency against another or a rise in the prices of gold. Today the risk is the collapse of the monetary system itself—a loss of confidence in paper currencies and a massive flight to hard assets." The whole world has known the financial intentions of the U.S. for a long time, so countries have to take self-defensive measures such as devaluing their own currencies, controlling their capital or jointly protesting against the U.S.. Additionally, another measure is to establish a regional financial cooperative mechanism.

After the 1997 Asian financial crisis, the countries in East Asian region have strengthened their financial cooperation, with a view to better preventing and responding to financial crises as well as reducing the harm caused by "the U.S. dollar hegemony."[35] With respect to the "financial regionalism" in East Asian regions, the American scholar Grimes thought that "financial regionalism...attempts to reduce currency volatility, to create frameworks to contain financial crises, and to develop local financial markets—in East Asia." "East Asian financial regionalism arises at least partially from the incentive to insulate regional economies from the dollar and from the U.S. economic policies." "If regional currency cooperation yields good results, it will greatly impact the global role of the U.S. dollar and the national strength of the U.S. [...] The desire to insulate regional economics from the U.S. macroeconomic policy choices and from the intrusiveness of global financial institutions and standards lies at the heart of East Asian financial regionalism."[36] Rickards further predicted, "Regional currency blocs could quickly devolve into regional trading blocs with diminished world trade."[37]

---

35 Some measures of cooperation that are worthy of attention are the establishment of the Chiang Mai Initiative by China, Japan, South Korea and other East Asian countries on May 6, 2000 and the Chiang Mai initiative Multilateralisation in December, 2009.

36 William W. Grimes (2009). *Currency and Contest in East Asia: The Great Power Politics of Financial Regionalism*. Ithaca: Cornell University Press, P2, 22 and 119 respectively.

37 James Rickards (2011). *Currency Wars: The Making of the Next Global Crisis*. New York: Portfolio/Penguin, P228-229.

Regarding China's intent to reform the international financial order which has long been dominated by the U.S., the American scholar Jonathan Kirshner commented as follows: "[...] China, a rising great power, will work for reform within existing international institutions where the status quo does not adequately reflect its growing importance. Because it is likely to meet with limited success on this front, given the entrenched interests of others, Beijing will also pursue its own international arrangements on a parallel track." The 2008 global financial crisis "accelerated the process of the RMB internationalisation, and it ended the project of converging with the American model." "The RMB internationalisation was regarded as a necessary step towards the process of establishing a multi-currency system, which would reduce the influence of the U.S. dollar, promote the stability of this system, amplify the voice of China, and reduce the harm caused by the U.S. dollar crisis."

Kirshner predicted that China would play a leading role in Asian currency and financial cooperation and "[all] this may be part of what rejecting the American model of financial governance looks like: putting the infrastructure in place for the Yuan to become more internationalised, promoting its use as a vehicle currency, and encouraging other central banks to hold Yuan as reserves while retaining some capital controls and other market-inhibiting devices. One way to encourage this would be through bilateral swap agreements, which China has quite actively pursued." "The bilateral currency swap facilitates the utilisation of and provides easy access to Yuan without requiring multilateral negotiations and without necessitating ambitious or comprehensive commitments to financial liberalisation." "No matter what the rate, however, regional monetary arrangements in Asia, anchored in Beijing, with features, practices, and norms recognisably distinct from the second U.S. postwar model [referring to the disconnection of the U.S. dollar from the gold in 1971], are very likely to emerge in the coming years."[38]

---

38 Jonathan Kirshner (2014). *American Power after the Financial Crisis*. Ithaca: Cornell University Press, P109, 114, 120, 121 and 123 respectively. See also Ronald I. Mckinnon (2013). *The Unloved Dollar Standard: From Bretton Woods to the Rise of China*. Oxford: Oxford University Press.

Fifth, China seeks to actively participate in global governance and pursue innovations in the mode of global governance. China, as a great power, has perforce to actively participate in international affairs, including maintaining peace, offering humanitarian aid, combating pirates, preventing proliferation of weapons of mass destruction, tackling global climate change, resolving regional conflicts, and promoting trade liberalisation and the development of the global economy. In fact, the international community also expects China to bear more international obligations and responsibilities, but China is only willing to do what is within its power and capabilities. China has devoted itself to strengthening the role and power of the global financial governance system in recent years. For example, China has repeatedly called for a greater representative status for itself as well as other developing countries in the International Monetary Fund and the World Bank.

The BRICS and the Shanghai Cooperation Organisation co-founded by China and other countries also aim to build a new global governance model. According to Rickards, "The BRICS' principal role has been to weigh in on global governance and the future of the international monetary system with one voice. The BRICS' leaders have begun to stake out radical new positions on five key issues: IMF voting mechanism, UN voting mechanism, multilateral assistance, development assistance, and global reserve composition. Their manifesto calls for nothing less than a rethinking or overturning of the post-World War II arrangements made at Bretton Woods and San Francisco that led to the original forms of the IMF, the World Bank, and the United Nations."[39] The Shanghai Cooperation Organisation also intends to become a global governance model for others' reference, particularly as it is diversifying and increasing its functions. "As geopolitics is increasingly played out in the realm of international economics rather than purely military-diplomatic spheres, the SCO's evolution from a security alliance to a potential monetary zone should be expected."[40]

---

39 James Rickards (2014). *The Death of Money: The Coming Collapse of the International Monetary System*. London: Portfolio/Penguin. P147.
40 Ibid, P152.

Sixth, China wishes to change the rules of global trade and investment. The Western countries had been focusing on promoting the liberalisation and marketisation of global trade and investment in the past, and were deadly opposed to the efforts of the state, state-owned enterprises and enterprises closely related to the state to seek profits with non-market methods that violated the rules of fair competition. In fact, however, governments of Western countries often help domestic enterprises and merchants to gain competitive advantages unfairly in foreign countries by means of political, economic, financial, diplomatic and military force. Two American scholars, Benn Steil and Robert Litan, pointed out the political power of "financial statecraft" in particular. The core of "financial statecraft" is to influence the trade volume and direction, currency exchange rate and the supply of funds by controlling the direction of the flow of funds, and to further impact the economic and political landscapes of other countries or even spark political unrest or the downfall of the government in those countries.[41]

In recent years, the economic power of China and some developing countries have been increasing, and their state-controlled funds are growing large. These countries strongly support their domestic enterprises and merchants in fighting for market share, merging and acquiring enterprises, obtaining resources, bidding for projects, concluding sales contracts and so on globally. Examples of such state actions are increasing unabated. More and more countries with relatively rich foreign exchange reserves establish sovereign funds for various types of investment and production around the world. Now, the competition among countries and market competition can be found in fields such as global trade, production, marketing, finance and commodities. The global economic order advocated by the West is an open, equal, fair and market competition-oriented system, but it has already undergone great changes. Enterprises, merchants and professionals

---

41 Benn Steil and Robert E. Litan (2006). *Financial Statecraft: The Role of Financial Markets in American Foreign Policy*. New Haven: Yale University Press; and Robert D. Blackwill & Jennifer M. Harris (2016). *War by Other Means: Geoeconomics and Statecraft*. Cambridge, MA: The Bekknap Press of Harvard University.

without assistance from their countries will be put in an unfavourable situation in the context of increasingly fierce international competition.[42]

China plays a significant role in promoting the transformation of the global economic landscape.[43] This has a lot to do with the fact that the state-owned economy constitutes a huge proportion of China's economic system and the state also actively participates in the economic affairs of the country. Some commentators in the West describe China's economic system a "state capitalism" which is highly successful. The Chinese model is also seen as a threat to the West not only because its basic tenets are in contradiction with the Western values of the free market and liberal democracy, but also because it is increasingly attractive to the developing nations, thus further eroding the ideological appeal of the West and changing the balance of power between China and the West. In the words of Kurlantzick, "The possibility that state capitalism will succeed as a model is a concern particularly in China, a state that is authoritarian, extremely powerful, possibly interested in promoting its economic model as a means of enhancing its overall power, and in some ways antagonistic to the United States and other leading democracies. Indeed, state capitalism's combination of economic strength and adaptability in China makes the Chinese model a genuine challenge to free-market capitalism. And if state capitalism succeeds in the long term, particularly in China, it could help Beijing— and to a lesser extent other authoritarian but efficient state capitalists— remake the international economic system, amass strategic power, and potentially use state companies as weapons."

---

42 See Ian Bremmer (2010). *The End of the Free Market: Who Wins the War Between States and Corporation?* New York: Portfolio; Ian Bremmer (2012). *Every Nation for Itself: Winners and Losers in a G-Zero World.* New York: Portfolio/Penguin; and David M. Smick (2008). *The World is Curved: Hidden Dangers to the Global Economy.* New York: Portfolio. Joshua Kurlantzick (2016). *State Capitalism: How the Return of Statism is Transforming the World.* New York: Oxford University Press.

43 The strategic function of China's national bank attracted much attention in the international arena. See Henry Sanderson and Michael Forsythe (2013), *China's Superbank: Debt, Oil and Influence–How China Development Bank is Rewriting the Rules of Finance.* Singapore: John Wiley & Sons.

In recent years, the West has also increased support for their domestic enterprises and businessmen so as to enable them to play a more prominent role in reforming and operating the financial system. In brief, the rise of China has pushed forward the emergence of a new global economic order that combines the participation of the state and market competition.

Seventh, to strenuously safeguard China's core interests. The rise of China has not only expanded its national interests around the whole world, but also increased the threats it faces. The previous policy of "hiding capacities and biding time" must be modified. However, though possessing numerous interests, China's core interests are limited, the meaning of which could change in response to the changing situation brought about by the rise of China itself. American scholars Elizabeth Economy and Michael Levi found that "[o]ver time, the number of core interests explicitly claimed by China has expanded. Originally, during the early 2000s, officials used the term to refer to Taiwan, when the territory's people appeared to be moving toward de jure independence. By 2006, it evolved to incorporate Tibet and Xinjiang, two regions in China with sizable and restive minority populations. In 2010, Dai Bingguo (戴秉國) reportedly told U.S. Secretary of State Hillary Clinton that the South China Sea was one of China's core interests. And in 2013, a spokesperson for the Ministry of Foreign Affairs claimed that the Diaoyu/Senkaku Islands in the East China Sea were a core interest."[44]

In recent years, China has become more and more determined to safeguard its core interests. Instead of just talking about taking actions as it did in the past, China actively resorts to diplomatic and military measures to defend its core interests. One obvious example is the decisive actions and determination shown in safeguarding China's interests in the East China Sea and the South China Sea despite the warnings and protestations of the U.S., Japan and some Asian countries.[45]

---

44 Elizabeth C. Economy and Michael Levi (2014). *By All Means Necessary: How China's Resource Quest is Changing the World.* New York: Oxford University Press, P145.

45 See David Shambaugh (eds.) (2005). *Power Shift: China and Asia's New Dynamics.*

Eighth, China intends to combat challenges posed by the U.S. and Japan in East Asia. The "pivot to Asia" and "Asia-Pacific rebalancing" strategies proposed by the U.S. are in fact the new "containment" strategy targeting China in the 21st century. The U.S. policy of "containment" targeted at China today can certainly not be compared to the containment strategies it implemented against China during the Cold War between the East and the West.[46] China and the U.S. were antagonistic towards each other at the time, without interaction in diplomacy or economics, but today they share many common interests. The aim of America's "containment" in the past was to destroy the Chinese communist regime. Today, however, the aim is not to end the Chinese communist regime or cripple its economy, but to contain the rise of China and prevent it from becoming a strategic threat to the U.S., especially in Asia. In addition, another aim is to force China to accept the international order crafted and led by the U.S. rather than attempting to establish another.

The core of America's "containment" strategy against China is to induce countries adjacent to China by every means to cooperate with the U.S. in economy and trade as well as on security, so that they can jointly form an economic/a trade and military network to encircle China and to constrain China's external expansion. In this "containment" strategy, the key roles are played by the U.S.-Japanese military alliance and the regional free trade system led by the U.S. In addition, the U.S. will also station additional military forces in the western Pacific. More will be said on this later. China's "counter-containment" strategy, on

Berkeley: University of California Press; Denny Roy (2013). *Return of the Dragon: Rising China and Regional Security*. New York: Columbia University Press; Toshi Yoshihara and James R. Holmes (2010). *Red Star over the Pacific: China's Rise and the Challenge to U.S. Maritime Strategy*. Annapolis: Naval Institute Press; Robert Haddick (2014). *Fire on the Water: China, America, and the Future of the Pacific*. Annapolis: Naval Institute Press; Bill Hayton (2014). *The South China Sea: The Struggle for Power in Asia*. New Haven: Yale University Press; Robert D. Kaplan (2015). *Asia's Cauldron: The South China Sea and the End of a Stable Pacific*. New York: Random House; and Bret Stephens (2014). *America in Retreat: The New Isolationism and the Coming Global Disorder*. New York: Sentinel.

46 See John Lewis Gaddis (2005). *Strategies of Containment: A Critical Appraisal of American National Security Policy during the Cold War*. New York: Oxford University Press; and Kurt M. Campbell (2016). *The Pivot: The Future of American Statecraft in Asia*. New York: Twelve.

the other hand, focuses on strengthening its strategic cooperation with Russia and actively establishing mutually beneficial relationships with its neighbouring countries by economic means. Apart from achievements in strengthening its military alliance with Japan, America's "containment" of China however has so far made little progress in other aspects.

Ninth, China wishes to strengthen its soft power globally. Compared with its increasingly strong hard power (namely military and economic power), China's soft power (namely cultural, institutional and intellectual power) lags far behind. There are many Chinese abroad nowadays, but some of them behave badly in public, which tarnishes the image of China abroad. One of the drawbacks of China's weak power is that the international community remains dubious and worried about China's intentions and actions, which to some extent impairs China's capability to lead and mobilize on a global scale, and also offers the opportunity for the rise of anti-China and anti-communist sentiments around the world. Obvious examples of this include the emergence of notion of "the China threat" in the West and the accusation of some developing countries against "Chinese neocolonialism."

Since mainstream global values are still defined as Western political values, it is difficult for China's political system, its ideals with respect to human rights, and its rule of law to be widely acknowledged or endorsed. Nonetheless, China's economic achievements, development model, achievements in alleviating poverty, capability for handling natural disasters, traditional culture, resistance to Western hegemonism, foreign aid as well as its advocacy of "a harmonious world" are to some degree accepted and praised globally.[47] It must be admitted that much more efforts are still needed to raise China's soft power to the level of its hard power. Currently, it is still extremely difficult for China to compete with Western countries in terms of soft power. However, with the increasing progress made by China, the inexorable decline of the

---

47 See Yao Yao (2014). *The New Chinese Propaganda History: Crafting Modern China's Rhetorical Authority Globally* (《新中國對外宣傳史：建構現代中國的國際話語權》). Beijing: Tsinghua University Press.

West, decreasing attractiveness of Western institutions and values in view of the financial crises and economic chaos created by the West's "irresponsible" monetary and fiscal policies, the exposure of the Western "double standard" in dealing with other countries, as well as the hard-to-handle economic hardships and social injustice in the West, it is possible to gradually narrow the soft power gap between China and the West.[48]

Tenth, China strives to advance the "One Belt, One Road" (一帶 一路) strategy proposed by Chinese President Xi Jinping at the end of 2013. The "One Belt, One Road" strategy is the most important national strategy devised by China since "Reform and Opening up". It is a grand strategy in the sense that it aims at integrating China's "going out" strategy with its international and foreign policy, taking into account both the domestic and the foreign situation on the one hand, and its hard power and soft power on the other. The "Belt" refers to the "Silk Road Economic Belt," while the "Road" refers to the "21st-century Maritime Silk Road."[49]

---

48 The competition between China (declaring "the peaceful rise") and Japan (advocating "the beautiful Japan") for soft power in East Asia is a good example. See Jing Sun (2012). *Japan and China as Charm Rivals: Soft Power in Regional Diplomacy*. Ann Arbor: The University of Michigan Press.

49 According to Wang Yiwei, the "Silk Road Economic Belt" includes three routes, namely the Eurasia Land Bridge-based North Belt (From Beijing through Russia and Germany to North Europe), the oil and gas pipeline-based Central Belt (From Beijing through Xi'an, Urumchi, Afghanistan, Kazakhstan and Hungary to Paris), and the transnational highway-based South Belt (From Beijing through southern Xinjiang, Pakistan, Iran, Iraq, Turkey and Italy to Spain). The focus of the Silk Road Economic Belt is to unblock the routes from China through Middle Asia and Russia to Europe (the Baltic Sea), through Middle Asia and West Asia to the Persian Gulf and the Mediterranean Sea, and to Southeast Asia, South Asia and the Indian Ocean. The skeleton of land bridges of the Silk Road Economic Belt mainly consists of the China-Pakistan Economic Corridor, Bangladesh-China-India-Myanmar Economic Corridor, the new Eurasia Land Bridge, and the China-Mongolia-Russia Economic Corridor. The China-Pakistan Economic Corridor is mainly for petroleum transportation; the Bangladesh-China-India-Myanmar Economic Corridor is to promote trade among ASEAN countries; the new Eurasia Land Bridge is the major logistics channel direct connecting China and Europe; and the China-Mongolia-Russia Economic Corridor has more on national security and energy development. The main direction of the "21st-century Maritime Silk Road" includes one from China's coastal ports through the South China Sea to the Indian Ocean and the Europe; from the coastal ports of China to the South Pacific Ocean through the South Sea. The Silk Road Economic Belt is a new economic development region based on the idea of the "ancient Silk Road." The Belt is basically an "economic belt," which demonstrates the idea of the centralised and coordinated development of cities along

President Xi Jinping once explained the aims of the "One Belt, One Road" strategy as follows: "The Silk Road Economic Belt is inhabited by nearly three billion people and represents a large market in the world with unparalleled potential. The potential for trade and investment cooperation among the relevant countries is enormous. We should discuss a proper arrangement for trade and investment facilitation, remove trade barriers, reduce trade and investment costs, increase the speed and quality of regional economic flows and achieve a win-win outcome in the region. [...] Southeast Asia has been the most important hub of the 'Maritime Silk Road' since ancient times. China would like to strengthen maritime cooperation with ASEAN countries, make full use of the China-ASEAN Maritime Cooperation Fund established by the Chinese Government, and foster a positive maritime cooperative partnership, so as to jointly build the 21st-century 'Maritime Silk Road.' China would like to expand pragmatic cooperation in various areas with ASEAN countries so that we can exchange needed goods and complement each other's advantages. We would like to share opportunities as well as face challenges with ASEAN countries for mutual development and prosperity."[50] He said further: "China will make joint efforts with

---

the Belt. The ocean is the natural area for economic, trade and cultural exchange among countries. The joint construction of the "21st-century Maritime Silk Road" provides a new trade road for China to connect with the world in view of the changing political and trade landscapes. The core values of the Maritime Silk Road are the value of thoroughfare and strategic security. Under the situation where China has become the second largest economies in the world, and the complex interrelationships between the global political and economic landscapes, the development and extension of the "21st-century Maritime Silk Road" can definitely strengthen China's strategic security. The 21st-century Maritime Silk Road, the Silk Road Economic Belt, the Shanghai Free Trade Zone and the high-speed rail strategy are all proposed under this background. See his (2015). *One Belt, One Road: Opportunities and Challenges* (《一帶一路：機遇與挑戰》). Beijing: People's Publishing House, P8-9. See also: Zhang Jie (eds.) (2015). *China's Regional Environment Review: "One Belt, One Road" and the Strategy on the Surrounding Areas* (《中國周邊安全形勢評估："一帶一路"與周邊戰略》). Beijing: Social Sciences Academic Press; Zou Lei (2015). *The Political Economic Study of China's "One Belt, One Road"* (《中國"一帶一路"戰略的政治經濟學》) Shanghai: People's Publishing House; Li Xiangyang (eds.) (2015). *2015 Report on the Development of Asia-Pacific Region: "One Belt, One Road"* (《亞太地區發展報告（2015）：一帶一路》). Beijing: Social Sciences Academic Press; Zhao Jianglin (eds.) (2015). *the 21st-Century Maritime Silk Road* (《21世紀海上絲綢之路》). Beijing: Social Sciences Academic Press; and Feng Bing. *"One Belt, One Road": The Chinese Logic for Global Development* (《一帶一路：全球發展的中國邏輯》). Beijing: China Democracy and Law Press.

50 CCCPC Party Literature Research Office (ed.) (2014). *Excerpted Expositions of Xi Jinping*

relevant countries to accelerate the interconnection of infrastructure and to successfully build the Silk Road Economic Belt and the 21st-century Maritime Silk Road. Using China's surrounding areas as the basis, we should speed up the implementation of the free trade zone strategy, expand the opportunities for trade and investment cooperation and establish the new landscape of an integrated regional economy. We should further cooperate in regional finance, prepare actively for the establishment of the Asian Infrastructure Investment Bank, and improve the network of regional financial security. We should also accelerate the opening up of regions along China's border and strengthen their mutually beneficial cooperation with neighbouring countries."[51]

From the economic perspective, the "One Belt, One Road" strategy is intended to maintain the long-term and continuous growth of the Chinese economy and promote it to a higher level. Additionally, it also aims to prevent China from falling into the "middle-income trap", to promote the transformation of China's industrial structure towards a more balanced, reasonable and advanced one, "digest" China's excess productive capacity and to promote the internationalisation of the RMB. From the perspective of diplomacy, through sharing achievements in economic development with countries in Asia, Africa and Europe, China purports to use the "One Belt, One Road" strategy to promote the development of those countries and connect their economies with China's more closely, so as to develop more intimate and harmonious relationships with one another.

In addition, the strategy can promote China's international status and soft power, strengthen its role in global governance and enhance its capabilities for formulating new rules, standards and norms for international affairs. These can help to build a fairer, more reasonable and peaceful international order and benefit humanity. In addition, the "One Belt, One Road" strategy will contribute to the "rebalancing"[52] of the

---

on *Comprehensively Deepening Reform* (《習近平關於全面深化改革論述摘編》). Beijing: Central Party Literature Press. P131.

51　Ibid, P133-134.

52　The "chief culprit" of global financial crisis is the imbalance of global economic system,

global economic system and the possibility of "rebalancing" between China's and America's economies, [53] namely changing the previous pattern where some countries accumulated excess savings, enjoyed trade surplus and increased foreign exchange reserves, while others were troubled by excess consumption, trade deficit and heavy debts.

Wang Yiwei (王義桅), a mainland scholar, clearly pointed out that, "As the paramount development strategy, the 'One Belt, One Road' strategy is devised to solve the problem of excess production capacity in China's market, procure resources, and extend strategic depth and strengthen national security."[54] "In brief, why should we build the 'One Belt, One Road'? The aim is to increase overseas direct investment, explore foreign markets, expand export of products, reduce excess production capacity, break down trade barriers and finally establish a global trade and currency system concordant with our nation's long-term interests."[55]

Other mainland Chinese scholars also made comments on the "One Belt, One Road" strategy from different perspectives. Zhang Yunling (張蘊嶺) believed, "From the perspective of 'opening up', it is obvious that the guiding idea of the great 'One Belt, One Road' strategy goes beyond the free trade zone and the multilateral trading system. It underscores the importance of the construction of an integrated development environment. The guiding idea also goes beyond China-centric interests and emphasizes joint construction and development. The cooperation projects stemming from the 'One Belt, One Road' strategy are determined through mutual consultation instead of negotiation, which is a new type of development and cooperation." "The 'One Belt,

particularly international trade. See Michael Pettis (2013). *The Great Rebalancing: Trade, Conflict, and the Perilous Road Ahead of the World Economy*. Princeton: Princeton University Press.

53 Compared with the U.S., China has made greater progress in restructuring its national economic structure and behaviour to make them more rational and sustainable. The current economic structure of the U.S. has little difference from that before the 2008 financial crisis. See Stephen Roach (2014). *Unbalanced: The Codependency of America and China*. New Haven: Yale University Press.

54 Wang, Yigui (2015). *"One Belt, One Road": Opportunities and Challenges* (《"一帶一路"：機遇與挑戰》). Beijing: People's Publishing House. P9-10.

55 Ibid, P11.

One Road' strategy is put forward at the right time, symbolising the significant strategic transformation of China's understanding of its relationships with neighbouring countries. The new national strategy is increasingly clear, and it is to promote the construction of a community of shared interests as well as a community of shared fate with our neighbouring countries based on the promotion of joint development." "Because of the 'One Belt, One Road', our neighbouring countries can take advantage of the 'fast train' of China's development to develop themselves more rapidly. Furthermore, China will also gain additional opportunities to expand from the development of neighbouring countries. And in turn, neighbouring regions can become a dependable zone for the extension of China's development." [56]

Piao Zhuhua (朴珠華), Liu Xiaomeng (劉瀟萌) and Teng Zhuoyou (滕卓攸), however, held the following views: "[...] The rise of China, the needs arising from the 'new [economic] normal [condition]', the upgrading of the industrial structure, and the continuous changes of the domestic economic structure together require a broader market to conduct reform. China has become a middle-income country in recent years. First of all, with the economic growth and the acceleration of overseas investment of our homeland, China must speed up the 'opening up' process so as not to fall into the 'middle-income' trap. [...] Second, at the current stage, China's production capacity in traditional industries and its foreign exchange reserves are excessive. [...] Under these circumstances, China should make use of its foreign exchange reserves to promote infrastructure construction in countries along the 'One Belt, One Road', as well as encourage Chinese enterprises to 'go out'. Next, although its economy is on the rise, China is still faced with the structural conflicts of development in the eastern and western regions. Developing the 'Silk Road Economic Belt' can effectively promote economic growth in the western regions." "[...] The establishment of the Asian

56 Zhang Yunling (2015). "How to Understand the Great Strategic Design of 'One Belt, One Road'?" ("如何認識 '一帶一路' 的大戰略設計"). In Zhang Jie (eds.) (2015). *China's Regional Environment Review: "One Belt, One Road" and the Strategy on the Surrounding Areas* (《中國周邊安全形勢評估：一帶一路" 與周邊戰略》). Beijing: Social Sciences Academic Press. P3-11, P7, 8, and 9 respectively.

Infrastructure Investment Bank provides financial support for the 'One Belt, One Road' strategy, promotes the construction of exemplary pilot projects such as the new Eurasia Land Bridge, the Bangladesh-China-India-Myanmar Economic Corridor and the China-Pakistan Economic Corridor, and expands the influence of China on countries along the 'Belt and Road'. Furthermore, the establishment of the Asian Infrastructure Investment Bank also symbolises China's entry into the period of capital export from merely exporting products as in the past."[57]

At present, China's international strategy has made some progress in promoting good neighbourliness as well as countering "containment". It is particularly worth mentioning that when the Asian Infrastructure Investment Bank was established in 2015, the number of the founding members from all continents reached 57, some of which (including Britain) joined over objections from the U.S.. On September 3, 2015, a grand military parade was held in Beijing to commemorate the 70th anniversary of the victory of the Chinese People's War of Resistance against Japanese Aggression and the victory of the Global War against Fascism. The South Korean president Park Geun-hye attended the parade regardless of the strong objections from the U.S. and Japan. All of these showcase the rise of China and the subtle changes in the international landscape.

Dramatic Changes in the International Landscape

After the problems related to "Hong Kong's future" were resolved, the international landscape changed dramatically and unexpectedly. As mentioned in Chapter 1, when the topic of "Hong Kong's

---

57 Piao Zhuhua, Liu Xiaomeng & Teng Zhuoyou (2015). "China's Political Risk Analysis of the Direct Investment Environment of 'One Belt, One Road'" ("中國對‘一帶一路’直接投資環境政治風險分析"). In Zhang Jie (eds.) (2015). *China's Regional Environment Review: "One Belt, One Road" and Strategy on the Surrounding Areas* (《中國周邊安全形勢評估：“一帶一路”與周邊戰略》). Beijing: Social Sciences Academic Press, P181-201, quoted from P183-184 and 185 respectively.

future" arose, the international landscape then offered a favourable international environment for smoothly and peacefully resolving the problems related to "the return of Hong Kong". It also helped to maintain the confidence of the West and the international community in Hong Kong's prospects, thus helping Hong Kong people to maintain their optimistic expectations for Hong Kong's future.

The international environment remained favourable until 1989 when the "June 4 event" broke out in Beijing. Later, in 1990, East Europe underwent gargantuan changes. In 1991, the Soviet Union collapsed, to everyone's surprise and amazement. The enterprise of communism around the world was dealt a heavy blow at that time. The enthusiasm and arrogance stemming from Western "triumphalism" suddenly soared. Many people believed that the values, democracy and market economy which the West, particularly the U.S., had all along articulated and ardently exported to other countries were the "end" and "peak" of the development of human history, and no better ideals or institutions were possible.[58] The dissolution of the Soviet Union and the dramatic changes in East Europe greatly strengthened the global primacy of the West. America's status as a global hegemon was more and more prominent, and its hard power of the economy and military forces as well as the soft power of values and institutions were definitely unparalleled in the world. The cocky Americans at that time were in high spirits and looked down upon others, which struck a sharp contrast with their dispirited and pessimistic outlook of the 1970s.

As the Americans became dizzy with success, a new stream of thought, neo-conservatism, rapidly emerged. It urged that, on the unprecedented occasion of its "unipolar" status, the U.S. should seize the moment and make full use of its absolutely unparalleled military forces to reshape the world based on its "self-image." In addition, the U.S. should also force other countries to reshape themselves according to American values and institutions, and operate in accordance with

---

58 The most typical work on this idea can be found in Francis Fukuyama (1992). *The End of History and the Last Man*. New York: Free Press.

the international order and rules of the game established by the U.S.. If everything proceeded as neo-conservatives expected, all countries would become American-style "democratic states" and the world would coincide with the "liberal international order" designed by the U.S.. As a result, the world peace would come naturally and easily.[59] The military operations in Iraq and Afghanistan by the U.S. were exactly the results of the dominance of this ideology.

The corollary of the dissolution of the Soviet Union and the dramatic changes in East Europe is the rapid decline of China's importance and strategic value in the calculation of the U.S., Japan and the Western world. The "June 4 event" elicited a strong response from the West, and the distinct differences in values, institutions and development paths between China and the West were starkly exposed to the world. The Western bloc led by the U.S. immediately imposed military, diplomatic and economic sanctions on China. China was suddenly isolated from the rest of the world. At the same time, Western countries put pressure on China regarding human rights and democratic issues and hoped to overthrow the CPC regime at one stroke, so as to end world history as well as human history's experimentation with communism. Guided by Deng Xiaoping's admonition of "keeping calm" and "hiding capacities and bidding time", however, the CPC has not only maintained its leading role in politics, but also carried on the development strategy of "Reform and Opening up" in the unfavourable international environment. Finally, China has succeeded in stabilising the national political landscape, promoting economic transformation and development, as well as proving the incorrectness of the Western prediction that the regime led by the CPC is "destined to collapse". After several years of political struggle, the relations between China and Western countries gradually went back to "normal" and satisfactory progress is being

---

59 Murray Friedman (2005). *The Neoconservative Revolution: Jewish Intellectuals and the Shaping of Public Policy*. New York: Cambridge University Press; Stefan Halper and Jonathan Clarke (2004). *America Alone: The Neo-Conservatives and the Global Order*. New York: Cambridge University Press; and G. John Ikenberry (2011). *Liberal Leviathan: The Origins, Crisis, and Transformation of the American World Order*. Princeton: Princeton University Press.

made in the fields of the economy, trade and finance. China joined the World Trade Organisation in 2001, Beijing hosted the Olympic Games in 2008, and Shanghai hosted the World Exposition in 2010. All these great events symbolise the sustained growth of China's status and influence around the world.

With regard to the implementation of "one country, two systems", the rapid rise of China has thoroughly changed the international landscape, but global changes in other aspects also make a difference. The interaction between all of these changes has had the following results. First, China's national strength is increasing rapidly and its influence is extending far and wide in the world. Second, the economic performance of the U.S. and Japan has been relatively lacklustre and their international influence is declining, so they are highly concerned about threats from China. Third, the European Union has arisen as a relatively independent group with growing clout. Fourth, Russia has gradually recovered from decadence and is reviving as a nation with a strategic status, but its relations with the West are strained and their conflicts of interests are still evident. Fifth, Japan's economy has been in a depressed state since the early 1990s.

China, Russia, the European Union, the U.S. and Japan have been the major powers in global competition, particularly among Asian countries, for the past decades, and they will remain so for quite some time into the future. The competition among great powers has not only a far-reaching impact on the world, but also a profound impact upon the implementation of the "one country, two systems" policy in Hong Kong. It is within this dynamic that China has carefully adopted measures that safeguard the viability of the policy while preserving mainland interests (particularly in the face of containment from the military and diplomatic alliance formed by the U.S. and Japan) vigorously promoting the China-centred strategies of "going west" and "developing Asia", deepening its strategic partnership with Russia, and striving for support from a number of important European countries.

Changes in the national landscape make the international environment of Hong Kong much more complicated and difficult for it to navigate. Before its return to China, Hong Kong benefited a great deal from both China and the West because of their mutually harmonious relations. In the new relations, however, conflicts and cooperation coexist and alternate. As a result, Hong Kong is placed in a dilemma. Since the central government and mainland compatriots are concerned about whether Hong Kong will become a tool for the West, particularly the U.S., the whole nation and the Hong Kong people together must face the problem of how to prevent Hong Kong from threatening national security. In addition, the rise of China will unavoidably entangle Hong Kong in Asia's political issues, so Hong Kong must change the previous attitude that shows preference for the West while neglecting the East. The changes in the geopolitical landscape place new but still unclear demands on the Hong Kong people as well as the Hong Kong SAR Government.

Essential Changes in America's Policy to China

Since its inception, Sino-U.S. diplomatic relations have undergone ups and downs, which can be generally described as "not that good yet not that bad". Although there are always disputes over trade and economic issues, let alone over human rights and democracy, the relations between the two countries are becoming increasingly close. By the end of the 20th century, the U.S. had become the major export market for China and a crucial source for China's foreign investment. A large proportion of China's revenues from export were spent on America's treasury bonds, which helped to maintain its low interest rates, low unemployment rates and high level of consumption. It was obvious that China and the U.S. were closely related and mutually dependent in terms of economy. In the field of politics, however, a series of issues such as the Taiwan issue, human rights issues, the dispute over the South China Sea and the issue of regional security disturbed the relations between China and the U.S.. Among those issues, the "June 4 event" made the greatest impact on their relations.

Although China's national strength had been increasing, the U.S. and international community paid little attention to it until the middle and late 1990s. They optimistically believed that even though China could make progress in modernisation, it wouldn't pose a threat to the West. On the contrary, they believed that the modernisation of China would bring it closer to Western institutions and ideology. As a result, China, in its path of development, would follow in the footsteps of the West, including adopting the Western liberal democracy and market economy. In the early 21st century, the world suddenly realised that the rise of China was irreversible and would bring along deep and profound repercussions. In particular, the development model underpinning China's economic wonder was different from the one that the West advocated, and China's model reached a level of recognition parallel to the Western one and became a valuable reference for some developing countries.

The increase in China's national strength has contributed to the expansion of its military forces and its influence in East Asia. Since Asia, particularly East Asia, would become the most dynamic region of the world economy in the future and a critical region over which the whole world would compete for influence and domination, the U.S. and its ally Japan have taken the "threat" from China as a serious matter. U.S. President Barack Obama announced the strategies of "pivot to Asia" and "Asia-Pacific rebalancing" in 2010, aiming to divert the nation's international strategic focus from the Middle East to East Asia. These were the greatest strategic changes in the U.S. in recent years.[60]

After World War II, the U.S. abandoned the previous doctrine referred to as "isolationism". Instead, it forcefully exerted its power and influence in international affairs. The U.S. crafted and imposed a "liberal international order" on those parts of the world that did not belong to the communist bloc. This new international order is based on concrete

---

60 In order to avoid irritating China, the U.S. seldom uses the phrase "returning to Asia" publicly. Instead, the term "pivot to Asia" is the more commonly used term.

and explicit rules of the game. It advocates free trade and open markets, the status of the U.S. dollar as the global core currency, democratic politics and protecting human rights. More importantly, it caters to the national interests of the U.S. and reflects America's values and national spirit. Under the "liberal international order," the U.S. has succeeded in setting up a cluster of key institutions and organisations, including the United Nations, the IMF, the World Bank, the Asian Development Bank, GATT (replaced by WTO in 1994), and so on.[61] In addition, the U.S. has also established a series of collective security institutions or treaties with its allies in all continents, so as to ensure regional security, particularly military security, and prevent security from being threatened by the communist bloc or the spread of communism. Among those institutions, NATO and the U.S.-Japan military alliance are definitely the most important.

Besides these formal and official organisations, a number of non-governmental, multinational organisations have been established and have performed an active role in the new international order. Although the aims of some non-governmental organisations are to a certain extent against the "liberal international order", particularly with respect to human rights, liberty, democracy, economic development, foreign aid, gender equality, racial equality, environmental protection, peace, nuclear weapons and poverty, most non-governmental organisations still basically recognize the "liberal international order", with the hope that the beliefs, ideals and principles which the order advocates come to pass. Generally speaking, under the "liberal international order", official and non-governmental organisations cooperate with each other, promote joint development and support the system.[62]

61 G. John Ikenberry (May/June 2014). "The Illusion of Geopolitics: The Enduring Power of the Liberal Order". *Foreign Affairs*. Vol. 93, No. 3, P80-90; James E. Cronin (2014). *Global Rules: America, Britain and a Disordered World*. New Haven: Yale University Press; Dan Plesch (2015). *America, Hitler and the UN: How the Allies Won World II and Forged a Peace*. London: I.B. Tauris; and Elizabeth Borgwardt (2005). *A New Deal for the World: America's Vision for Human Rights*. Cambridge, MA: The Belknap Press of Harvard University Press.

62 See Anne-Marie Slaughter (2004). *A New World Order*. Princeton: Princeton University Press.

In the U.S., there are several particularly significant ideas under-girding its post-war strategy. First, American strategists believe that only under the "liberal international order" led by the U.S. could the peace and development of the world be ensured. The U.S. is able to "maximize" national interests and security through this international order. For the sake of the effective operation of this order, the U.S. is willing to bear the costs unilaterally and provide its allies and interna-tional community with "public goods" free of charge, which include the security of navigation and uninhibited passage, regional peace, global governance, the U.S. dollar's status as the currency for global trade and investment, the open international market, humanitarian intervention and guarantee of human rights.[63] The American conservative strate-gist Robert Kagan even asserted, "[After World War II, perhaps] the era of peace we have known has something to do with the enormous power wielded by one nation [here referring to the U.S.]."[64] As a result, "Contrary to what one often hears, multi-polar systems have historically been neither particularly stable nor particularly peaceful."[65] In conclu-sion, "International order is not an evolution; it is an imposition [by some power]."[66] From the perspectives of many Americans, if the inter-national order is not designed and led by the U.S., it would not be able to bring peace, development and benefits to human beings. A multi-polar world couldn't be a stable and effective international political order.

Second, although the U.S. has long been the hegemon of the Americas, from the perspective of security and national interests, the U.S. wouldn't allow another great power to control the destiny of other continents, particularly Eurasia with its large population and rich resources. Zbigniew Brzezinski, the then National Security Advisor of President Carter, emphasized that, "America's global primacy is

---

63 Michael Mandelbaum (2005). *The Case for Goliath: How America Acts as the World's Government in the 21st Century*. New York: Public Affairs.
64 Robert Kagan (2012). *The World America Made*. New York: Alfred A. Knopf, P4-5.
65 Ibid, P83.
66 Ibid, P97.

directly dependent on how long and how effectively its preponderance on the Eurasian continent is sustained.... Eurasia is the globe's largest continent and is geographically axial. A power that dominates Eurasia would control two of the world's three most advanced and economically productive regions. A mere glance at the map also suggests that control over Eurasia would almost automatically entails Africa's subordination, rendering the Western Hemisphere and Oceania geographically peripheral to the world's central continent."[67] The American conservative strategist Mearsheimer held a similar opinion, asserting that, "regional hegemons attempt to check aspiring hegemons in other regions because they fear that a rival great power that dominates its own region will be an especially powerful foe that is essentially free to cause trouble in the backyard of the fearful great power. Regional hegemons prefer that there will be at least two great powers located next to each other in other regions, because their proximity will force them to concentrate their attention on each other rather than on the distant hegemon."[68] He demonstrated with a historical example, "The United States began making serious moves to get militarily involved in Asia during the summer of 1941, not because American leaders were determined to bring peace to the region, but because they feared that Japan would join forces with Nazi Germany and decisively defeat the Red Army, making hegemons of Germany in Europe and Japan in Northeast Asia. The United States fought a war in the Far East between 1941 and 1945 to prevent that outcome. As in Europe, American troops were stationed in Northeast Asia during the Cold War to prevent the Soviet Union from dominating the region, not to keep peace."[69]

Third, according to the analysis of American strategists, although the U.S. has significant advantages over other countries in terms of national strength, its leading role is constantly being challenged due to

67 Zbigniew Brzezinski (1997). *The Grand Chessboard: American Primacy and Its Geostrategic Imperatives.* New York: Basic Books, P30-31.

68 John J. Mearsheimer (2014). *The Tragedy of Great Power Politics.* New York: W.W. Norton, P41-42.

69 Ibid, P266.

the rise of other great powers. Charles Kupchan, an American scholar of international politics, pointed out that, "The emerging landscape is one in which power is diffusing and politics diversifying, not one in which all countries are converging toward the Western way. [...] The twenty-first century will not be America's, China's, Asia's, or anyone else's; it will belong to no one. [...] A global order, if it emerges, will be an amalgam of diverse political cultures and competing conceptions of domestic and international order."[70] Accordingly, the "task [of the U.S.] at hand is not guiding rising powers into the Western harbours. Rather, it is establishing a new order whose fundamental terms will have to be negotiated by Western powers and newcomers alike. The West will have to give as much as it gets as it seeks to fashion a new international order that includes the rest [referring to emerging powers]."[71] "Accordingly, emerging powers will want to revise, not consolidate, the international order erected during the West's watch. They have different views about the foundations of political legitimacy, the nature of sovereignty, the rules of international trade, and the relationship between the state and society."[72] He warned that, "Equating legitimacy with responsible governance rather than liberal democracy, tolerating political and ideological diversity, balancing between global governance and devolution to regional authorities, fashioning a more regulated and state-centric brand of capitalism—these are the types of principles around which a new order is likely to take shape."[73]

Fareed Zakaria, a famous American political commentator, described the modern world as a "post-American world." In this modern world, the U.S. still has the most powerful national strength, but other countries are rising and have some strengths as well. As a result, the U.S. cannot dominate the world according to its own will. Instead, it should cooperate with other countries to deal with rising global issues

---

70 Charles A. Kupchan (2012). *No One's World: The West, the Rising Rest, and the Coming Global Turn*. New York: Oxford University Press, P3.

71 Ibid, P5.

72 Ibid, P7.

73 Ibid, P11-12.

which were becoming more and more complicated.[74] Zbigniew Brzez-inski also expressed a similar opinion in his recent work, suggesting that in an increasingly complicated and changing world, rather than "acting on its own," the U.S. should seek more cooperation and understanding from other countries.

The American strategic analyst Ian Bremmer also believed that the time when G7 joined hands to take charge of international affairs is gone, and even G20 which was established after the 2008 global finan-cial crisis cannot take on this responsibility. Accordingly, he believed that the modern world is in fact controlled by G0, which means that no country can lead the whole world, and every country competes with others based on its own interests.[75] Ian Bremmer further asserted that the role of the free market model, as a new economic model proposed by the U.S. after the war, is declining, and global economic affairs are gradu-ally led by states and large enterprises. As a result, economic behaviour is less likely to be influenced by the economy or profits. Rather, poli-tics is more and more taken into consideration.[76] Accordingly, the U.S. should make wise decisions based on changes in the international land-scape as well as its own interests and strengths, and give up the previous diplomatic policy which costs a great deal but provides few benefits.[77] Lyle Goldstein, another American scholar, proposed that the U.S. should regard China as an important rival, respect its interests as well as ideas, and satisfy its needs and requirements in various areas, and must avoid disastrous wars between the two countries.[78] Some other scholars also offered advice on reducing disagreements and conflicts between the two countries.[79]

74 Fareed Zakaria (2011). *The Post-American World*. New York: W.W. Norton & Co.

75 Ian Bremmer (2012). *Every Nation for Itself: Winners and Losers in a G-Zero World*. New York: Portfolio/Penguin.

76 Ian Bremmer (2010). *The End of the Free Market: Who Wins the War Between States and Corporations*. New York: Portfolio.

77 Ian Bremmer (2015). *Superpower: Three Choices for America's Role in the World*. New York: Portfolio/Penguin.

78 Lyle J. Goldstein (2015). *Meeting China Halfway: How to Defuse the Emerging U.S.-China Rivalry*. Washington, DC: Georgetown University Press.

79 Michael Swaine (2011). *America's Challenge: Engaging a Rising China in the 21st Century*.

Fourth, according to a popular theory in the U.S. and the West, there are low chances of war among democratic countries. The first reason is that the decision-making power of the governments of democratic countries is restricted by the people, who normally object to launching wars or using military forces against other countries. Second, there is no need for governments of democratic countries to start wars for the purpose of reinforcing or strengthening their political legitimacy.[80] China is not a democratic country (at least from the perspective of Westerners) and the regime of the CPC is dictatorial. Accordingly, with its increasingly powerful strength as well as intense and severe domestic conflicts,[81] there are plenty of opportunities for China to launch wars against other countries.

Apparently, however, only a few American political elites are willing to admit and accept the inevitable arrival of a "multi-polar world". Instead, more and more of them believe that China and the U.S. are embroiled in a "zero-sum" game, so it would be difficult to achieve "win-win"[82] results. Although quite a few people are pessimistic about the prospects of America's national strength,[83] most people continue

Washington DC: Carnegie Endowment for International Peace; James Steinberg and Michael O'Hanlon (2014). *Strategic Reassurance and Resolve: U.S.-China Relations in the 21st Century.* Princeton: Princeton University Press; Thomas J. Christensen (2015). *The China Challenge: Shaping the Choices of a Rising China.* New York: W.W. Norton; and Lyle J. Goldstein (2015).

80 See Spencer R. Weart (1998). *Never at War: Why Democracies Will Not Fight One Another.* New Haven: Yale University Press. However, some scholars think that the situation is complicated and it is more likely for emerging democratic countries to go to war at the early stage of democratization. See Edward D. Mansfield and Jack Snyder (2005). *Electing to Fight: Why Emerging Democracies Go to War.* Cambridge, MA: MIT Press.

81 See Gerard Lemos (2012).*The End of the Chinese Dream: Why Chinese People Fear the Future.* New Haven: Yale University Press; Timothy Beardson (2013). *Stumbling Giant: The Threats to China's Future.* New Haven: Yale University Press; Ho-fung Hung (2016). *The China Boom: Why China Will Not Rule the World*; and David Shambaugh (2016). *China's Future.* Cambridge: Polity.

82 Michael D. Swaine (May/June 2015). "The Real Challenge in the Pacific: A Response to How to Deter China." *Foreign Affairs*, Vol. 94, No. 3, P145-153.

83 This topic has drawn the attention of quite a few scholars. See David S. Mason (2009). *The End of the American Century.* Lanham: Rowman & Littlefield; George Packer (2013). *The Unwinding: An Inner History of the New America.* New York: Farrar, Straus and Giroux; Alfred W. McCoy, Josep M. Fradera, and Stephen Jacobson (eds.) (2012). *Endless Empire: Spain's Retreat, Europe's Eclipse. America's Decline.* Madison: The University of Wisconsin Press; and Glenn Hubbard and Tim Kane (2013).*Balance: The Economics of*

to insist that the U.S. still has the capacity to lead international affairs on its own without sharing its leadership with other great powers, and should actively strengthen its leadership by various means.[84]

American strategic researchers, who are reluctant to see other great powers lead the world in partnership with the U.S., generally tend to believe that the rise of China is a mortal threat to the U.S. and therefore must be contained, otherwise the U.S., its allies as well as the whole world will suffer. According to Samuel Huntington, a famous American political scholar, the inevitable clash between Western civilisation and the Chinese Confucian civilisation would affect the survival and development of the Western civilisation. The Western world must be led by the U.S. if it is to win such a war. He predicted that, "The balance of power among civilisations is shifting: the West is declining in relative influence; Asian civilisations are expanding their economic, military, and political strength."[85] He believed that the Confucian secular culture would be revived as the Asian values to be recognised by the world. Generally speaking, powerful societies tout the universalistic essence of their features as suitable for all societies. Weak societies, however, have only some peculiar features which could be applied in only a few areas. "The mounting self-confidence of East Asia has given rise to an emerging Asian universalism comparable to that which has been characteristic of the West."[86] At last, Huntington warned that, "Chinese hegemony will reduce instability and conflict in East Asia. It also will reduce American and Western influence there and compel the United States to accept what it has historically attempted to prevent: domination of a key region of the world by another power [meaning China]."[87]

Many American strategists believe that China is the biggest threat

Great Powers from Ancient Rome to Modern America. New York: Simon & Schuster.

84 Bruce Jones (2014). *Still Ours to Lead: America, Rising Powers, and the Tension between Rivalry and Restraint*. Washington, D.C.: Brookings Institution Press.

85 Samuel P. Huntington (1996). *The Clash of Civilisations and the Remaking of World Order*. New York: Simon & Schuster, P20.

86 Ibid, P109.

87 Ibid, P237.

to the hegemony of the U.S., but they have different opinions about how big the threat is and what the international landscape would become in the end.[88] For example, Andrew J. Nathan, an American "China hand", thought that China's domestic issues are so tricky that China has neither time nor energy for outward expansion. "China is too bogged down in the security challenges within and around its borders to threaten the West unless the West weakens itself to the point of creating a power vacuum.... The main tasks of Chinese foreign policy are still defensive: to blunt destabilizing influences from abroad, to avoid territorial losses, to moderate surrounding states' suspicions, and to create international conditions that will sustain economic growth. What has changed is that these internal and regional priorities are now embedded in a larger quest: to define a global role that serves Chinese interests but also wins acceptance from other powers."[89]

Another American "China hand" David Shambaugh warned that the West shouldn't overestimate China's international influence because China is yet to be an all-around world power. It is at most a "partial power," whose soft power cannot be compared to that of the West at all.[90] Recently, Shambaugh wrote a widely read essay, predicting that China would "crack" due to all kinds of domestic conflicts and problems.[91] The American scholar Joseph Nye, who had come up with the concepts of "soft power" and "smart power", stated that even if China's overall economic strength is becoming comparable with that of the U.S., its hard power and soft power are still far from parallel to that of the U.S. in the coming decades. Accordingly, "the American Century" will still continue.[92] Jonathan Fenby, who has long studied the modern

---

88 Wang Jisi (2015). *Relations between the Big Powers: Will China and the U.S. Go Their Separate ways or Share the Same Destination?* (《大國關係：中美分道揚鑣，還是殊途同歸？》). Beijing: China CITIC Press.

89 Andrew J. Nathan and Andrew Scobell (2012). *China's Search for Security.* New York: Columbia University Press, Pxi-xv.

90 David Shambaugh (2014). *China Goes Global: The Partial Power.* New York: Oxford University Press.

91 David Shambaugh (March 6, 2015). "The Coming Chinese Crackup". *The Wall Street Journal.*

92 Joseph S. Nye, Jr. (2015). *Is the American Century Over?* Cambridge: Polity, P46-70.

history of China, along with other scholars, also asserted that it is impossible for China to dominate the 21st century. [93]

A few American scholars of international relations even inferred optimistically that the rise of China doesn't indicate that it would over-throw the U.S.-led "liberal international order". Instead China would integrate itself into that system and strive to promote reform of the order so as to render it more beneficial to Chinese interests. For example, John Ikenberry believed, "China, despite its ascent, has no ambitious global agenda; it remains fixated inward, on preserving party rule. [...] China now wants a larger role in the International Monetary Fund and the World Bank, a greater voice in such forums as the G-20, and wider global use of its currency. [...] Across a wide range of issues, China and Russia are acting more like established great powers than revisionist ones. [...] They do not have grand visions of an alternative order. For them, international relations are mainly about the search for commerce and resources, the protection of their sovereignty, and, where possible, regional domination. They have shown no interest in building their own orders or even taking full responsibility for the current one and have offered no alternative visions of global or political progress."[94]

However, the mainstream views of the U.S. and the West seemed to be different. Instead, the spread of the "China threat theory" is rampant. Westerners assert that China is planning to build a new world order shaped and led by itself so as to replace the one established by the U.S. after World War II. The new order would not only reject Western values and institutions, but also threaten the interests and security of the West.[95]America's and Japan's policies towards China unmistakably

93 Jonathan Fenby (2014). *Will China Dominate the 21st Century?* Cambridge: Polity; Josef Joffe (2014). *The Myth of America's Decline: Politics, Economics, and a Half Century of False Prophecies.* New York: Liveright Publishing Corp.; and George Friedman (2009). *The Next 100 Years: A Forecast for the 21st Century.* New York: Doubleday.

94 G. John Ikenberry (May/June 2014). "The Illusion of Geopolitics: The Enduring Power of the Liberal Order". *Foreign Affairs,* Vol. 93, No. 3, P80-90, P88-90. Quite a few scholars didn't think that China would become a threat to the West. See for example Lionel Vairon (2014). *China Threat? The Challenges, Myths, and Realities of China's Rise.* New York: CN Times Books.

95 See David Shambaugh (11 June 2015). "Race to the Bottom". *South China Morning Post,* PA15.

show that China is regarded as their powerful counterpart, so they strive for policies to contain the rise of China. The Chinese Government uses the phrase "China's peaceful development" instead of "China's peaceful rise" to demonstrate its benevolent intentions. However, this attempt to pacify the West and others has yielded few positive results and failed to change the sceptical attitude of the West towards China's "striving for success despite all difficulties" and "revanchism."[96]

Since many American and Western strategists are convinced that "the rise of China as a great power" is irreversible, the urgent task therefore becomes how to assess the threat from China and figure out how to deal with it. Martin Jacques' prediction is the most frightening. He is sure that China will "rule" the world and forcibly apply their traditional principle of "the tribute system" to other countries so as to establish an unfair and hierarchical China-centred international order and subject other countries to Chinese hegemony. As a result, the international order established by the West in the past centuries would finally come to an end.[97]

Generally, American strategists believe that China could definitely become a mortal threat to the U.S. hegemony, so they must pay sufficient attention to China. Some even think that China has long planned to replace the U.S. as the global hegemon in one hundred years.[98] They believe that China is a threat to the U.S. in many aspects, including replacing the U.S. as the global hegemon, weakening America's say in the world, contending with the U.S. for its interests and primacy in Asia, Africa or other regions, eroding the primacy of the

---

96 See Zheng Bijian (2012). *China's Peaceful Development and Building a Harmonious World* (《中國和平發展與構建和諧世界》). Beijing: People's Publishing House; and Zheng Bijian (2014). *China's Development Strategy: China's Peaceful Rise and the Cross-Strait Relations* (《中國發展大戰略：論中國的和平崛起與兩岸關係》). Taipei: Commonwealth Publishing Group.

97 Martin Jacques (2009). *When China Rules the World: The End of the Western World and the Birth of a New Global Order*. New York: Penguin. See also Christopher A. Ford (2010). *The Mind of Empire: China's History and Modern Foreign Relations*. Lexington: The University Press of Kentucky.

98 Michael Pillsbury (2015). *The Hundred-Year Marathon: China's Secret Strategy to Replace America as the Global Superpower*. New York: Henry Holt and Co.; and Peter Navarro (2015). *Crouching Tiger: What China's Militarism Means for the World*. New York: Prometheus Books.

U.S. dollar through the RMB internationalisation, blocking the navigation in the Western Pacific through military expansion, particularly strengthening the navy and air force, excluding the U.S. from the China-led regional economic cooperation, striving to change international politics and the international rules of the game, as well as rejecting the Western standpoint that human rights take precedence over national sovereignty and substituting that with the Chinese traditional views on sovereignty and national security. Considering the security and interests of the U.S. and the West, the rise of China therefore has to be contained by all means available.

American scholars James Steinberg and Michael O'Hanlon gave a summary of these views as follows: "As China's power has grown, there has emerged a more explicit school of thought [in the U.S.] that China's rise poses a direct threat to American primacy and thus to American security and that therefore the United States needs to check China's capabilities." From [the values] perspective, actively supporting democratic change and human rights not only is consistent with the U.S. values but also can help bring about 'peaceful evolution' of the Chinese political system that would make China's rise less threatening. They therefore champion direct support for activists in China, including measures to undermine China's information control system, as well as a forceful role in challenging China's human rights policy in international fora." "[...] maintaining Taiwanese de facto (if not de jure) independence is a major barrier to China's military expansion beyond the so-called first island chain and therefore key to continued maritime supremacy."[99]

In the following discussion, views on the "China Threat" from several influential Western professors and scholars are quoted to illustrate some typical standpoints. According to John Mearsheimer, a champion of political realism, all great powers tend to be aggressive. In order to increase its own strength, a great power would not only seek

---

99 James Steinberg and Michael E. O'Hanlon (2014). *Strategic Reassurance and Resolve: U.S.-China Relations in the Twenty-First Century.* Princeton: Princeton University Press, quoted from P65-66, 68 and 70 respectively

to damage the interests of other countries, but also defeat countries that decrease its strength to increase their own strengths. This is the inevitable result of the lack of a central authority above nations to protect some countries from aggression. All nations have offensive military forces, and they are all ignorant of the others' intentions. As a result, in order to maintain its hegemonic status, every great power strives to surpass others in national strength, so as to protect itself from threats. He asserted pessimistically that, "Unfortunately, a policy of engagement [with China] is doomed to fail. If China becomes an economic powerhouse, it will almost certainly translate its economic might into military might and make a run at dominating Northeast Asia. Whether China is democratic and deeply enmeshed in the global economy or autocratic and autarkic will have little effect on its behaviour, because democracies care about security as much as non-democracies do, and hegemony is the best way for any state to guarantee its own survival. Of course, neither its neighbours nor the United States would stand idly by while China gained increasing increments of power. Instead, they would seek to contain China, probably by trying to form a balancing coalition. The result would be an intense security competition between China and its rivals, with the ever-present danger of great-power war hanging over them. In short, China and the United States are destined to be adversaries if China's power grows."[100]

John Mearsheimer believed that a regional hegemon will attempt to check aspiring hegemons in other regions because it fears that a rival great power that dominates its own region would become an especially powerful foe that will create trouble in its backyard. Regional hegemons prefer that there be at least two great powers located together in other regions, because their proximity would force them to concentrate their attention on each other rather than on the distant hegemon. As a result, "A much more powerful China can also be expected to try to push the United States out of the Asia-Pacific region, much as the United States

---

100 John J. Mearsheimer (2014). *The Tragedy of Great Power Politics*. New York: W.W. Norton, P4.

pushed the European great powers out of the Western Hemisphere in the nineteenth century."[101] "The optimal strategy for dealing with a rising China is containment. [...] Toward that end, American policymakers would seek to form a balancing coalition with as many of China's neighbours as possible. The ultimate aim would be to build an alliance structure along the lines of NATO, which was a highly effective instrument for containing the Soviet Union during the Cold War. The United States would also work to maintain its domination of the world's oceans, thus making it difficult for China to project power reliably into distant regions like the Persian Gulf and, especially, the Western Hemisphere."[102]

Political commentator Geoff Dyer is sure that a war in the 21st century would certainly happen between China and the U.S. He thinks that China has the ambition of replacing the U.S. as the global hegemon. He has stated that since the negotiation between President Nixon and Mao Zedong in 1972, the most peaceful and prosperous period in the modern history of Asia was the following forty years. According to the "China-U.S. agreement," the U.S. recognised China's return to the international community, while China agreed tacitly on America's primacy in Asia. However, neither written documents nor official announcements about this agreement could be found. As a result, this agreement, which specified the role of the U.S. in Asia, is gradually collapsing. Now, China intends to change the military and political landscapes in the Asian region and restore its primacy in Asia. The dispute over naval power in the Western Pacific between China and the U.S. is the prime example of the contest between these two great powers. According to his estimate, the relations between China and the U.S. might develop in two directions. The first possibility is that China and the U.S. would become more and more interdependent, so their military or other contests would be suppressed. The second one is that the contest would result in two different global economic systems in the world, namely the

---

101 Ibid, P371.
102 Ibid, P384-385.

U.S.-centred Western system and the system serving primarily the interests of Beijing. Regarding the U.S.-driven TPP (Trans-Pacific Partnership)[103] and TTIP (Transatlantic Trade and Investment Partnership), he believed that those two negotiations on free trade zones share the same theme—to counter the threat from China's "state capitalism" together with Japan and the European Union.[104]

The agendas of both TPP and TTIP cover topics that are unacceptable to China and that are not given prominent place by the "China Model". The topics include the protection of intellectual property rights, the prohibition against government subsidies to enterprises, foreign investment protection and labour rights. The U.S. believes that if negotiations on these two free trade zones succeed and are able to entice many more countries to participate, the U.S. can then install new game rules to make international trade more open, thus forcing China to accept and abide by the rules. Consequently, the U.S. can reinforce its status as the rule makers for the world trade system, while China has no alternative but to seek survival in a U.S.-dominated system.[105]

Christopher Coker, a British scholar of international relations, also agreed that it is not impossible for an all-out war between China and the U.S. to break out because in the past three hundred years, all wars among great powers were caused by issues regarding the rules and norms of the international system. Both China and the U.S. are "exceptional" powerhouses, and their values are diametrically opposed to each other. Beijing is challenging the international order which Washington

---

103 See Jeffrey J. Schott, Barbara Kotschwar, and Julia Muir (January 2013). *Understanding the Trans-Pacific Partnership.* Washington, D.C.: Peterson Institute for International Economics; and Peter A. Petri, Michael G. Plummer, and Fan Zhai (November 2012). *The Trans-Pacific Partnership and Asia-Pacific Integration: A Quantitative Assessment.* Washington, D.C.: Peterson Institute for International Economics.

104 See Richard Rosecrance (2013). *The Resurgence of the West: How a Transatlantic Union Can Prevent War and Restore the United States and Europe.* New Haven: Yale University Press. The author believed that "the progressive merger of the economies of Europe and America would lead to the creation of a new and magnetic bloc in world politics, one that would be irresistible to others as well. Japan, Canada, Southeast Asian countries, and eventually China would be drawn into the Atlantic vortex, to their great economic and ultimately political betterment." P92.

105 Geoff Dyer (2014). *The Contest of the Century: The New Era of Competition with China— and How America Can Win.* New York: Alfred Knopf.

has long viewed as being completely under its control. Now, these two powers are falling into the so-called "Thucydidean Trap," and unable to get out of it. The so-called "Thucydidean Trap" refers to the situation where a conservative powerhouse that would like to maintain the current situation is challenged by an emerging powerhouse. Because of this trap, a war between China and the U.S. seems unavoidable.[106]

According to the prediction of the American strategist Edward Luttwak, the expansion of China's national strength in Asia would definitely provoke fear in neighbouring countries, which would enable the U.S., together with those countries, to contain China. He asserted that China shouldn't expand its military forces while promoting economic development, because the former would lead to fear and countermeasures from other countries, which would ultimately harm the interests of China's economic development. One of the fears of China's neighbouring countries is that it would take away their precious marine resources by relying on its increasingly powerful national strength, so the threat from China to the Pratas Islands and the Spratly Islands cannot be ignored. Another fear is that they would be forced to sign agreements with China that would benefit China's bilateral trade. As Luttwak sees it, the Chinese leaders are deliberately pursuing several incompatible aims—rapid economic growth, rapid military expansion, with an attendant increasing influence globally. In order to restrict China's expansion, "Since 2010, the United States Department of Defence, together with other different services, has effectively supported measures related to the 'containment' of China in policies related to China made by the State Department."[107]

Aaron Friedberg, a young American scholar of China-U.S. relations, holds more extreme and explicit views, thinking that only by overthrowing the CPC regime could the U.S. be relieved of the threat

106 Christopher Coker (2015). *The Improbable War: China, the United States & the Logic of Great Power Conflict.* New York: Oxford University Press.

107 Edward N. Luttwak (2012). *The Rise of China vs. the Logic of Strategy.* Cambridge, MA: The Belknap Press of Harvard University Press. P238. See also Jakub J. Grygiel & A. Wess Mitchell (2016). *The Unquiet Frontier: Rising Rivals, Vulnerable Allies, and the Crisis of American Power.* Princeton: Princeton University Press.

from China. He has stated, "If through inadvertence, error, or deliberate decision we permit China as presently constituted to dominate
Asia, our prosperity, security, and hopes of promoting the further
spread of freedom will be seriously impaired. Our businesses could
find their access to the markets, high-technology products, and natural
resources of some of the world's most dynamic economies constricted
by trade arrangements designed to favour their Chinese counterparts.
[...] Control over the vast oil and gas reserves believed to lie beneath
the South and East China Seas, plus assured access on favourable
terms to energy imports from Central Asia and Russia, could greatly
reduce Beijing's vulnerability to a possible American (or Indian) naval
blockade. With the United States gone from East Asia, China would be
able to bring Taiwan to terms, and it would most likely be able to block,
neutralise, or preempt the emergence of serious military challenges
from Japan or South Korea. Freed of the necessity of defending against
possible threats from its maritime periphery, Beijing would be able to
devote more resources to setting terms with its continental neighbours
and it would be able to more easily project military power to defend or
advance its interests in other parts of the world, including the Middle
East, Africa, and Latin America. Before it can hope to compete with
the United States on a global scale, China must first establish itself as
the foremost power in its own region. If Asia comes to be dominated
by an authoritarian China, the prospects for liberal reform in any of
its non-democratic neighbours will be greatly diminished. Even the
region's established democracies could find themselves inhibited from
pursuing policies, foreign and perhaps domestic as well that might
incur Beijing's wrath. With its enhanced global reach and influence,
China would also be able to more effectively support non-democratic
regimes in other parts of the world and to present some variants of its
own internal arrangements as a viable alternative to the liberal democratic capitalism of the West." [108] "[To fundamentally transform the

108 Aaron L. Friedberg (2011). *A Contest for Supremacy: China, America, and the Struggle for Mastery in Asia*. New York: W.W. Norton, P7-8.

U.S.-China relationship], a change in China's domestic regime will be necessary. While they might induce some initial instability, liberal-ising reforms would eventually ease or eliminate ideology as a driver of competition, enhance the prospects for cooperation in dealing with a variety of issues, from trade to proliferation, and reduce the risk that future disputes might escalate to war." [109] "Stripped of diplomatic niceties, the ultimate aim of the American strategy is to hasten a revo-lution [in China], albeit a peaceful one, that will sweep away China's one-party authoritarian state and leave a liberal democracy in its place."[110]

Jonathan Holslag, a young German scholar, even asserts that a war between China and some Asian countries is unavoidable because China's core interests, along with its demands as a great power, are incompatible with its proposal of "peaceful development" in the polit-ically complicated Asia. "The rise of China as a great power" would certainly impair the interests and security of other Asian countries and even the U.S.. China is a "revisionist" power with the determination to alter the current international order and distribution of powers and to become the new global hegemon. [111] In addition, American military scholar Navarro also believed that with the constant advance of China's military technology and the constant reduction in America's defence budgets, a China-U.S. war is doomed to happen. [112]

New International Landscape in the Contest between China and the U.S.

To sum up, an intense game between the U.S., a "conserva-

---

109 Ibid, P57.
110 Ibid, P184.
111 Jonathan Holslag (2015). *China's Coming War with Asia*. Cambridge: Polity Press.
112 Peter Navarro (2015). *Crouching Tiger: What China's Militarism Means for the World.* New York: Prometheus Books.

tive" and "status-quo" power, and China, a rising great power, is an unavoidable international event. The intensity of the game, however, is increasing due to America's insufficient confidence in its institutions and prospects, its fear of Chinese ideas and the notion that "Someone not from our clan must have a different mind," as well as the rising and strengthening Chinese nationalism and patriotism. The U.S. is well aware that it can no longer maintain its hegemony using only its own strength as in the past, so it must form alliances with other countries. Similarly, although China refuses to forge military alliances with any other countries, it is inevitable for it to develop strategic partnership with other countries due to the coercion and "containment" of the U.S..

Although many people in the U.S. believe that China is a threat to the U.S., only a few of them explicitly advocate a war with China. Considering the close and inseparable economic and financial link-ages between China and the U.S.,[113] China's power to "take revenge" and "counteract," the interests and anxieties of America's allies, the war-weariness of the Americans, their potential tendency toward "isola-tionism" as well as the huge cost of opposing China, most American strategists propose to balance or constrain China's strength by various means and prevent the over-expansion of China's strength. They also suggest the indispensability of a favourable environment and proper incentives for China to accept the U.S.-led international order and be willing to pursue national interests within it.

At the beginning of his tenure, President Obama and his coun-selors hoped that Washington and Beijing could cooperate in order to become senior partners in charge of international affairs (under the so-called G-2 arrangement). However, American decision-makers soon concluded that China would regard this shared leadership as a compro-mise on the part of America. They thought that, obviously, a more effective policy should be based on the balance of strength, so a more

---

113 Zachary Karbell (2009). *Superfusion: How China and America Became One Economy and Why the World's Prosperity Depends on It.* New York: Simon & Schuster; and Stephen Roach (2014). *Unbalanced: The Codependency of America and China.* New Haven: Yale University Press.

powerful diplomatic policy was needed. In November, 2009, Obama met with leaders of the Association of East Asian Nations (ASEAN) for the first time. Later in June, 2010, Obama agreed to attend the newly-organized East Asia Summit in order to prevent it from being led by China. "Barack Obama was the first [American] president who had to worry that America might actually be overtaken by a rival since the end of the Cold War."[114]

In 2011, President Obama proclaimed with much fanfare to "pivot to Asia" and pursue the "rebalancing" strategy, with the aim of renewing and strengthening the U.S. diplomatic and military relations with the countries adjacent to China, particularly ASEAN countries, and pledging support from the U.S., to encourage these countries to contest with China for interests in the East China Sea and the South China Sea, including the islands, reefs, waters, maritime and submarine resources. The paramount aim of the U.S. is to weaken China's influence in East Asia as well as to erode the relations between China and its neighbouring countries.[115] America's "rebalancing" policy is its first response to the challenges taken up by China in East Asia. "Kurt Campbell, the State Department's senior official for Asia, thought that the United States had staked its claim to a dominant role in the Asia-Pacific region for the next forty, thirty, fifty years."[116]

In order to bolster its "pivot to Asia" strategy, the U.S. is strengthening its military forces in the Western Pacific on a large scale, deploying nearly half of its naval and air forces there. Furthermore, it is using the new "air-sea" battle strategy to enhance its military strength against China, with the aim of confining China's naval and air forces within the first island chain and preempting a greater threat to the Western Pacific.[117]

---

114 Stephen Sestanovich (2014). *Maximalist: America in the World from Truman to Obama.* New York: Alfred A. Knopf, P314.

115 Kurt M. Campbell and Ely Ratner (May/June 2014). "Far Eastern Promises: Why Washington Should Focus on Asia." *Foreign Affairs.* Vol. 93, No. 3, P106-116.

116 Ibid, P314.

117 The U.S. underwent a "military revolution" in 1980s, during which advanced technology and mobility were developed to enhance the effectiveness of air-land battle to fight against

The aim of America's proposal for TPP is to establish a regional free trade system which would apply the stringent principles of free trade in accordance with America's program so as to reinforce its economic primacy in Asia-Pacific regions and weaken China's economic momentum. China will be excluded from this system because it would be difficult for China to abide by its rules and regulations. Some American scholars clearly point out that it is difficult to separate trade from politics in the contest between China and the U.S.. "Even trade agreements, like the U.S.-Korean Free Trade Agreement and the proposed Trans-Pacific Partnership (TPP), are sometimes promoted in terms of their broader impact on U.S.-China relations."[118]

Chinese strategists are well aware of America's plots. Qi Dapeng (綦大鵬) for example believed that countries around the world are currently in an unprecedented stage of fierce competition to shape the international political and economic order. The first item in the competition is the contest for free trade arrangements. In 2013, the U.S. strived to complete the TPP negotiation and initiated the TTIP negotiation with the European Union. "The conclusion of both agreements will lay the foundation for the establishment of the Asia-Pacific and

---

the Soviet Union. In the past few years, in order to respond to China's increasing "military threat," the U.S. also made innovations in its strategy of air-sea combat capabilities which, if successful, would result in the complete blockade of China by the U.S. navy and air force. The aim was to respond to China's improvement in the quality and quantity of its naval and air forces, and to prevent China from forcing America's military forces far away from the coast of China. Nevertheless, air-sea battle requires a total war with China, and it is not clear whether the U.S. has enough determination to launch it. See Aaron L. Friedberg (2014). *Beyond Air-Sea Battle: The Debate Over U.S. Military Strategy in Asia*. London: The International Institute for Strategic Studies. Also see Carnes Lord and Andrew S. Erickson (2014). *Rebalancing U.S. Forces: Basing and Forward Presence in the Asia-Pacific*. Annapolis: Naval Institute Press; and Peter Navarro (2015). *Crouching Tiger: What China's Militarism Means for the World*. New York: Prometheus Books, P187-201.

118 James Steinberg and Michael E. O'Hanlon (2014). *Strategic Reassurance and Resolve: U.S.-China Relations in the Twenty-First Century*. Princeton: Princeton University Press, P69-70. Bruce Jones analysed it in a similar way: "In 2011 and 2013 respectively, President Obama announced U.S. engagement on two more major trade initiatives—the Trans-Pacific Partnership (TPP) and the Transatlantic Trade and Investment Partnership (TTIP). Both initiatives have a double logic: driving economic growth and reinforcing core alliances." See Bruce Jones (2014). *Still Ours to Lead: America, Rising Powers, and the Tension between Rivalry and Restraint*. Washington, D.C.: Brookings Institution Press, P105-106. See also Jeffrey A. Bader (2012). *Obama and China's Rise: An Insider's Account of America's Asia Strategy*. Washington, D.C.: Brookings Institute Press.

U.S.-European free trade zones. As a result, the global trade map will be greatly transformed, and the U.S. will again become the centre of the global economy. [...] Based on geopolitical calculations, Japan is giving priority to the TPP negotiation while initiating negotiation on the China-Japan-South Korea free trade zone at the same time. Japan is attempting to make new rules for free trade together with the U.S., and force China to obey the rules once it joins the trade bloc in the future, so as to gain head start advantage over China in development. In addition, the initiation of negotiation on the formation of a Japan-EU free trade zone symbolises Japan's completion of its free trade strategic layout in three major economically prosperous regions in the world—East Asia, the U.S. and Europe. The European Union is also developing free trade zone strategies. In 2013, Europe concluded free trade zone agreements with Canada and opened a series of negotiations on free trade zones with the U.S., Japan, India and ASEAN. The free trade zone arrangements of developed countries in the West reflect their attempts to prolong their prerogative of setting the guidelines on international trade when faced with the simultaneous rise of a group of emerging powers. Since the U.S.-led TPP might weaken the leadership of ASEAN countries in East Asian economic cooperation and produce divisions within ASEAN, it therefore also started the "RCEP (Regional Comprehensive Economic Partnership)" negotiation with China, Japan, South Korea, Australia, New Zealand and India in 2013. Since the U.S. is excluded from RCEP, TPP and RCEP are in fact competitors. Not to be outdone, China also initiated the China-South Korea and China-Japan-South Korea negotiations on free trade zones. With respect to some middle and small countries such as South Korea, Singapore and Vietnam, they also didn't lag behind in the competition of arranging free trade zones. They also took advantage of competition among great powers and concluded a series of free trade zone agreements with them."[119] "It

---

119 Qi Dapeng (2014). "An Exploration of the Historic Trend of the Great Transformation of the International Order: an Overview of this Year's International Strategic Situation" ("探索國際大變局的歷史走向—本年度國際戰略形勢總論"), in Institute of Research, National Defence University. *The International Strategic Situation and China's National Security* (《國際戰略形勢與中國國家安全》). Beijing: National Defence University Press. P2-3.

seems that competition for free trade arrangement won't attract as much attention as traditional large-scale wars do. In essence, however, the competition lays the foundations for the countries' status and power in the future international system under the new historical circumstances. [...] In the long term, results of the competition will not only determine the trend of the global economic order, but will also influence trends of global politics and security."[120]

Sun Zhe (孫哲), a mainland Chinese scholar of international relations, expounded his views on America's Asia-Pacific strategy: "America's comprehensive policy of 'strategic refocusing' in the Asia-Pacific, particularly Southeast Asia, since 2011 was already made one year ago or even earlier. The prime consideration behind this policy was the influence of the rise of China on these regions. In fact, during the administration of George Walker Bush, the U.S. had already made major changes in its Asia-Pacific strategy twice in response to the rise of China. The first change was made when Bush took office. He eschewed the policy on China made by the Clinton administration and instead regarded China as a 'strategic competitor' rather than a 'strategic partner.' He declared that the U.S. would strengthen its military presence in the Western Pacific to 'prevent the re-emergence of a new rival.' The core of the first change was to 'preventively' contain the rise of China with the help of America's allies like Japan and South Korea. The second change took place about five years later. According to the National Security Strategy Report released by the White House in 2006, the focus of America's strategy on East Asia shifted from security alone to security with economy and trade. In particular, the U.S. wanted to expand the trade and investment in East Asia and become more engaged in its existing regional organisations and diplomatic affairs. As far as China was concerned, the U.S. emphasised the idea of the 'responsible stakeholder,' attached more importance to the expansion of cooperation with China, and held a cautiously optimistic attitude toward the future of China. This change focused on improving relations with China and

---

120 Ibid, P4.

engaging itself in cooperation in East Asia, particularly in the areas of trade and economic integration. However, with most of its strategic strength trapped in Iraq and Afghanistan, the U.S. could hardly involve itself fully in Asia-Pacific affairs as it had planned. As a result, in the cooperation between the U.S. and the Asia-Pacific countries, the dominant topic was still security, and the major participants in security cooperation were America's allies such as Japan, South Korea and Australia as well as countries such as Indonesia and the Philippines, both of which played a comparatively greater role in the anti-terrorist wars launched by the U.S. In brief, until the end of Bush's presidency, the extent of America's involvement in the trade and economic integration in East Asia had been relatively limited. On the whole, the U.S. didn't pay enough attention to this region. In the wake of 2009, America's anti-terrorism efforts decreased, and China successfully showed its powerful political and economic clout during the global financial crisis. After that, American strategists become increasingly worried about the security of America's status in the future Asia-Pacific geopolitical landscape, afraid that the rapid rise of China would make a great impact on the current regional situation. Due to this concern, the *National Security Strategy Report* released by the White House in 2010 put forward the strategy of 'revitalising the U.S. to lead the world', which mainly focused on the Asia-Pacific regions. Furthermore, according to the *Annual Report on China's Military Power* released by the United States Department of Defence in August, 2011, China's rapid development of military technology and its 'maritime ambitions' were emphasised in particular, and the U.S. again 'showed their anxiety about Beijing's tougher military stance.'"

"In addition to the doubts and worries about China's military development, the U.S. is also afraid that as China's economic strength grows rapidly and particularly as ASEAN-centred 'ASEAN+X' mechanism is being reinforced, the U.S. will gradually be excluded from China-led trade and the economic integration of East Asia in the near future."

"Does America's new round of strategic changes in Southeast

Asia mainly target China? The answer is definitely yes. However, we should be aware that the fundamental reason for the strategic changes is neither the replacement of the 'Sinologists' by the 'China bashers' inside the U.S. Government nor China's tough stance on the South China Sea issues. Instead, the reason is that the rapid rise of China becomes a challenge to America's medium- and long-term primacy in the Asia-Pacific security, trade and economic systems, a psychological challenge in particular. As long as the U.S. continues to be have such kind of psychological hang-ups, it will constantly adjust its strategy for Southeast Asia and the Asia-Pacific in response to the changing situation there, so as to slow down the declining trend of its power as far as possible, and continue to play a leadership role in these regions."

"In the long run, as China is rising, the U.S. will certainly increase its strategic involvement in the Asia-Pacific, with the focus on Southeast Asia, so as to recover its dominance in affairs related to security, trade and economic cooperation in the Asia-Pacific region. As to the means to be used to pursue its Southeastern Asia strategy, the U.S. will deploy diplomacy, military force, economics and trade, as well as American values in different ways in line with its own needs, adapting to changes in the regional situation. Among these various means, the most crucial is strengthening military deployment and security partnerships in the region."[121]

Tang Yongsheng (唐永勝) and Pang Hongliang (龐宏亮), scholars from mainland China, had similar observations. According to their analysis, "There are still many problems that cannot be ignored between China and the United States, which are mainly reflected in the fact that the United States is steadily promoting the 'Asia-Pacific rebalancing' strategy, including increasing investment in the armed forces in order to establish strategic prepositioning to defend against China, and promoting the Trans-Pacific Partnership Agreement (TPP) to establish a US-led trade mechanism. [...] The United States continues to

---

121 Sun Zhe (eds.) (2012). *Changes in the Asia-Pacific Strategy and New Great Power Relations between China and the U.S.* (《亞太戰略變局與中美新型大國關係》). Beijing: Current Affairs Press, quoted from P265-267, 267, 268 and 273 respectively.

enhance the strength of the U.S. Pacific Command while strengthening the traditional military alliance with Japan. [...] TPP is a new type of trade model carefully designed by the United States in order to remake the FTA [Free Trade Agreement] traditional trade pattern. It sets strict standards on labour, environmental products, intellectual property, government procurement, state-owned enterprises and other issues. For China, it is nearly impossible to meet these standards under the existing political and economic systems. From this perspective, the TPP agreement will eventually become a 'high standard' trade agreements which excludes China. [...] Once the TPP agreement is implemented in the countries mentioned above, it will certainly weaken the existing '10+1', '10+3' and the China-Japan-ROK dialogue mechanisms in East Asia. Thus it will weaken the role of China in the promotion of economic integration of the Asia-Pacific region, and even marginalise China."[122]

Another important part of America's "rebalancing strategy" is to tilt Southeast Asian countries toward itself, creating obstacles or barriers in China's surrounding areas to prevent China's expansion. To this end, the United States has encouraged some Southeast Asian countries to fight China for the territories and resources in the South China Sea, shaped international opinion in their favour, provided them with weapons, promoted military cooperation, and sent the U.S. troops to their strategic areas while condemning China from time to time for its "illegal" actions or "aggression". From the U.S. strategists' perspective, the strategic value of the South China Sea to China is comparable with that of the Caribbean to the United States. Kaplan, who specializes in America's Asia strategies, offered the following description in his book: "China's position vis-à-vis the South China Sea is akin to America's position vis-à-vis the Caribbean Sea in the nineteenth and early twentieth century. The United States recognised the presence and claims of European powers in the Caribbean, but sought to dominate the region,

---

122 Tang Yongsheng & Pang Hongliang (2014). "Analysis of Sino-U.S. Relation" ("中美關係評析"), in Institute for Strategic Studies, National Defence University (2014). *The International Strategic Situation and China's National Security* (《國際戰略形勢與中國國家安全》). Beijing: National Defence University Press, P16-33, 27-29.

nevertheless. Moreover, it was domination of the Greater Caribbean Basin that gave the United States effective control of the Western Hemisphere, which, in turn, allowed it to affect the balance of power in the Eastern Hemisphere. Perhaps likewise with China in the 21st century."[123]

If the South China Sea becomes a "Chinese lake", it will not only affect maritime trade due to the obstruction of the channel, but also pose a threat to the U.S. allies in East Asia, especially to Japan's survival and interests. American strategist Kagan believes that "The fact that China is trying to use its growing naval power not to open but to close international waters offers a glimpse into a future where the U.S. Navy is no longer dominant. [...] The move from American-dominated ocean to a collective policing by multiple great powers—even if it occurred— might turn out to be a formula for competition and conflict rather than a bolstering of the liberal economic order."[124]

Zhao Yi (趙毅), a Chinese scholar of strategy, described the Southeast Asia strategy of the U.S. as follows: "Since the United States announced its 'pivot to Asia' in 2010, ASEAN has become the main target of America's strategic adjustment. ASEAN was at that time favoured by Japan as Sino-Japanese relations deteriorated. As a result, in 2013, ASEAN's geopolitical importance improved significantly. It was ardently sought after by the United States and Japan. In order to restore its power in the Western Pacific region, the United States readjusted its strategies, setting the alliance with ASEAN as an important goal while strengthening relations with its traditional allies in the Asia-Pacific region. Militarily, the United States is expanding its forces in Southeast Asia. Economically, it is trying to draw the Southeast Asian countries to join the TPP. And politically, it employs democracy and human rights as the rationale to export Western values, through official and non-governmental channels, to Southeast Asia and Indochina. These efforts

---

123 Robert D. Kaplan (2014). *Asia's Cauldron: The South China Sea and the End of a Stable Pacific.* New York: Random House, P13. He believes further that the game between China and the U.S. in South Asia will start in the near future. See Robert D. Kaplan, Monsoon (2010). *Monsoon: The Indian Ocean and the Future of American Power.* New York: Random House.

124 Robert Kagan (2012). *The World America Made.* New York: Alfred A. Knopf, P77.

have paid off as most Southeast Asian countries have moved noticeably closer to the U.S. as a whole in recent years. [...] The United States and Japan's association with ASEAN has remarkably changed the geo-strategic situation in the Western Pacific region. ASEAN's status has been on an upward trajectory since the Cold War. Its geopolitical importance has also enjoyed constant improvements as large countries have paid more attention to it. This can be shown in two aspects: not only have the powers in the [Western Pacific] region including China, Japan, the United States, Russia, India, Australia and Canada maintained close contact with ASEAN, but countries and organisations outside the region have also attached great importance to cooperation with it."[125]

Long before the United States announced its "return to Asia" strategy, China had already worked hard to improve its relations with neighbouring countries, Southeast Asian countries in particular, reinforcing their economic, trade and financial ties, and had achieved positive results. Despite the disputes over territorial waters and seabed resources, China has claimed its sovereignty over the South China Sea and the East China Sea with a strong commitment and concrete actions, including military actions in recent years. Notwithstanding this, except for Japan, East Asian and Southeast Asian countries still maintain reasonable relationships with China, and cooperate with China on security matters despite the disputes some of them have with China over territorial waters. In order to cope with the challenges from the U.S., China is bound to further strengthen its relations with neighbouring countries. China's rising economic power and willingness to share the dividend so fits development under the principle of "good neighbourliness" and "mutual benefits", make it quite competitive in the contest with the United States in East Asia and Southeast Asia.

China's neighbouring countries are committed to their economic development, and are absolutely reluctant to be involved in war. So

---

125 Zhao Yi (2014). "Assessment of the Internal and External Relations of ASEAN" ("東盟內外關係走向評估"), in Institute for Strategic Studies, National Defence University (2014). *The International Strategic Relations and China's National Security* (《國際戰略形勢與中國國家安全》). Beijing, National Defence University Press. P69-87, 69-74.

they prefer to stay "neutral", thus making it very difficult for the U.S. to forge an "anti-China" alliance with them. Besides, the U.S.'s looming economic difficulties and rising economic protectionism also have hindered TPP negotiation. The prospects of TPP are still uncertain despite its official endorsement by the governments of the countries involved. Steve Chan, a scholar of international study, saw the situation as follows: "East Asian elites have collectively pivoted to a strategy of elite legitimacy and regime survival based on economic performance rather than nationalism, military expansion, or ideological propagation." "I argue that governing elites in East Asia have increasingly staked their legitimacy and regime survival on their economic performance and that this decision is consequential in moving their regional relations in a cooperative direction." "An awareness of the severe opportunity costs of armament and alliance has restrained East Asian countries from behaving according to balance-of-power reasoning. The shadow of the future disposes them to eschew balancing policies and inclines them to cooperate. Interlocking international and domestic bargains, involving self-restraint and mutual restraint, buttress East Asia's regional stability and cooperation. The longer these bargains are sustained, the greater the prospect of continued cooperation."[126]

Liu Aming (劉阿明), a strategist from mainland China, had a similar opinion: "In building a regional order, the initiatives of East Asian countries are largely demonstrated by their action of building a policy to balance the powers within the region. ...Generally, East Asian countries have adopted a 'hedging' strategy to address the changes of power in the international system since the 1990s. Instead of only allying with or excluding a certain country, they work to include all the major powers in all the regional affairs. The purpose is to absorb these powers into a single body with ASEAN to form a new regional order through closer economic and political relations. In this way, the big

---

126 Steve Chan (2012). *Looking for Balance: China, the United States, and Power Balancing in East Asia.* Stanford: Stanford University Press. P4, 13, and 17. Also see Sun Zhe (2012). *Strategic Transformation in the Asia-Pacific Region and U.S.-China Relations* (《亞太戰略變局與中美新型大國關係》). Beijing: Current Affairs Press.

powers will take greater responsibility for the security and prosperity of the region, and be more interested in maintaining regional stability through political and diplomatic means. Small and medium-sized countries will thus receive greater respect for their role. In other words, the East Asian countries neither approve of a regional order of dominance by one country, nor are they in favour of a so-called 'balance of power' order dispersing the power of small countries into different groups led by big countries. What they aspire for is a balanced world where major powers can 'monitor each other' to prevent adventurism on the part of the major powers while small and medium-sized countries are assured a unique role in regional systems."[127]

The election of nationalist populist Donald Trump as the new president of the U.S. in November 2016 has resulted in the exit of the U.S. from TPP, signifying that TPP is no longer seen by the new U.S. Government as a cost-effective strategy to contain China. As far as TTIP is concerned, the rise of political forces against globalisation in both the U.S. and the European Union as well as the gross difficulty of reconciling their divergent interests would make it impossible to seal a comprehensive trans-Atlantic trade and investment deal in the near future. Consequently, the attempts of former President Obama's to forge a more open and high-grade global free trade system and thereby compel China to subscribe to an international system dominated by the West, instead of trying to build an alternative system, have not succeeded. Nevertheless, despite the appearance of some "isolationist" sentiments in the U.S. and the failure of TPP and TTIP to come about, in view of the intense geopolitical fight for primacy between the U.S. and China in Asia and the huge stakes involved, it is unlikely that the U.S. will give up its longstanding strategy to contain China and prevent China from becoming the hegemonic power in Asia. Accordingly, Sino-American contest in Asia will continue unabated. The South China Sea and the East China Sea will continue to be flashpoints between the

---

127 Liu Aming (2012). *The Changing East Asia and the Relations between China and the United State* (《變動中的東亞與中美關係》). Beijing: Intellectual Property Publishing House. P29-30.

two countries, with Japan siding strongly with the U.S.

Japan is the U.S.'s most important political and military ally in Asia. In recent years, China has been rising but the grievances stemming from the history between China and Japan have not yet been satisfactorily settled. In addition, Japanese nationalism and right-wing forces have also emerged. Fearful of China, Japan hopes that the U.S. will help it counterbalance China as well as help it regain its position as a pivotal "normal country" in the international arena. All these considerations have tilted Japan toward the United States in recent years. From another point of view, in the past, China, Japan and the United States all faced a common threat from the Soviet Union. However, their strategic partnerships have been greatly fractured after the breakdown of the Soviet Union lifted the threat.[128] In response to the rise of China, the United States has actively roped Japan in to develop a U.S.-Japan military alliance,[129] and coaxed Japan to strengthen its cooperation with the U.S. in East Asia, and to assume more responsibilities for maintaining the "security" and "peace" in the region (including Taiwan). The United States not only strengthens military cooperation with Japan to counterbalance China, but also favours Japan in the Sino-Japanese territorial dispute. Both actions exacerbated Sino-Japanese relations. At the same time, Japan stepped up to draw the Southeast Asian countries in with the intention of strengthening its influence on those countries, which inevitably intensified the contest between China and Japan in Southeast Asia.

Over the past decade or so, China and Japan clashed frequently over a number of issues. The most severe issues included Japan's attitude towards its historical crimes, the visit paid by Japanese politicians to Yasukuni Shrine with its Class-A war criminals, the poisoned dumplings case, the sovereignty of the Diaoyu Islands (known as Senkaku

---

128 Thomas J. Christensen (2011), *Worse than a Monolith: Alliance Politics and Problems of Coercive Diplomacy in Asia.* Princeton: Princeton University Press.

129 Michael J. Green and Zack Cooper (eds.) (2015). *Strategic Japan: New Approaches to Foreign Policy and the U.S.-Japan Alliance.* Lanham: Centre for Strategic & International Studies. Qiao Linsheng (2006). *The Foreign Policy of Japan and ASEAN.* Beijing: People's Publishing House.

Islands in Japan) and the seabed resources in the East China Sea.[130] The newly signed Treaty of Mutual Cooperation and Security with the United States and the revised domestic laws enables Japan to further expand its military forces and send troops for non-combat activities overseas to support the U.S. military actions in other countries.[131] In short, all the changes in the evolving situation has worsened the relations between China and Japan.

With regard to the evolution of Sino-Japanese relations, Liu Jiangyong (劉江永), an expert in Japanese studies in mainland China, observed: "It was not until Japan was defeated in 1945 that Japan started the transition of its state model under external forces, promulgated the Constitution of Japan, and embarked on the road of peaceful development. The strategy of 'cutting down armaments and prioritising economic development' implemented by the Prime Minister Shigeru Yoshida was a policy of state-building based on trade. It resulted in the spectacular economic growth of Japan. However, Nakasone and other Japanese rulers pursued another change of the character of the Japanese state beginning in the late 1980s. Domestically, they put economic growth aside and dedicated themselves to amending the Constitution. Externally, they sought to become a major 'political power' and change the post-war order with the U.S. support." "The main content of Japan's China strategy had shifted from economic cooperation in the 1980s to counterbalancing China on security issues as proposed by the Abe Cabinet nowadays. Japan's national development model is heading even further towards the right. [...] In East Asia, there is an alarming tendency: for one thing, multilateral joint military exercises on the Korean peninsula continue; for another, the Japanese Government is trying to lift the ban on collective self-defence and enable itself to

---

130 Sheila A. Smith (2015). *Intimate Rivals: Japanese Domestic Politics and a Rising China.* New York: Columbia University Press; Richard C. Bush (2010). *The Perils of Proximity: China-Japan Security Relations.* Washington, DC: Brookings Institution Press; and Claude Meyer. (2011). *China or Japan: Which Will Lead Asia.* New York: Columbia University Press.

131 Kent E. Calder (2009). *Pacific Alliance: Reviving U.S.-Japan Relations.* New Haven: Yale University Press; and Richard J. Samuels (2007). *Securing Japan: Tokyo's Grand Strategy and the Future of East Asia.* Ithaca: Cornell University Press.

engage in overseas joint operations with the U.S. and other countries in the name of 'self-defence'. The Abe Cabinet, on the one hand, proposes the doctrine of 'active peace'. On the other hand, Japan attempts to rely on the United States, and to call on the Philippines to undertake joint strategic efforts to counterbalance China in East Asia. Moreover, Japan even endeavours to forge security partnerships with NATO. All these actions on the part of Japan are in fact, paving the way for promoting 'violent multilateralism' in East Asia with the US-Japan alliance at its core…. The conflicts between China and Japan will become increasingly acute as Japan moves to the right politically and as the character of the Japanese state changes. Issues concerning history, the Diaoyu Islands and Taiwan will continue to haunt both countries."[132]

According to the observation of American scholar Denny Roy, for Japan and the United States, "China is not only a rising great power. It is a returning great power, arguably the first in history."[133]A resurgent great power may harbour the thought of revenge, so it is particularly dangerous. "From Tokyo's point of view, China is usurping Japan's previous position in Southeast Asia. The race between Tokyo and Beijing for free trade agreements with Southeast Asian countries is as much about political leadership as it is about enlarging economic opportunities." "[…] many retired or nongovernment Japanese strategists argue that the PRC possession of Taiwan would seriously threaten Japan by placing an unsinkable Chinese aircraft carrier amid the sea lane that carries most of Japan's energy supply." "[…] Controlling Taiwan would provide the PRC with a massive platform for air and naval bases, an extension of the Chinese coastline two hundred to 250 miles eastward, and unfettered access to the Pacific Ocean from Taiwan's east coast. This would offer PLA [People's Liberation Army] forces a commanding position from which to control the Western Pacific and to threaten the

---

132 Liu Jiangyong (2014). "A Great Change is Coming to the Sino-Japan Relations after Sixty Years" ("中日關係正迎來新甲子大變局"), in Institute for Strategic Studies, National Defence University (2014). *International Strategic Relations and China's National Security.* Beijing: National Defence University Press. P34-25, 42-45.
133 Denny Roy (2013). *Return of the Dragon: Rising China and Regional Security.* New York: Columbia University Press. P3.

U.S. bases on Okinawa or a U.S. Navy task force sailing in from the east."[134]

Apart from the contest in East and Southeast Asia, the Sino-American contest in Central Asia is heating up rapidly. Many countries in this region have emerged after the disintegration of the Soviet Union. Some are still in a state of political instability. However, Central Asia not only enjoys an important strategic status, but is also rich in oil, gas and natural resources. The Sino-American contest in Central Asia includes both cooperation and competition. Central Asian countries deal with the two sides in accordance with the interests of their countries or those of their rulers. "Chinese interests in Central Asia mainly involve the field of economy, followed by the construction of the belt of good-neighbourliness. Geographically, Central Asia is a transportation hub for China to reach faraway Russia and Ukraine and then move on to Europe. Using Central Asia as the hub, China can also cross the Caspian Sea and go to the Caucasus region, and from there reaching Iran and the Persian Gulf region in the Middle East. As the stable home front of China, Central Asia is also a stable source of resources and strategic support."[135]

China is a strong competitor to the United States. Nevertheless, the United States is willing to let China play a role in Central Asia for the stability of the region. "To avoid Russia's regaining of control over Central Asia, the United States encourages more powers to conduct competition in this area. The European Union, Japan, ROK [Republic of Korea], Turkey and other countries also actively promote their influence by all means among Central Asian countries. The United States hopes that China will shoulder more responsibilities in Afghanistan and Central Asia to maintain regional stability and impede Russia at the

---

134 Ibid. P89, 95 and 78.
135 Yang Lei (2014). "The Strategic Situation of Central Asia and the Evaluation of the Utility of the Shanghai Cooperation Organisation", ("中亞戰略形勢與上海合作組織作用評估"), in Institute for Strategic Studies, National Defence University. *International Strategic Relations and China's National Security* (2013-2014) (《國際戰略形勢與中國國家安全》(2013-2014)). Beijing, National Defence University Press. P125-145, 139.

same time."[136] American scholar Alexander Cooley believed that the situation in Central Asia has become a miniature of the global games between great powers, through which we can see the future of the international political situation. "Central Asia had become a natural experiment for observing the dynamics of a multi-polar world, including the decline of the U.S. authority, the pushback against Western attempts to promote democratisation and human rights, and the rise of China as an external donor and regional leader."[137]

China's "One Belt, One Road" strategy actually elevates the Sino-American contest around the world to a more comprehensive, diverse and intense level. At present, this strategy can be seen as an active response on the part of China, to a certain degree, to the U.S.'s "Pivot to Asia" and "Contain China" strategies, representing a strategic "anti-containment" move and a shift "from defence to offence". Interestingly, China's "One Belt, One Road" strategy is an active response to the American "eastward" strategy (that is the "return to Asia", "pivot to Asia" or "Asia-Pacific rebalancing" strategy). It does so by countering the American offensive with China's "westward" strategy. China has been committed to promoting regional economic cooperation with neighbouring countries, firstly with ten ASEAN countries, then with the ROK and Japan. Cooperation with ASEAN has obtained remarkable results in particular.[138] The "westward" strategy, to some extent, benefits from America's overreach and failures which have led to huge losses for U.S.'s strategic interests in the Middle East and Central Asia due to the overestimation of its capabilities.[139] In the past, the United States exces-

---

136 Ibid, P137.
137 Alexander Cooley (2012). *Great Games, Local Rules: The New Great Power Contest in Central Asia*. New York: Oxford University Press. Pxiv.
138 See Alice D. Ba (2009). *[Re] Negotiating East and Southeast Asia: Region, Regionalism, and the Association of Southeast Asian Nations*. Stanford: Stanford University Press; and Gilbert Rozman (2004). *Northeast Asia's Stunted Regionalism: Bilateral Distrust in the Shadow of Globalization*. Cambridge: Cambridge University Press.
139 See Michael MacDonald (2014). *Overreach: Delusions of Regime Change in Iraq*. Cambridge, MA: Harvard University Press. MacDonald thought that although the U.S. overthrew Saddam Hussein's regime in Iraq through violence, a pro-Iranian Shia regime was established thereafter, thus leading to the worsening of the U.S.'s geopolitical situation in the Middle East.

sively intervened in the military and political affairs of the Middle East and Central Asia, not only bringing irreparable disasters to the local people but also hurting the feelings of the local people. The chaotic situation in Syria and Iraq engendered the rise of ISIS, an even more deadly terrorist force posing threat to the whole world. But it is this situation that provides a favourable international environment (with little inference from the U.S.) for China's peaceful rise. America's dilemma in the Middle East and Central Asia provides just the opportunity and conditions for China's "westward" strategy. As American scholar Vali Nasr stated: "The story of the price the United States will pay for its failure to understand that the coming geopolitical competition with China will not be played out in the Pacific theatre alone. Important parts of that competition will be played out in the Middle East, and we had better be prepared for the jousting and its global consequences. [...] Just as we pivot east, China is pivoting further west. And it is doing so through its close and growing economic and diplomatic relationships with the Arab world, Pakistan, Iran, and Turkey."[140]

Although the eastward strategy of China's "Belt and Road Initiative" faces a lot of difficulties, and its future is hard to predict, it certainly bothers the U.S. and its allies and even causes a certain degree of panic. The United States must believe that the intention of China's grand strategy is to dominate Eurasian Affairs, becoming the "hegemon" of the Eurasian continent. Preventing any great power from dominating Eurasia has always been the goal of the U.S. strategies. There is no wonder why the "Belt and Road Initiative" strategy will inevitably become a strategic nightmare for the United States and intensify the Sino-U.S. contest throughout the world. And it explains why President Obama did not allow the U.S. to join the Asian Infrastructure Invest-

---

140 Vali Nasr (2012). *The Indispensable Nation: American Foreign Policy in Retreat*. New York: Doubleday. P3-4. Marc Lynch even despaired that except for Jordan, the U.S. relationship with its allies in the Middle East has weakened. Marc Lynch (2016). *The New Arab Wars: Uprisings and Anarchy in the Middle East*. New York: Public Affairs. On Sino-U.S. strategic competition in Central Asia, see Alexander Cooley (2012). *Great Games, Local Rules: The New Great Power Contest in Central Asia*. New York: Oxford University Press; and Thomas Fingar (Ed.) (2016). *the New Game: China and South and Central Asia in the Era of Reform*. Stanford: Stanford University Press.

ment Bank sponsored by China.

To compete with the United States around the world and counter the "contain China" strategy, China is forging a new strategic partnership with the resurgent Russia. After its foundation in 1949, China has been pursuing an independent foreign policy, except for the military alliance with the Soviet Union at the onset of the Cold War that lasted only for a few years. Sino-Soviet relations have been in a hostile state for a long time and did not turnaround until the eve of the breakdown of the Soviet Union. The opportunity to improve Sino-Russian relations arose from concurrent changes on the part of both the United States and China, specifically "the advent of a long-term Western presence in Central Asia (because the U.S. and its allies sent troops in Afghanistan), the Iraq war, the erosion of American political and normative influence, skyrocketing energy prices that have enabled Russia's resurgence, and the spectacular rise of China. However, "The real shift occurred [...] in July 1986, during [Gorbachev's] speech in the far eastern port city of Vladivostok, home of the Soviet Pacific Fleet. Vladivostok marked the point when the Soviet Union moved from its previous policy of containment to one of engagement."[141]

Most post-Soviet countries have maintained friendly relations with China since the breakdown of the Soviet Union. Russia, however, at one time hoped to strengthen its relations with the West in order to revive the country. Unfortunately, the West does not see Russia as an "ally", but instead carries out a series of provocative and "containment" measures which Russia sees as unfriendly and threatening.

Russian President Vladimir Putin and the Russian people have a long list of grudges against the United States and the West in many aspects, including NATO's eastward expansion, the exclusion of Russia from Europe-Atlantic collective security organisation, Russia's reluctant cooperation with NATO in the Balkan wars, censure of Russia by the West for starting wars in Chechen, the West's support for Russia's

---

141 Bobo Lo (2008). *Axis of Convenience: Moscow, Beijing, and the New Geopolitics.* Washington, DC: Brookings Institution Press. Quotations come from P15 and 27-28 respectively.

domestic opposition, the West's attempt to reduce European countries' dependence on Russian oil and natural gas, the West-engineered "colour revolutions" in Russia's neighbouring countries (Ukraine, Georgia and the Middle East countries), and the West's encouragement of some neighbouring countries to provoke and defy Russia.

Russia does not share the West's position that the international community has the right to send troops and conduct humanitarian intervention in a given country as taking on the responsibility to protect the people from "humanitarian disasters", and instead insists on the sanctity of the classical theory of absolute national sovereignty formed in the nineteenth century and the principle of non-interference in the internal affairs of other countries. From Russia's perspective, the United States not only refuses to treat it as an equal but also regards it as a loser of the Cold War. Even before coming fully into power, Putin had been determined to reverse the situation and erase the humiliation of the 1990s. He has not altered his resolve to protect the territorial integrity of Russia, reinvent Russia's position as a great nation, strengthen Russia's influence in the former Soviet republics nearby, enhance the decision-making role of the United Nations and its Security Council in international affairs, and preserve the framework and bilateral arms-control accords that the Soviet Union signed with the United States during the Cold War. All these actions are to bolster Russia's status as a great power in the world.

After 2002, U.S.-Russian relations become more intense, especially when Western countries had achieved their goal of regime change in Kosovo through military invasion and when the West employed military forces in Serbia without the authorisation of the U.N. Putin's speech at the international security conference in Munich, Germany, in February 2007 pushed the relations between Russia and the West into a new stage. His speech indicated that Russia would not accept the agenda on international affairs set by the United States. In order to counter the threat from the West, Russia plans to form a Russia-led Eurasian Union and strengthen its strategic cooperation with China. The most shocking move for the West, however, was Russia's invasion of Georgia in 2008 and Russia's annexation of Crimea in 2014. These actions not only have a great impact on the existing international situation, but also introduce a measure of uncertainty to international politics.[142]

The economic sanctions and diplomatic isolation imposed by the West on Russia turned out to strengthen the strategic partnership between China and Russia, and their cooperation in the fields of economy and trade is particularly prominent. The China-Russia strategic partnership is neither a military alliance nor a partnership against the West. Instead, its main function is to shatter the West's "containment" strategy, and prevent the West from dominating international affairs. The Shanghai Cooperation Organisation and the BRICS group are the most important platforms for the cooperation between the two countries. Nonetheless, a closer strategic cooperation between China and Russia will inevitably give rise to serious concerns and misgivings in the West.[143]

Some Western scholars argue that the real intention of China and Russia is to radically change the existing international situation. Mead, an American scholar of international politics, analysed the situation as follows: "China, Iran, and Russia were never bought into the geopolitical settlement that followed the Cold War, and they are making increasingly forceful attempts to overturn it. That process will not be peaceful, and whether or not the revisionists succeed, their efforts have already shaken the balance of power and changed the dynamics of international politics." "In Europe, the post-Cold War settlement involved the unification of Germany, the dismemberment of the Soviet Union, and the integration of the former Warsaw Pact states and the Baltic Republics into NATO and the EU. In the Middle East, it entailed the dominance of

---

142 Angela E. Stent (2014). *The Limits of Partnership: U.S.-Russia Relations in the Twenty-First Century.* Princeton: Princeton University Press; Jeffrey Mankoff (2014). "Russia's Latest Land Grab: How Putin Won Crimea and Lost Ukraine". *Foreign Affairs.* Vol. 93, No. 3 May/June, P60-68.

143 Gibert Rozman (2014). *The Sino-Russian Challenge to the World Order: National Identities, Bilateral Relations, and East versus West in the 2010s.* Washington, DC: Woodrow Wilson Centre Press; and Ivan Krastev and Mark Leonard (2015). "Europe's Shattered Dream of Order: How Putin Is Disrupting the Atlantic Alliance". *Foreign Affairs.* Vol. 94, No. 3 (May/June), P48-58.

Sunni powers that were allied with the United States (Saudi Arabia, its Gulf allies, Egypt, and Turkey) and the double containment of Iran and Iraq. In Asia, it meant that uncontested dominance of the United States, embedded in a series of security relationships with Japan, South Korea, Australia, Indonesia, and other allies."[144]

Two other American scholars also offer similar views. Douglas Schoen and Melik Kaylan believed that "Russia and China now cooperate and coordinate to an unprecedented degree—politically, militarily, economically—and their cooperation, almost without deviation, carries anti-American and anti-Western ramifications. Russia, China, and a constellation of satellite states seek to undermine American power, dislodge America from its leading position in the world, and establish a new, anti-Western global power structure. And both Russia in Eastern and Central Europe and China throughout Asia are becoming increasingly aggressive and assertive, even hegemonic, in the absence of a systematic U.S. response—notwithstanding the Obama administration's 'strategic pivot to Asia.'"[145]

Germany is gradually adopting an "independent" international strategy after the reunification. Instead of depending on the United States as it did in the past, it has in recent times even drawn a clear line with the United States in terms of its foreign policy. Germany is able to do so due to the constant rise of its national strength, the divergence of its national interests from those of the United States, and disagreements between the European Union and the United States over some major geopolitical issues such as Iraq. From the perspective of geopolitics, there is a need for Germany to strengthen its relations with Russia. From the perspective of drawing on each other's economic strength, there is a need for Germany to tighten economic ties with China. In recent years, Germany has actively improved its ties with Russia and China, representing another development in the new international land-

---

144 Walter Russell Mead. "The Return of Geopolitics: The Revenge of the Revisionist Powers". *Foreign Affairs*. Vol. 93, No. 3 (May/June 2014), P69-79. Quotations come from P69-70 and 70 respectively.

145 Douglas E. Schoen and Melik Kaylan (2014). *The Russia-China Axis: The New Cold War and America's Crisis of Leadership*. New York: Encounter Books. P3.

scape and offering a "wedge" for China to break through America's "containment".[146]

Yan Xuetong (閻學通), a Chinese mainland international politics scholar provided a neat summary of China's overall international strategy under the new international situation: "The structural conflicts between China and the U.S. will enter a critical period as China's 'pole' in a bi-polar world continues to strengthen over the next ten years. The foreign policy of China will have completed the transition from that of a regional great power to that of a global power by 2023. This transition is characterised by three main aspects. First, China will be more assertive in its foreign policy, shifting from passive diplomacy to active diplomacy. Second, China will enhance its leading role in the world, take more responsibility for international security and promote new international standards and global cooperation. Third, China's foreign strategy will be politically oriented and focus on improving its international strategy, and China may revive the strategy of forming alliances.

The centre of the world is moving eastward, which will urge China to deal with other countries and regions in different ways. Located in East Asia itself, China will need to take East Asia as its core arena and classify the countries there in line with its East Asian policy, allying with the friendly and the neutral while isolating enemies, thereby expanding its strategic partners. In order to build a safe rear area, China will put Russia, Central Asia and South Asia as its strategic buffer zone, rely on the SCO [Shanghai Cooperation Organisation] to deepen the strategic partnerships, and pursue South Asia sub regional economic cooperation and stabilise the strategic relationships with South Asian nations. All these are done to reduce the pressure [on China] from the international system during China's rise. Since there are far fewer common interests than conflicts concerning the strategic interests of China and the great powers in Europe and the U.S., China should be

---

146 Hans Kundnani (2015). *The Paradox of German Power*. New York: Oxford University Press; Hans Kundnani and Jonas Parello-Plesner (2012). "China and Germany: Why the Emerging Special Relationship Matters for Europe". European Council on Foreign Relations Policy Brief, May 2012; and Stephen F. Szabo (2015). *Germany, Russia, and the Rise of Geo-Economics*. London: Bloomsbury Academic.

cautious in dealing with Europe and North America, preventing the conflicts of interest from intensifying and elevating. In Europe, China needs to maintain good relations with friendly countries to ensure the neutrality of the European Union. Africa, Latin America and the Middle East are far away from China geographically. They have neither strategic conflicts of interest nor any common interests with China. So China should unite these areas by adopting a politics-oriented strategy in Africa, paying attention to the major powers in Latin America and retaining an economically-based strategy in the Middle East. As for North America and Oceania, China should focus on the United States and Australia. Increasing intra-regional conflicts will be the major trend for the next decade. Therefore, China should focus on bilateral strategy in the region, to be supplemented by multilateral strategy.

The strategic competition between the United States and China is inevitable in a context of polarisation. The policy toward the United States remains the priority of China's bilateral policies. The key of the U.S. policy will be to ensure that the competition remains peaceful, while the key of foreign policy in general is to strive for more high quality strategic friendships. The objective of the Russia policy will be to deepen strategic cooperation and seek alliances. As for Japan, China will through continued engagement urge it to avoid confrontation and to adopt a balanced position between the U.S. and China. Regarding Germany, China's goal is to make it a strategic economic partner, which calls for frequent Sino-German dialogues on human rights so as to mitigate any politically negative incidence. As for France, it is vital to respect its international status and strengthen consultations with it on global affairs, prompting France to cherish its dialogue with China and ensuring that France stays neutral with respect to the Sino-U.S. competition. Regarding India, China needs to adopt the policy of separating politics from economics to ensure that India will not involve itself in the Sino-U.S. strategic competition in East Asia. Given that British foreign policy is moderate, China can adopt a policy of 'non-action' toward

Britain."[147]

In short, the eastward shift of the centre of gravity of global economy, the United States's "pivot to Asia" strategy, the strengthened U.S.-Japan military alliance, the rapid rise of China, the resurgence of Russia, the enhancement of the Sino-Russian strategic cooperation, the Sino-American strategic competition in different parts of Asia, and the decline of economic and soft power in the West are the main causes of the drastic changes of international situation in the past two decades or so. The changes of the international landscape will inevitably lead to the changes of Hong Kong's international situation as well, bringing along a new context for the practice of "one country, two systems".

The Changes of Hong Kong's International Situation

The reason why I have gone to great lengths to elaborate in detail on the current international situation and China's situation in the world is to provide those Hong Kong people without an international vision to understand the changing international situation facing Hong Kong in a more comprehensive and profound way. In the rapidly changing international situation, Hong Kong should clearly and properly position itself in the international environment and in the development of China, as this is very much related to the long-term interests, existence and development of Hong Kong. There is no time for complacency or indifference. As I have mentioned earlier in this book, when the topic of "Hong Kong's future" appeared in the early 1980s, the international environment was favourable for Hong Kong. China, the United States, Japan and the West at the time worked together to combat the military and diplomatic expansion of the Soviet Union. The tensions between China and the Soviet Union pushed China to seek support from the West. The relatively friendly relations between China and the West contributed to

---

147 Yan Xuetong (2013). *Inertia of History: China and the World in the Next Ten Years* (《歷史的慣性：未來十年的中國與世界》). Beijing: CITIC Press. P214-215.

the smooth return of Hong Kong. The Western countries also pledged to continue their support for the development of Hong Kong after the handover, as it would be conducive to the great economic interests of the West in Hong Kong. Of course, the United States and the West sincerely hoped that Hong Kong, after its handover, would play a role in nudging China toward the road of "pro-West" and "peaceful" evolution, and eventually ending the rule of the Chinese Communist Party in China. With support from both China and the West, Hong Kong's situation at that time was really favourable indeed.

From an international point of view, even if other things do remain unchanged, there would inevitably have been a qualitative change in the relations between Hong Kong and the West after the handover. To a certain degree, Hong Kong served the interests of the West's policy toward China before the handover because Hong Kong was part of the Western Bloc, both politically and economically. In the eyes of the West, however, even though Hong Kong continues to maintain close relations with the West after its reunification with China, Hong Kong has from then on irreversibly left the Western Bloc and become part of China politically. Thus, the subject of Hong Kong's loyalty should be and would be to China and has to serve China's interests and security. The West will neither see Hong Kong as a political ally nor extend a "helping hand" to Hong Kong when it is in need, regardless of Hong Kong people's aspirations and their attitudes toward the West.

When frictions occur between China and the West, Hong Kong has no choice but to stand by the side of China. However, the Western powers are still able to take advantage of the open and liberal environment of Hong Kong to propagate Western politics and ideology, and support the anti-government forces and various "subversive" activities in the mainland China. Of course, the United States will continue to support the prosperity, stability and development of Hong Kong, and curtail its aggressive activities against China in Hong Kong when Sino-U.S. relations are good, but also vice versa.

After its reunification with China, Hong Kong's economic relations with the mainland China have grown increasingly close while the mainland China enjoys ever greater economic influence on Hong Kong as compared with the influence of the West. As the economies of both sides integrate more and more with each other and Hong Kong participates more actively in the "Five-year Plan" of the mainland, we can conclude that Hong Kong has by now become an "integral" part of China economically.

Overall, the strategic importance of Hong Kong for the West has decreased noticeably since the handover.[148] Hong Kong is an integral part of a powerful China. The Chinese Government firmly opposes any foreign intervention in Hong Kong affairs, which would be regarded as an unfriendly attempt to intervene in China's internal affairs. The West has realized its limited influence in Hong Kong, and thus no longer pays much attention to Hong Kong. The quantity of reports on Hong Kong by the Western media also shows a downward trend. In the past, the Western media often praised the development and achievements of Hong Kong and criticised Singapore, particularly for its authoritarian governance and the restrictions on freedoms. However, the reverse is true today as Singapore's strategic partnership with the United States has strengthened and its strategic importance to the West is increasing. The attitude of "disparaging Hong Kong while promoting Singapore" has become more obvious in Western media. Nevertheless, when anti-Beijing actions and movements erupt in Hong Kong, the Western media will give them wide coverage and use them as propaganda materials against China.

Western countries are fully aware that the governance of Hong Kong will never fall into the hands of the opposition forces who identify with Western values. Moreover, they do not want to "over-antagonise" China because of their support for the opposition in Hong Kong. So they will not make too great a "political investment" in the opposition. However, this does not mean that the Western powers will not offer

---

148 I have read a lot of papers and books on Sino-U.S. relations. Not a single American expert or scholar mentioned Hong Kong in the research on the Sino-U.S. game. This testifies indirectly to the insignificance of Hong Kong in the U.S. strategy towards China.

any support or encouragement to the opposition. While Hong Kong was a part of the Western camp, there were basically no attempts by the Western powers to oppose or upset the colonial government. A prosperous and stable Hong Kong under British colonial rule was compatible with the strategic interests of the West.

Since its return to China, the opposition in Hong Kong has always gained frequent spiritual and material support from the West when it undertakes to challenge the central authority or the Hong Kong Government, including support from Western governments, politicians, media, semi-official organisations and NGOs in particular. The West has done so for several reasons. First, Westerners always feel that they have a moral and religious responsibility to promote their "universal values" and particularly democracy in other places. Second, now that Hong Kong is a part of China, the effectiveness of governance and the political stability of Hong Kong are no longer as important to the West as they were. Third, since China has committed itself to "socialism with Chinese characteristics" and has been quite successful in this project, Hong Kong's role in nudging China toward "peaceful evolution" has become irrelevant. On the contrary, if Hong Kong contributes to the "peaceful rise" of an authoritarian, wealthy and powerful China, a prosperous and stable Hong Kong may not accord with the strategic interests of the West in the context of intensifying contest between China and the West. Fourth, although it is impossible for the opposition in Hong Kong to seize power, they however can still make political troubles for the Chinese Government, thus distracting the Chinese leaders' attention from the mainland's development and slowing down China's rise. Fifth, if "one country, two systems" encounters difficulties in Hong Kong due to the disruptions by the opposition, the exemplary role of Hong Kong for Taiwan will disappear, hence hindering the process of China's reunification and permanently postponing the time when Taiwan will become China's "unsinkable aircraft carrier". Sixth, if the central authority is forced to quell the political struggle waged by the opposition through tough means or even the use of force, China's international image and soft power will be damaged, which will weaken China's position in the global contest between East and West.

In the second half of 2014, the "Occupy Central" movement broke out in Hong Kong and lasted up to 79 days. By occupying several busy areas in Hong Kong, the protestors, mostly young people, vented their anger against Beijing for refusing to accede to their democratic demands. This movement has made a great impact on the international reputation of both Hong Kong and China. Apparently Western forces have played a role in this movement. I personally can't by any means accurately measure the role of the Western forces in the "Occupy Central" movement. But, according to my friends in mainland China, the central authority is firmly convinced that the West had a hand in the movement. When two countries enter into a fierce strategic contest, neither side will assume that the other side is a fair player nor harbour goodwill toward the opposing nation. Instead, each side will try to defend its own interests and security by imagining the worst scenario. It might be quite difficult to convince the central authority that the United States and other external forces (including even Taiwan) do not intend to do harm to Hong Kong at present when China is threatened by the U.S.'s "containment" policy. The central authority is assuming that the United States and other external forces plan to include Hong Kong in the grand "containment" strategy against China, and believe that Hong Kong is likely to become one of the battlefields of the Sino-U. S. contest. Therefore, "national security" becomes an increasingly important consideration of the central authority when formulating their Hong Kong policy.[149]

China's Hong Kong policy has always been closely related to its international strategy. "One country, two systems" does not stand alone but is always an important part of China's incessant strategic adjustments to the changing international landscape[150] At present, the

---

149 A friend of mine who is able to influence the Hong Kong policy of the central government, told me that Hong Kong is and will be the battlefield of the Sino-U.S. game. National security becomes a more and more important consideration of the central government in making Hong Kong policy.

150 Lau Siu-kai (2013). "Hong Kong's Role in China's International Strategy", ("香港在中國國際戰略中的角色") in Lau Siu-kai, *The Politics of Hong Kong after Its Return* (《回歸後的香港政治》).

new international situation shows that the global competition between the camp led by the United States and Japan and that led by China and Russia has become increasingly intense, while East and Southeast Asia have become two important regions hotly contested by all sides.[151] The gist of China's new international strategy is to firmly safeguard its national sovereignty, security and development interests in East and Southeast Asia, establish a stable, fair and peaceful order of politics, security, economy and finance for national development, and promote China's power and leadership in East Asia. Specifically, the objectives include defending China's territorial waters, maritime rights and interests in the East China Sea and the South China Sea, ensuring the security and smooth navigation of the shipping sea lanes that connect East Africa and China, and ensuring the supply of energy and resources.

America's "return to Asia" policy is aimed at impeding China's international strategy in order to restrain China's rise in East Asia. It is natural that the central authority requires Hong Kong to support and coordinate with mainland China to carry out the new international strategy and endeavour to protect national security at the same time. China will naturally draw on Hong Kong's special status and advantages to strengthen its relations with neighbouring countries and regions, trade and financial relations in particular, and establish a new order in East Asia, especially in Southeast Asia. Therefore, China's investments and interests in Hong Kong will multiply, be more diversified and will become more strategically valuable. Any changes that would induce Hong Kong to deviate from "one country, two systems", or result in Hong Kong to become "an independent political entity" would constitute an obstacle for the implementation of China's international strategy, or even worse, a threat to China's political and financial security. Currently faced with a serious international situation, the Chinese central authority will absolutely not allow Hong Kong to

Hong Kong: The Commercial Press. P77-109.

151 Bruce Gilley & Andrew O'Neil (eds.) (2014). *Middle Powers and the Rise of China*. Washington, DC: Georgetown University Press; and Parag Khanna (2008). *The Second World: Empires and Influence in the New Global Order*. New York: Random House.

challenge its authority, or change itself into "a base for subversion" utilised by foreign forces against China. The Chinese central authority will intervene in any action taking place in Hong Kong which smacks of "collusion with external forces". Most importantly, the HKSAR regime cannot fall into the hands of anti-communist or anti-China forces.

The rise of China and Sino-American contest have also changed the perceptions of China and the United States toward Hong Kong. The Chinese central authority suspects that the United States is working to turn Hong Kong into a detrimental factor for the Communist regime without regard to its interests in Hong Kong. In the interest of national security, the new National Security Law approved by the Standing Committee of the National People's Congress in 2015, emphasises Hong Kong's obligation and responsibility for national security in two places. Article 11 of the law stipulates: "The sovereignty and territorial integrity of China shall not be infringed upon or divided. Maintaining the sovereignty, unity, and territorial integrity of the nation shall be the common obligation of all Chinese people including the Hong Kong and Macao compatriots and the Taiwan compatriots." Article 36 requires that "The Special Administrative Region of Hong Kong and the Special Administrative Region of Macao must fulfill their responsibility to safeguard national security." At a time when Hong Kong has not completed the local legislation of Article 23 of the Basic Law and China is faced with a grim international environment, the central authority's requirement and expectation for Hong Kong to protect national security is of great importance, indicating that China will not allow Hong Kong to become a threat to national security.

Even if China adopts some safeguarding measures to ensure that it would not be threatened by Hong Kong, such as by introducing the national laws concerning national security to Hong Kong, there's actually no need for it to be too worried about the possibility of retaliation against it from the United States or other Western countries because China is strong enough today to withstand such attacks. Moreover, doing harm to Hong Kong does not accord with the interests of the U.S. or the West. After all, Hong Kong still holds some economic and strategic value for them.

Hong Kong inevitably tilts toward China in the new international situation; otherwise the people of Hong Kong will not be tolerated by the central authority and mainland compatriots. At the same time, the intentions of the U.S. and the West have become more elusive. It is suspected that they intend to get more involved in Hong Kong affairs. External forces have all along been supportive of the mainstream opposition parties, mainly those who attempt to use Hong Kong to influence the mainland political situation, including ending the communist regime. It is very likely that all those "nativist" and "separatist" groups that have appeared in recent years will received the attention, encouragement and support of the external forces. As a result, China and the West, especially the United States, shall engage in contests in Hong Kong. Some Western politicians and officials have already expressed their support for selecting the Chief Executive of the HKSAR through universal suffrage several times in the past two years. They also counselled the opposition in Hong Kong on protest strategies and by doing so exerted political pressure on the central authority. Some U.S. politicians even threaten to punish Hong Kong if freedom and human rights are violated there. These are some examples of how the West has actively involved itself in Hong Kong's politics.

The relations between Hong Kong and the West will be more complicated in the future. As an international metropolitan city, how Hong Kong shoulders its responsibility for safeguarding national security while maintaining a proper relationship with the West will be a great test for the Hong Kong people and the HKSAR Government in particular. On the one hand, Hong Kong hopes to maintain close economic and trade ties with the West, as the Hong Kong people continue to venerate Western culture. On the other, Hong Kong opposes any interference in Hong Kong's internal affairs or creating political chaos in the territory by the West. Hong Kong is particularly unwilling to be used by the West to do harm to China. I believe that the discussions, negotiations and disputes concerning Hong Kong between the Chinese Government and the West will increase in the future. The

Chinese Government will warn Western countries not to interfere in China's internal affairs from time to time while the Hong Kong Government will also "wrestle" with the West under the guidance of the central authority occasionally.

The contest between China and the West in Hong Kong will certainly engender political polarisation in the territory. After all, almost all the opposition forces in Hong Kong in varying degrees identify with the Western values, rely on Western support and maintain an anti-communist or even anti-China attitude. Any act from the central authority to prevent the West from meddling in Hong Kong affairs will give rise to dissatisfaction and opposition from them. On the contrary, the "people who love China and Hong Kong" would certainly stand by the central authority and the HKSAR Government. The gap between the political differences of both sides will be enlarged by the contest between China and the West. The conflict between the opposition and the "people who love China and Hong Kong" will also increase the opposition's hostility toward the central authority.

Hong Kong will become increasingly integrated in the regional economic cooperation framework dominated by China in East and Southeast Asia. As the "One Belt, One Road" strategy is carried out, the relations between Hong Kong and the rest of Asia will grow closer. For most Hong Kong people, this is a difficult trend to adapt to. Since they have grown accustomed to being close to the West since the Second World War, and they have very little knowledge of Asia or even look down upon Asia, falling prey to a mentality of blindly worshiping the West. Hong Kong people are increasingly able to be aware of the fact that the world economic centre is moving eastward and that the future development of Hong Kong will depend more heavily on Asia.[152] However, to establish a closer connection between Hong Kong and other Asian countries, the supporting infrastructure needs either to be improved or to be rebuilt, with a focus on the transportation and

---

152 Wendy Dobson (2009). *Gravity Shift: How Asia's New Economic Powerhouses Will Shape the 21st Century*. Toronto: University of Toronto Press.

communication network, verbal and written communication, personnel who are familiar with Asian affairs, the connections between the HKSAR Government and the governments of other Asian countries, and Asian countries' awareness of Hong Kong.

How to make the best use of the powers to conduct external affairs endowed by "one country, two systems" to develop extensive relationships with Asian countries will be a major challenge for Hong Kong in the future. If it can be handled successfully, the strategic value of Hong Kong to China will rise greatly, and the practice of "one country, two systems" in Hong Kong will receive new impetus. In addition, Hong Kong can elevate its international status if it develops itself into a hub of Asia, serving the interests of the economic integration in Asia and the economic ties between Asia and Europe.

CHAPTER FIVE

The Critical Moment in the Practice of "One Country, Two Systems"

In the eyes of the international community, of the Chinese Government, of mainland compatriots, of the Hong Kong people and of overseas Chinese, the last twenty years has demonstrated that "one country, two systems" in Hong Kong has by and large succeeded. The facts have proved that "one country, two systems", put forward by Deng Xiaoping, is instrumental to the peaceful resolution of the problems related to "Hong Kong's future", preserving Hong Kong's property and stability as well as Hong Kong's economic value to China. There is no doubt that Hong Kong has experienced a great many of ups and downs since the handover, and its future would remain full of difficulties. At present, there is no comprehensive research on, not to mention any convincing explanations of, the connection, if any, between the problems that Hong Kong has encountered after the handover and "one country, two systems", and the possibility that Hong Kong might be better off it had remained a British "colony".

In order to "clarify" which problems are and which problems aren't relevant to the practice of "one country, two systems" in Hong Kong following 1997, I would like to speculate on a series of circumstances and events which have not actually happened. In other words, I shall conduct a "counterfactual analysis", as sociologists sometimes do. My starting point is: if Hong Kong had remained a British "colony", what would have taken place in Hong Kong in the last nineteen years? Many people do believe that numerous changes of post-1997 Hong

Kong are caused by the departure of the British and the onset of "one country, two systems" and blame the policy for Hong Kong's post-handover "plight". My goal is to dig out the facts related to that opinion. That is, what changes in Hong Kong transpiring over the past twenty years are actually "caused" by "one country, two systems"?

In my opinion, even if Hong Kong had not returned to China, the following phenomena would still appear:

(1) The globalisation of trade and finance would weaken the international competitiveness of Hong Kong due to the intensification of competition among countries.

(2) As a mature economy, Hong Kong's economy would only grow at a moderate rate.

(3) Hong Kong would see more economic and financial volatility than before.

(4) Hong Kong's manufacturing sector would continue to move northward. The hollowing out of industries is unavoidable.

(5) China's "reform and opening up" policies would continue to push forward and deepen.

(6) China's rise would maintain a strong momentum.

(7) The economic relationship between Hong Kong and mainland China would get closer.

(8) China would draw on Hong Kong's advantages to carry out its development strategy.

(9) Hong Kong's economic development would be more dependent on China's development.

(10) Hong Kong would continue to play the role of the bridge between China and the world.

(11) The income gap and the gap in living standards between Hong Kong and mainland China would narrow.

(12) Hong Kong's wealth would be increasingly concentrated in the hands of the rich while the gap between the rich and the poor would continue to widen. The opportunities for social mobility would not be able to meet the needs of the middle class and the younger generation.

(13) Hong Kong people would lament the growing social injus-

tice and the narrowing of its industrial base.

(14) A range of deep-seated conflicts would become salient.

(15) The colonial government would be more active and interventionist in the provision of social welfare, land and housing, and public services to ease various social conflicts.

(16) The colonial government would maintain its longstanding economic policy. It would continue to abstain from intervening in the economy; the colonial government would not take actions to diversify the industrial structure of Hong Kong.

(17) The public consensus on the direction of development in Hong Kong, on the government's role in economy and social development and on other important public policies would continue to erode.

(18) Hong Kong people would be pessimistic about Hong Kong's economic prospects and their confidence in Hong Kong would drop gradually.

(19) The British would not undertake democratic reform but at the most would open additional channels for those who are more educated to participate in political activities. Hong Kong's political power would still be firmly held by the colonial government.

(20) China's economy would grow rapidly; the living standard of the mainland Chinese would rise; numerous state-owned enterprises and private enterprises would invest in Hong Kong; greater numbers of mainland compatriots would buy properties, shop, travel, study and work in Hong Kong. All these phenomena would dampen Hong Kong people's sense of superiority, undermine their self-confidence and lead to an "identity crisis" among the Hong Kong people and friction between the Hong Kong people and mainland compatriots.

(21) Apart from drawing on Hong Kong's advantages, China would not include Hong Kong in the national development strategy. The British would not allow that either. Even if Hong Kong were embroiled in economic difficulties, China might do something to help but would not be extremely eager to come to Hong Kong's "rescue".

(22) Some form of "nativism" which is prejudiced against the mainland compatriots would come into being. Some Hong Kong people

would turn certain Western values into Hong Kong's "core values", glorifying colonial rule with the purpose of building a "new identity" for the Hong Kong people that would elevate their sense of pride and self-confidence. However, calls for "self-determination" and "Hong Kong independence" would not erupt.

(23) The British would work hard to restrain acts which might challenge, provoke or offend the CPC and the Chinese Government in order to maintain good relations with China.

(24) The colonial government would face more and more serious problems in its rule as various social conflicts and political discontent accumulated and became entangled.

(25) Hong Kong would remain part of the Western bloc politically, but its economic relationship with the West will inexorably weaken over time. The West would continue to use Hong Kong to influence China's development even though the effectiveness of those acts would be limited.

According to my speculation, many political, economic, social and livelihood problems that disappoint the Hong Kong people and bring about public discontent, in fact, have nothing to do with the handover or the implementation of the "one country, two systems" policy. The main reason for the appearance of these problems undoubtedly is the rapid rise of China and its enormous impact on Hong Kong. Supposing that Hong Kong were still governed by the U.K., the colonial government would still be faced with as many knotty problems as the HKSAR Government is today. Moreover, if Hong Kong were still a British "colony", I surmise that the Chinese Government would hardly rack its brain in order to help Hong Kong.

My further speculation is that the obstacles to the practice of "one country, two systems" are mainly political since Hong Kong's return to China and that "one country, two systems" will engender some inevitable and irreversible political changes in Hong Kong. The political changes are as follows:

(1) Although the central authority holds considerable power, Hong Kong people are endowed with unprecedented autonomy under

"one country, two systems".

(2) The development of democracy in Hong Kong will move ahead, with "mass democracy" gradually replacing "elite democracy". The collision of these two "democracies" during the transitional period is bound to lead to an unstable and confused political situation.

(3) Slowed by Hong Kong political leaders' limited wisdom, political experience, and capability of handling crises, the building of a new and authoritative Hong Kong regime is a slow and difficult process. Hong Kong people's unwillingness to support and trust the "pro-Beijing" HKSAR Government and the constant challenges instigated by the opposition are the principal obstacles to Hong Kong's governance.

(4) Checked and challenged by the Legislative Council and the judiciary, whose powers have expanded after the handover, the "executive-led" political system does not function in line with expectations. Due to the lack of determination, courage and capability, the Chief executive of the HKSAR has not, as a rule performed his duties well enough to protect national security, the interests of the central authority, and ensure the successful implementation of "one country, two systems" and the Basic Law.

(5) Under "one country, two systems," the Hong Kong bourgeoisie gains the political power and status that they never would have attained during the "colonial" period. They use these powers to protect their own interests, which to some extent constrains the ability of the HKSAR Government to promote the transformation of Hong Kong's economic structure, to improve people's livelihood, to narrow the gap between the rich and the poor, and to reform land and housing policies. As a result, it fosters in society worries about the "collusion between government and business", grudges against the rich and discontent with the HKSAR Government.

(6) A series of "explosive" political issues pop up due to different and hard to reconcile understandings of the "one country, two systems" policy and interpretations of the Basic Law, triggering intense political conflicts and political instability.

(7) Irrespective of their preferences, the question of how to

comprehend and deal with their new identity as "citizens of the People's Republic of China" unavoidably arouses speculation and anxieties among the Hong Kong people. People of Hong Kong struggle over whether they should maintain their weak and vague identity as "Chinese people" as they did in the past while manufacturing an identity as "Hong Kong people" that would "separate" themselves from mainland compatriots, or seriously face up to relations with the People's Republic of China and mainland compatriots. (The "identity" issue is a prominent political cleavage in Hong Kong today. It is also the source of political conflicts).

(8) The Chinese Government is willing and even keen to offer strong support to Hong Kong's economic development as Hong Kong is now a part of China, or even generously "conceding benefits" to Hong Kong. Of course, the relevant measures are also in line with the development-related interests of the state.

(9) The central authority deeply incorporates Hong Kong into the national development strategies, making Hong Kong an important part of national development, sharing dividends brought by national development with Hong Kong.

(10) As a result of active promotion by the central authority, the economic exchanges between Hong Kong and the mainland China cover a wider range of areas and become more frequent. However, these exchanges also generate more conflicts between the Hong Kong people and the mainland compatriots and become a catalyst for the rise of "nativism" and the construction of a "new identity" among a minority of Hong Kong people. (Even though these phenomena would still appear without Hong Kong's handover, they would not be as acute and divisive as they are even at present).

(11) After the departure of the British, all kinds of social conflicts that had been "contained" or "controlled" by the colonial government in the past, are now fully exposed, and they are further heated by democracy, producing a great many political conflicts, many of which were unforeseen. Before the handover, the British pampered the opposition, divided the pro-establishment camp, and adopted "divide and rule"

political tactics, creating even more political conflicts, and all of these have added difficulties to the future governance of the HKSAR.

(12) Some Hong Kong people, the opposition and anti-communists in particular, believe that they have the responsibility and right, as Chinese citizens, to actively and directly participate in mainland politics, with the goal of "ending the one-party dictatorship," of promoting China's "peaceful evolution", of supporting the anti-communist and anti-government forces in the mainland, of setting Hong Kong up as a "model" of democracy for China and of allying themselves with external forces in order to criticise the human rights record and status of the rule of law in China. These people continue to believe that Hong Kong's return to China means that they can exploit the political privileges endowed to Hong Kong by "one country, two systems" to interfere in mainland politics or change the political condition in China. (By doing so, these people have in effect violated the "prescription" of "well water not interfering with river water" [井水不犯河水]).

(13) Having lost the barrier that prevents it from becoming an "anti-communist base" or a "base for subversion", Hong Kong proves difficult to govern and directly provocative to the central authority, leading to direct confrontations and frictions between the central authority and the Hong Kong people.

(14) Following ceaseless provocations on the part of the opposition, the central authority perforce takes action in order to protect national security, uphold the authority of the central authority, and to ensure that the implementation of "one country, two systems" remains in line with the original blueprint. (The British, to some extent, had limited Chinese Government's interference in Hong Kong politics during "colonial" rule; after the handover, however, anti-communists and the opposition have basically refused to identify with "one country, two systems" and the Basic Law).

(15) The absence of Britain's involvement makes room for some international organisations, foreign powers, and even Taiwan forces to interfere in Hong Kong politics. The alliance between part of the external forces and the opposition of Hong Kong exacerbates the

worries of the central authority. Thus the central authority sees a greater urgency to participate in the affairs of Hong Kong.

(16) As Hong Kong has left the Western camp politically, the West grows increasingly integrated into China economically. The value and importance of Hong Kong to the Western world continues to diminish. (All along, the Western world expected that Hong Kong would guide China to follow the path "peaceful evolution", only to find that China instead had taken up "a socialist path with distinct Chinese characteristics", which posed a serious political and economic threat to the West. Accordingly, the "usefulness" of Hong Kong to the West has declined. On the other hand, if the West believes that Hong Kong will contribute to the success of China's "socialist path with distinct Chinese characteristics", then the continued prosperity and stability of Hong Kong might not be in the strategic interests of the West. How would the West then deal with Hong Kong? What is Hong Kong's role in the strategic game between China and the West? How does China perceive Hong Kong's role in China's collision with the West? These are some of the unavoidable new issues that might affect the implementation of "one country, two systems").

From my point of view, although the purpose of "one country, two systems" is to maintain the "status quo" of Hong Kong as of the mid-1990s, this policy by itself brings long-term and profound changes to Hong Kong. These changes would take place mainly in the polit-ical field, which has to do with Hong Kong's position in the world and within China, the relationship between the central authority and the HKSAR, Hong Kong's political "ecology", and Hong Kong people's identity. I have done a detailed analysis of these changes in Chapter 3. These changes, and the derivative conflicts and problems, are the main reasons for the discrepancy between the "ideal type" of "one country, two systems" and the reality. In short, the major cause of the deviation of the actual state of "one country, two systems" from the "blueprint" is political. The reason that the actual "one country, two systems" does not meet expectations is that, in the absence of effective exercise of the powers at the disposal of the central authority, many controversial political issues remain unresolved, for instance, formulating solid local governance in the HKSAR and settling a number of essential judicial issues. Consequently, the policy of "one country, two systems" has yet to be comprehensively and accurately carried out.

In fact, both the central authority and the opposition in Hong Kong are not happy with the current state of "one country, two systems" in Hong Kong. They both feel that the current "stalemate" should not continue, and should be overcome as soon as possible. In fact, the political struggle over the arrangements for the popular election of the chief executive reflects that both sides are eager to break the dead-lock in their favour by entrenching their own understanding of "one country, two systems" in Hong Kong. Hong Kong is now at a critical moment as the central authority and the opposition are embroiled in a "showdown", each trying to make its understanding of "one country, two systems" prevail forever. And Hong Kong's future will depend on the outcome of this fierce battle. As I mentioned earlier, the central authority and the opposition have totally different interpretations of "one country, two systems" and the Basic Law. Both of them are not happy with the actual state of "one country, two systems" as it is seen as not meeting their respective expectation or goals. The central authority is dissatisfied because the implementation of the policy does not proceed in accordance with the original "blueprint", and consequently, the national sovereignty, security and development interests have not been effectively safeguarded; the authority and responsibility of the central authority has not received due recognition and respect in Hong Kong; the central authority's interpretation of "one country, two systems" and the Basic Law have encountered skepticism and resistance on the part of the opposition and their followers; and the opposition's irrationality in their all-out and unconditional political struggles continues against the central authority and the HKSAR Government, etc.

The opposition is dissatisfied because it takes for granted that Hong Kong is an "independent political entity". "From that point of view, the opposition is of the view that the central authority is exer-cising powers which they don't have and taking on responsibilities that do not belong to them. The opposition accuses the central authority of being biased in favour of the "people who love China and Hong Kong", interfering in Hong Kong's internal affairs and democratic development, undermining Hong Kong's high degree of autonomy, and establishing a HKSAR Government that is out to destroy the original institutions, the original way of doing things, the proper procedures, the "core values" and the social cohesion of Hong Kong.

Both the central authority and the opposition have thus stepped up efforts to change the situation of Hong Kong, striving to push Hong Kong to develop and operate in accordance with their own understanding of "one country, two systems". In this case, the struggle and confrontation between the central authority and the opposition will be difficult to avoid.

The Strategy and Deployment of the Central Authority

The outbreak of a large-scale demonstration against the local legislation of Article 23 of the Basic Law in the middle of 2003 has compelled the HKSAR Government to shelve the legislative work indefinitely. Since then, the central authority has decided to change its policy of "non-intervention" or "the best way to deal with Hong Kong is to do nothing ("不管就是管好")" with regard to Hong Kong. Otherwise, the operation of "one country, two systems" would head toward the wrong direction, and the security and interests of the country and the central authority would suffer. The central authority is beginning to realise that the supporting conditions for successfully carrying out "one country, two systems" in Hong Kong are fully available, that the authority of the HKSAR Government is limited and its capacity for governance needs improvement, that the opposition still holds a recalcitrant attitude, and that many Hong Kong people still are resistant toward the central authority and the motherland.

Moreover, the investments and interests of the mainland enterprises (both state-owned enterprises and private enterprises) in Hong Kong are increasing rapidly. More and more mainland enterprises are listed on the Hong Kong Stock Exchange. The total market capitalisation of these companies accounts for more than a half of the market value of all the listed companies in Hong Kong. China's "going out" strategy is moving forward rapidly. As an important platform for China's "going out" strategy, Hong Kong's interests will become increasingly tied with national interests. As the strategic competition between China and the West—and the Sino-American competition in particular—intensifies, America's Hong Kong strategy will also be the concern of the Chinese Government. All of these facts will spur the central authority to pay more attention to Hong Kong's political situation and will interfere in it when necessary for the good of China as well as for the good of Hong Kong. In this context, the "hands-off" approach to deal with Hong Kong affairs is no longer viable.

Indeed, in the first six years after the handover, the central authority has all along strictly or even over-prudently abided by the "non-intervention" policy on Hong Kong affairs. The most obvious evidence for that is that the central authority's officials and mainland people abstain from talking about Hong Kong, and rarely have contacts with people from different sectors of Hong Kong. Most research institutions in the mainland for Hong Kong affairs that had been quite active before the handover have been abolished after 1997. The only piece of evidence for intervention by the central authority is that it continues its support for the "people who love China and Hong Kong" in many aspects, especially during the election of the chief executive, the Legislative Council and the District Councils. The reason for that is to prevent the opposition from controlling the HKSAR Government, the Legislative Council and the District Councils and to ensure that the "one country, two systems" is on the right track. The practice of the "non-intervention" policy amply shows that the central authority was too optimistic about Hong Kong's politics and economy after its return to China, had overestimated the capacity for governance and political calibre of the HKSAR Government, and had seriously underestimated the political clout of various opposition forces and their determination to confront Beijing. Needless to say, the central authority also worries that intervening in Hong Kong affairs would incur criticism, weaken the still fragile public confidence in the central authority, in Hong Kong, and in the "one country, two systems" policy. Equally importantly, the central authority has been well aware of the dearth of people on the mainland who were familiar with Hong Kong or had done research on Hong Kong's conditions and development. In consequence, the central authority has not been confident that they might grasp and master the situation of post-handover Hong Kong.

Although the central authority appropriately devised the "one country, two systems" policy to solve the issue of "Hong Kong's future", it lacked a conception of the changes the policy will bring forth. The designers of the "one country, two systems" could not predict all the changes and difficulties that might affect the implementation of the policy, not to speak of preparing for all the possible scenarios before and during implementation. Without doubt, the best method to deal with the challenges newly-arising is to make appropriate adjustments to the "one country, two systems" policy in an open, pragmatic, flexible and investigative manner. The "non-intervention" policy should be an intrinsic requirement for the central authority stipulated by the "one country, two systems" policy. Unless exceptional cases occur in Hong Kong, the central authority will not easily change the "non-intervention" policy. Furthermore, a major national policy change cannot be achieved in a single effort but demands repeated argumentation and meticulous arrangements. In this way, it is easier to understand that why the central authority has "sat on its hands" though Hong Kong has undergone troubles after its return to China.

Obviously, the large-scale demonstration in 2003 has prompted the central authority to shift its policy toward Hong Kong from "non-intervention" to "non-intervention, but doing something positive" (不干預，但有所作為) since then. In recent years the central authority has further amended the policy, which can be loosely described as "non-intervention, but doing something effectively" (不干預，但善於作為). Simply speaking, the central authority has already got rid of the previous hesitation about interfering in Hong Kong affairs and is more prepared to exercise their authority and statutory powers for the sake of performing their duties. The national interests and the faithful implementation of "one country, two systems" is now more important than the perception and feelings of the Hong Kong people. After all, the central authority believes that the country, the central authority and the Hong Kong people share common interests under "one country, two systems". Only some Hong Kong people are not able to see it clearly for the time being because of various ideological and psychological barriers.

The evolving policy of the central authority toward Hong Kong policy purports to correct deviations in the implementation of "one country, two systems", to resolve conflicts and problems arising during the implementation of the policy, the political conflicts and problems in particular, and to create a favourable environment for the successful implementation of "one country, two systems". The central authority has noticed that quite a few Hong Kong people have an understanding of the Hong Kong policy that is different from the central authority's interpretation, and that is because they have been "poisoned" by the opposition's "alternative interpretation" of "two countries, two systems" and the Basic Law.

The *White Paper* (a short term for the document issued by the Information Office of the State Council, the People's Republic of China entitled *The Practice of the "One Country, Two Systems" Policy in the Hong Kong Special Administrative Region* [2014]) describes the above situation as follows: "The practice of "one country, two systems" has come to face new circumstances and new problems. Some people in Hong Kong have yet felt comfortable with the changes. Still some are even confused or lopsided in their understanding of "one country, two systems" and the Basic Law. Many wrong views that are currently rife in Hong Kong concerning its economy, society and development of its political structure are attributable to this. The continued practice of "one country, two systems" in Hong Kong requires that we proceed from the fundamental objectives of maintaining China's sovereignty, security and development interests and maintaining the long-term stability and prosperity of Hong Kong to fully and accurately understand and implement the policy of "one country, two systems", and holistically combine upholding the principle of "one country" with respecting the differences of "two systems", maintaining the power of the central authority with ensuring the high degree of autonomy of the HKSAR, and letting the mainland play its role as a strong supporter of the HKSAR with improving the competitive edge of Hong Kong. In no circumstance should we do one thing and neglect the other."[1]

Under the new situation, the basic idea of the central authority is that, in maintaining the national interests and the implementation of "one country, two systems" in Hong Kong, Beijing cannot entirely rely on the HKSAR Government and the forces of "the people who love China and Hong Kong". The central authority must "come forward" at critical moments and exercise their powers properly to perform their duties, working with the HKSAR Government and the forces of "the people who love China and Hong Kong" to make the practice of "one country, two systems" a success.

The new thinking of the central authority's Hong Kong policy has been brewing for years, becoming more mature and can be implemented in step-by-step manner. In summary, the new Hong Kong policy has several aspects:[2]

(1) Working with "the people who love the country and Hong Kong" to make greater efforts to explain the historical background, domestic and international situation, Hong Kong's situation at the time and the main objectives behind the policy of "one country, two systems" and the Basic Law. It would highlight that "one country, two systems" and the Basic Law are a major national policy which takes the interests of the country, Hong Kong and Western countries into account, seeking "mutual benefits and win-win solutions". The publication of the *White Paper* has enormous significance in the current political atmo-

---

1   *The White Paper* (《白皮書》). P30.
2   Part of the following discussion comes from Lau Siu-kai (2013). *Hong Kong's Politics after Its Return to China* (《回歸後的香港政治》). Hong Kong: The Commercial Press. P2-76.

sphere. Although the wording of the *White Paper* is slightly different from that of the Chinese leaders before Hong Kong's return, the content is largely the same. The publication of the *White Paper* and the more frequent explanation of the "one country, two systems" policy and the Basic Law by national leaders, mainland experts and scholars in recent years are also geared to provide the central authority with more say in "one country, two systems." It should be emphasised that the central authority affirms the importance of Hong Kong for national development and are determined to whole-heartedly support the prosperity, stability and development of the HKSAR. These statements are helpful in removing the self-doubts among the Hong Kong people by refuting those views that "belittle Hong Kong," asserting that Hong Kong is no longer valuable to the country. These views are propagated by people who intend to undermine the self-confidence of the Hong Kong people and stir up discontent with the central authority, the HKSAR Government and Hong Kong society.

(2) Inculcating and strengthening the concept of "one country." Considering that quite a few Hong Kong people tend to highlight the "two systems" while neglecting the "one country", the central authority will make more efforts to clarify the superior-subordinate relationship between the "one country" and the "two systems", in which the country is prior to the HKSAR, the mainland's socialism is prior to Hong Kong's capitalism, the central authority is prior to the HKSAR Government, and the Constitution of People's Republic of China is prior to the Basic Law of Hong Kong. The *White Paper* clearly points out that "The 'one country' is the premise and basis of the 'two systems', and the 'two systems' is subordinate to and derived from 'one country'. But the 'two systems' under the 'one country' are not on a par with each other. The fact that the mainland, the main body of the country, embraces socialism will not change."[3] The central authority hopes that the Hong Kong people would always bear the national sovereignty, security and development interests in mind, giving top priority to the overall situ-

---

3   *The White Paper* (《白皮書》). P31.

ation, establishing the perspective of the country and of the nation, as well as thinking about the interests and prospects of Hong Kong. If the Hong Kong people can develop a better understanding of the importance of national development to the development of Hong Kong, and appreciate the support and assistance that the central authority has all along offered Hong Kong, they will realise the importance and relevance of the country to Hong Kong.

(3) Making great efforts to clarify to the Hong Kong people the powers and responsibilities of the central authority under "one country, two systems". In order to eliminate the misunderstandings of "one country, two systems" in Hong Kong spread by the opposition, the national leaders, central government officials, mainland experts and scholars have taken the trouble to explain the powers and responsibilities of the central authority under "one country, two systems" to the Hong Kong people over the past few years. To my understanding, the main purpose of the *White Paper* is to encourage Hong Kong people to think about issues from the perspective of "one country," and to be fully aware of and respect the powers and responsibilities of the central authority. The *White Paper* clearly emphasises that "the system of the special administrative region, as prescribed in the Constitution of the People's Republic of China and the Basic Law of the HKSAR, is a special administrative system developed by the state for certain regions. Under this system, the central authority exercises overall jurisdiction over the HKSAR, including the powers directly exercised by the central authority, and the powers delegated to the HKSAR by the central authority to enable it to exercise a high degree of autonomy in accordance with the law. The central authority has the power of oversight over the exercise of a high degree of autonomy in the HKSAR."[4] It is expected in the days ahead that the central authority will widely proclaim its powers and responsibilities under "one country, two systems" among the Hong Kong people to pave the way for actively employing its powers to ensure that the practice of "one country, two

---

4   Ibid, P7.

systems" in Hong Kong is on the right track, thus lessening the obstacles that might appear when exercising power.

(4) Declaring solemnly the central authority's (new) Hong Kong policy under the new situation. The Resolution of the CPC Central Committee on Certain Major Issues Concerning Comprehensively Advancing the Law-Based Governance of China, approved at the 4th Plenary Session of the 18th Central Committee of the Chinese Communist Party on October 23, 2014, is an extremely important policy document, wherein a declaration of the central authority's policy concerning Hong Kong and Macao is made: "[The Communist Party of China] will guarantee the practice of 'one country, two systems' according to the law and move forward the unity of the motherland; insist on the supreme legal status of the Constitution and its highest legal effect; comprehensively and accurately implement the policies of 'one country, two systems', 'the people of Hong Kong governing Hong Kong', 'the people of Macao governing Macao', and a high degree of autonomy; conduct affairs strictly according to the Constitution and the Basic Laws; perfect systems and mechanisms connected to the implementation of the Basic Laws; exercise the powers of the central authority according to the law; guarantee a high degree of autonomy according to the law; support the Special Administrative Region's chief executives and governments in governing according to the law; guarantee the development of economic and trade relationships between the mainland, Hong Kong and Macao, and of exchange and cooperation in all areas; prevent and oppose interference by foreign powers in Hong Kong and Macao affairs, guarantee that Hong Kong and Macao remain prosperous and stable in the long run."

President Xi Jinping's speech at the celebration of the 15th Anniversary of Macao's Return to the Motherland and the Inauguration of the Fourth-term Government of the Macao Special Administrative Region on December 20, 2014 is also noteworthy. Though President Xi's speech is about Macao, it also has practical implications and significance to Hong Kong. He said: "We are glad to see that the principles of 'one country, two systems', 'Macao people governing Macao', and a high degree of autonomy as well as the Basic Law of the Macao SAR have won massive support from the people of Macao and have been implemented in real earnest. The constitutional order of the Macao SAR as prescribed by the Constitution and the Basic Law is respected and upheld. The central authority's overall governing power is effectively exercised while the high degree of autonomy enjoyed by the SAR is fully guaranteed."

He also said: "'one country, two systems' is a basic national policy. To firmly uphold this policy is vital to ensuring the long-term prosperity and stability of Hong Kong and Macao. It is also an important part of our endeavour to fulfil the Chinese dream of great national renewal. It conforms to the fundamental interests of the country and the nation, the overall and long-term interests of Hong Kong and Macao, and the interests of international investors. To continue to advance the cause of 'one country, two systems', we must stay committed to the fundamental purpose of 'one country, two systems', jointly safeguard national sovereignty, security and development interests, and maintain the long-term prosperity and stability of Hong Kong and Macao. We must continue to govern Hong Kong and Macao and uphold the practice of 'one country, two systems' according to law. We must both adhere to the 'one China' principle and respect the differences of the two systems, both uphold the powers of the central authority and ensure a high degree of autonomy in the SARs, both give play to the role of the mainland as the staunch supporter of Hong Kong and Macao and increase their competitiveness. At no time should we focus only on one side to the neglect of the other. This is the only way leading to sound and steady progress. Otherwise, a misguided approach from the beginning, just like putting one's left foot into the right shoe, would lead us nowhere."

The political statement above is to highlight the powers and responsibilities of the central authority under "one country, two systems", and show the sincere and serious attitude of the central authority towards their powers and responsibilities. The central authority is determined to protect the national security and interests on its own.

(5) Insisting that Hong Kong's administrators must be the "people who love China and Hong Kong". In response to the fierce struggle on selecting the chief executive by universal suffrage, the national leaders and central government officials constantly reiterate the importance of the "people who love China and Hong Kong" to the success of "one country, two systems". The *White Paper* reminds the Hong Kong people: "The fact that Hong Kong must be governed by patriots is well grounded in laws. Both the Constitution and the Basic Law provide for the establishment of the HKSAR, which works for China's national unification, territorial integrity and maintaining Hong Kong's long-term stability and prosperity."[5]

The *White Paper* also admonishes: "loving the country is the basic political requirement for Hong Kong's administrators. If they are not consisted of by patriots as the mainstay or they cannot be loyal to the country and the HKSAR, the practice of 'one country, two systems' in the HKSAR will deviate from its right direction, making it difficult to uphold the country's sovereignty, security and development interests, and putting Hong Kong's stability and prosperity and the wellbeing of its people in serious jeopardy."[6]

Li Fei (李飛), Director of the Basic Law Committee, gave the most candid and appropriate exposition of this important point: "The controversy over the universal suffrage has lasted twenty-eight years counting from the controversy over the direct elections in 1986, and at least seventeen years counting from the return of Hong Kong to China. The controversy is a major political issue distressing Hong Kong for nearly three decades. What is the essence of the problem?"

"As the National People's Congress Standing Committee members pointed out in the discussion of the Report of the Chief Executive of HKSAR and the Decision (draft), the essence of the problem does not lie in the decision whether to adopt universal suffrage or not, or in the decision whether to practise democracy or not, but in the

---

5   Ibid, P35.
6   Ibid, P35.

controversy over the jurisdiction of the HKSAR, which reflects various longstanding political issues of Hong Kong. Even though seventeen years have passed after Hong Kong's return to China, a few Hong Kong people are still unwilling to accept the fact that China has resumed exercising sovereignty over Hong Kong, and unwilling to accept the central authority's powers to govern Hong Kong. Those people have an 'alternative interpretation' of 'one country, two systems' and the Basic Law, and work with external forces to provoke political disputes constantly. They direct the attack at the central government and attempt to turn Hong Kong into an independent political entity."

"In summary, the opposition's argument and aspiration on the issue of universal suffrage are to allow their representative to serve as the chief executive. This of course cannot be allowed. If an opposition figure becomes the chief executive, the sovereignty, security and development interests of the country would inevitably be undermined and the prosperity and stability of Hong Kong damaged. In that case, we will not only be difficult to face our predecessors who strived arduously for the return of Hong Kong to China, but also hard to explain to all the Chinese people, including those Hong Kong people who 'love China and Hong Kong' as well as our progeny."

"Some Hong Kong people hope that the central authority will make concessions based on the proposals put forward of a minority of Hong Kong people [meaning the opposition]. These aspirations are understandable. However, the current controversy over universal suffrage is a matter of principle, and therefore one upon which the central authority cannot compromise. The key spirit of the decision of the NPC Standing Committee this time can be summarised in two sentences which have been repeated over the past year. The first is that, 'The universal suffrage system must comply with the Basic Law of Hong Kong and the NPC Standing Committee's decision'; the second is that, 'The chief executive has to be a person who loves China and Hong Kong. A person who confronts the central authority can never be the chief executive.'"

"In these two sentences the central authority has clearly made known their position on the core issue of universal suffrage, which is also the most controversial problem in Hong Kong society. In other words, these two sentences explicitly tell those Hong Kong people who persist in confronting the central authority the fact that they would have no chance to become the chief executive, be it in the past, present, or future. It sounds very 'tough' but it contains the ardent expectation that those who harbour unrealistic ideas will return to the path of loving the country and loving Hong Kong instead of spending their lives in the street, and that they will contribute to building a peaceful Hong Kong, adding a little positive energy to the community."[7]

(6) Bringing the major political issues back to the central authority (把重大政治課題收歸中央) and letting the central authority make the decisions, and removing those issues from Hong Kong over the medium and long-term. A good example of giving the central authority giving the "final word" (一錘定音) is that the central authority has power of determination over the arrangements for the popular election of the chief executive which take precedence over the freedom of expression of the Hong Kong people, in particular, the opposition. Another good example is that the central authority claims the power to ensure that the political system of the HKSAR is an "executive-led" system with the chief executive at its core as opposed to a system with a "separation of powers", to be divided among the executive, the legislature and the judiciary.

Yet another example is the interpretation of article 104 of the Basic Law by the Standing Committee of the National People's Congress in November 2016 on its own initiative to prevent those elected Legislative Council members who attempt to advocate "Hong Kong independence" during the oath-taking process from taking office.

If the major political issues are "withheld" by the central authority, Hong Kong will only need to handle the less politically controversial and the practical issues. The pan-democrats (the most

---

7   Li Fei (September 1, 2014). "The Political and Legal Connotation of the Decision of the Standing Committee of the National People's Congress—Speech at the Briefing Session with the Senior Officials of the HKSAR".

important segment of the opposition) can no longer embarrass or lambast the HKSAR Government with "hot" political issues, and their ability to mobilise the masses to challenge the central authorities will also be weakened. On the other hand, with the major political issues now in the hands of the central authority, the pro-Beijing camp may have more opportunities to work with the opposition on practical issues, and by doing so, gradually mitigate mutual suspicion. Indeed, the establishment of a set of stable and reasonable criteria to guide the relations between the "two systems" cannot be accomplished overnight. Even the "ultimate" set of rules may not be able to satisfy all sides, but it can nevertheless foster a decent relation between them, promote friendly interactions and reduce tensions and collisions.

(7) The central authority will exercise their powers actively, seriously and prudently under "one country, two systems", and "institutionalise" the processes whereby the powers are exercised, including "legitimising", "standardising", and "concretising" the processes, making them more "transparent" and "reasonable". Because the central authority is determined to ensure that the practice of "one country, two systems" falls in line with the original "blueprint" with the powers at their disposal, there is hence the need to institutionalise the processes of power exercise. The *White Paper* describes the matter as follows: "The fact that the Standing Committee of the NPC exercises the power of interpretation of the Basic Law in accordance with the law is aimed at maintaining the rule of law in Hong Kong, as it oversees HKSAR's implementation of the Basic Law and protects the high degree of autonomy of the region."[8]

This means that the central authority will throw off the shackles which have inhibited them from interpreting the Basic Law in the past, and actively adopt NPC Standing Committee's interpretation as the means to correct the deviations from the Basic Law. The "normalisation" or "routinisation" of NPC Standing Committee's interpretation of the Basic Law will become a trend. The *White Paper* also mentions:

8   *White Paper* (《白皮書》). P34.

"We should improve the systems and mechanisms related to implementing the Basic Law, which will help enhance its authority. [...] As the practice of "one country, two systems" continues and the Basic Law is further implemented, it is imperative to further improve the systems and mechanisms in relation to the implementation of the Basic Law. In particular, it is necessary, with an eye to the lasting peace and order in Hong Kong, to exercise well the power invested in the central authority as prescribed in the Basic Law and see to it that the relationship between the central authority and the HKSAR is indeed brought onto a legal and institutionalised orbit."[9] Undoubtedly, the most important thing here is the supervisory powers over the chief executive and the Legislative Council.

(8) Strengthening the leadership of the central authority in researching, formulating and implementing Hong Kong policy. To better exercise their powers and perform their duties, the central authority has to better understand the nature, architecture, mode of operation and framework of Hong Kong's "one system" and Hong Kong people's ideas and values, and hence master Hong Kong's situation more effectively. Since the central authority will involve itself in Hong Kong affairs more frequently, their Hong Kong policy should be more explicit, transparent, institutionalised, predictable, reasonable, and scientific. The central authority needs to discuss with Hong Kong people when developing their policy toward Hong Kong and make the policy public to the international community regularly in order to boost public and international confidence in Hong Kong and China.

The central authority also needs to establish a mechanism for extensive communication with all sectors in Hong Kong, instead of totally relying on the HKSAR Government to do the job. That would help to increase the Hong Kong people's identification with the country and to enlarge the ranks of "the people who love China and Hong Kong". To achieve the goals mentioned above, the central authority would improve its personnel, institutions and capability in the area of

---

9   Ibid, P34.

Hong Kong research as soon as possible. The Chinese Association of Hong Kong and Macao Studies was founded at the end of 2013 with support of the central authority. This organisation is committed to promoting research on specific Hong Kong issues among Hong Kong and mainland scholars and providing political and policy advice. Apart from serving as a think tank for the central authority on Hong Kong policy, the association is also committed to training elect specialists in Hong Kong studies.

(9) Furthering the economic integration of Hong Kong and the mainland China with particular emphasis on sharing the dividends of economic integration with Hong Kong people from all social sectors. To this end, Hong Kong would be encouraged to position itself strategically in the national development process and formulate a development strategy that would be to the benefit of all social sectors and all social classes. The *White Paper* affirms Hong Kong's status and role in the national development in the new era. "By consolidating and enhancing its existing advantages, Hong Kong can better play its role in introducing external investment and talents, in absorbing internationally advanced technologies and managerial expertise, in serving as a bridge for implementing China's 'go global' strategy, and in helping to quicken the shift of the mode of growth on the mainland. In addition, Hong Kong's experience can be of reference for the mainland with respect to pursuing innovative methods of social and economic management."[10] Actively encouraging Hong Kong to take part in the "One Belt, One Road" strategy and in the development of the Guangdong-Hong Kong-Macao Greater Bay Area are good examples. (粵港澳大灣區)

(10) Helping Hong Kong to handle its social conflicts. The central authority asserts a role in easing and solving conflicts relating to the "one country, two systems" policy, particularly with respect to coordinating the interests of various sectors and overcoming and counteracting the impediments posed by the conservative forces in Hong Kong. In fact, as conflicts and challenges occur under the "one country, two

---

10 Ibid, P38.

systems" policy, the central authority will inevitably have to intervene in Hong Kong's affairs, aiming to promote internal stability, resolve social conflicts and overcome difficulties in governance.

After years of observation and study, the central authority has deepened its understanding of Hong Kong's capitalist system, understanding that a good business environment is needed to maintain Hong Kong's prosperity and stability. While there is no need to over-prioritise the interests of the capitalists, balancing, coordinating and reconciling the interests of various social classes and generations is an urgent matter. Just as important, the Government of the HKSAR should keep some distance from the bourgeoisie, and by doing so enhance its political autonomy.

It remains a difficult task in the foreseeable future to make significant changes in Hong Kong's political system and to enlarge the political influence of the middle class and working class, as the central authority will still have to focus on dealing with the opposition in Hong Kong. As a result, the most feasible method is that the central authority mandates that the HKSAR Government adjust its policies, particularly in such areas as economic development, land, housing, welfare and labour, so as to better protect the interests of the middle class, labour the youth as well as other disadvantaged groups, and to reduce the social conflicts and the destabilising factors. As far as implementing the policies that promote "social justice" is concerned, the central authority should assist the HKSAR Government in overcoming the objections and impediments posed by the bourgeoisie and conservative forces, and persuade the capitalists and vested interests to bear in mind the overall interests of Hong Kong, to assume more social responsibilities and to make some "sacrifices" in exchange for political stability and effective governance.

In addition, with the rise of China, the central authority is less and less dependent on Hong Kong's bourgeoisie. Before the return of Hong Kong, "disinvestment" was the most powerful weapon with which Hong Kong capitalists could "blackmail" the Chinese Government. Currently, the central authority no longer remains vulnerable to such threats. For one thing, large-scale withdrawal of capital from Hong Kong is unlikely to happen, and if it does, it is very likely that most of the capital would be withdrawn to the mainland. For another, as China now has huge foreign exchange reserves and financial resources, its dependence on Hong Kong's capital becomes lower and lower. Even if the Hong Kong's capitalists bypass the HKSAR Government to appeal to the central authority with the intention of changing the public policies of Hong Kong, the effect will be negligible, as the HKSAR Government has increased its autonomy with respect to the bourgeoisie.

Another important development is that the state-owned enterprises and private enterprises from the mainland make up an increasing proportion of Hong Kong's economy, but they have little influence in Hong Kong's politics due to the central authority's "non-intervention" policy. To promote social equity and justice, the central authority should encourage enterprises from the mainland, state-owned enterprises in particular, apart from profit-making, to take an active part in Hong Kong's affairs, especially in charity, national education, and technical training in order to create more job opportunities for the youth, to help Hong Kong people better utilise the development opportunities springing from the development of the country, and to break the monopolies in some industries. It is not difficult to imagine how mainland enterprises can strengthen the Hong Kong people's sense of belonging to China if they are able to play a positive role in Hong Kong's social and economic development.

(11) Strengthening the efforts to build up the forces of "the people who love China and Hong Kong". The goal is to ensure that Hong Kong is administered by patriots; therefore, instead of waiting for patriots to come forward, the central authority should work together with the HKSAR Government to actively nurture "the people who love China and Hong Kong" via a multitude of policies and measures. According to the *White Paper*, "The central government will continue to encourage the people of Hong Kong to carry forward their fine traditions of inclusiveness, mutual support and respect for the rule of law and order. It calls on the Hong Kong people to seek common ground

while reserving differences, be tolerant and help each other in the fundamental interests of the nation and the general and long-term interests of Hong Kong, to achieve the broadest unity under the banner of loving the country and Hong Kong with strengthened social harmony and stability through compromise and mutual assistance."[11] The most important way to nurture the "people who love China and Hong Kong" is to provide opportunities for talented individuals from a broad spectrum of society to gain experience in various social sectors or to work in the councils (legislative and district), the HKSAR Government, as well as various kinds of independent organisations and advisory bodies, so that they can be trained as talented persons to "govern Hong Kong".[12]

(12) Reinforcing the leadership and guidance of the HKSAR Government and the forces of "the people who love China and Hong Kong". The central authority should provide the necessary leadership and guidance to enable the HKSAR Government to engage fully in a cooperative process with the central authority in order to accurately implement the "one country, two systems" policy. The chief executive of the HKSAR should follow and implement the recommendations and executive orders from the central authority in order to safeguard national security and national interests in Hong Kong. Additionally, the chief executive and principal officials should be selected prudently, and the offices of the chief executive and principal officials accountable to the central authority should follow formal procedure.

It is stated in the *White Paper* that "The chief executive reports his/her work to the central government on an annual basis, on the implementation of the Basic Law and other items for which he/she is accountable to the central government; and the state leaders give guidance to the chief executive on major matters related to the implementation of the Basic Law."[13] It is imperative that the high-level civil servants of

---

11 *White Paper* (《白皮書》). P37-38.

12 Lau Siu-kai (2013). "Preliminary Discussion on the Construction of Governing Coalition" ("構建管治聯盟芻議"), and "On the New Political Doctrine in Hong Kong" ("關於香港的新政治主張"). *Hong Kong's Politics after Its Return to China* (《回歸後的香港政治》). Hong Kong: The Commercial Press. P330-354 and P355-363 respectively.

13 *White Paper* (《白皮書》). P8. Also in Lau Siu-kai (2013). "The Central Government's Power

the HKSAR fully understand the central authority's "one country, two systems" policy, and be aware that they should be loyal to the country and the central authority, firmly support the "one country, two systems" policy and the chief executive's administrative policies and philosophy. The high-level civil servants should neither evade their responsibilities on grounds of "political neutrality" which didn't exist in the colonial period, nor secretly collude with the opposition, impeding the governance of the HKSAR Government.

(13) Nurturing the Hong Kong people's (particularly with respect to the youth) understanding of and respect for the country. In recent years, many young people have been highly alienated from and antagonistic towards the country, the Chinese nation and the central authority, becoming organisers of and participants in all kinds of radical actions. The hard core of the 2014 "Occupy Central" movement was, surprisingly enough, made up of students. Ironically "veteran" opposition activists have been largely induced to participate by radical students. In view of the fact that the opposition is losing support from the masses, they turn their attention to the youth as their new supporters and possible successors. In order to attract, recruit and train the youth and turn them into opposition activists, the opposition in varying degrees cater to their "nativist" and "separatist" sentiments and radical tendencies, particularly those radical opposition figures who are themselves sympathisers with these sentiments. Schools, universities in particular, thus become the new battlegrounds for political disputes and struggles.

In the past decade, the central authority had identified the young people as a major problem, and worked hard to promote the young people's awareness of and identification with the country and the nation via national education. But it remains a difficult task, because such efforts are often challenged and rejected by the opposition, teachers and students, and its implementation by the HKSAR Government has

---

of Appointment, Removal and Supervision of the Principal Officials of the SAR Becomes the New Normal" ("中央對特區主要官員的實質任免權和監督權將成為新常態"). Hong Kong and Macao Journal (《港澳研究》), 2 (Vol. 7), P15-16.

been clumsy and ineffective.[14] After re-evaluation, the central authority still considers national education as the very means of enlightening the youth. But by now the central authority is able to look at national education from a broader perspective and have a more realistic understanding of the possible goals to be achieved through national education. They are also aware that the youth are discontented with the central authority for many reasons, including dissatisfaction and concern about their less than ideal existential and development prospects, disaffection toward the HKSAR Government and particularly chief executive Leung Chun Ying (梁振英). The youth also think that their interests are infringed upon due to the growing relations between Hong Kong and the mainland, and they are worried that the original "superior" institutions and values of Hong Kong risk destruction at the hands of the central authority. In addition, the young people are quite influenced by the "populism", anti-authority sentiments and the "philosophy" of struggle coming from foreign countries and the opposition in Hong Kong.

In my opinion, instead of directing efforts toward producing "patriots" among the youth as soon as possible, the central authority's policy should focus on deepening the young people's awareness of the historical background and strategic objectives of the "one country, two systems" policy, while strengthening their understanding of the interdependent relationship between the mainland and Hong Kong, as well as the concept of "community of common fate" held in common by the Hong Kong people and the mainland compatriots. The central authority hopes that the young people are fully aware the consequences for Hong Kong of confronting the country and the central authority, particularly the high price that will be paid by Hong Kong and themselves. The central authority also knows that the implementation of national education cannot be accomplished overnight or solely relying on schools and teachers. Instead, national education has to be promoted through a variety of means and methods, and that the central authority, the main-

---

14　Lau Siu-kai (2013)."The Development of the Central Government's Policies toward Hong Kong after the Handover" ("回歸後中央政府對港政策的發展"), in Hong Kong's Politics after Its Return to China (《回歸後的香港政治》). Hong Kong: The Commercial Press. P39-42.

land, the HKSAR Government and all sectors of the Hong Kong society have different roles to play, and their efforts should be complementary with one another so as to obtain the best results. Additionally, although schools, teachers and public examinations are necessary to promote national education, it also requires the robust economic development in Hong Kong so that the young people can better enjoy the benefits brought by the growing economy and economic integration of Hong Kong with the mainland. Furthermore, the serious inequality and injustice in Hong Kong have to be reduced so that more opportunities of political participation are made available to the young people.

(14) Engaging in "a decisive battle" with the opposition. The central authority will take tougher and bolder measures in response to the opposition's fiercer onslaught against and challenges to the "one country, two systems" policy and the central authority. On the one hand, the central authority will adopt the strategies of "no appeasement", forceful counteroffensive, and decisive implementation of the law. On the other hand, the central authority will try to gain the goodwill from the "moderate" opposition activists, and gradually convert them into the "loyal opposition". The "loyal opposition" represent those who are willing to cooperate with the central authority and acknowledge the political systems of the country and Hong Kong. The near future will be the crucial moment of "the decisive battle" between the central authority and the opposition, and the central authority, with a panoply of powers, mechanisms and influences, is bound to win this battle. The eventual victory of the central authority would mean that Hong Kong will follow the path of "one country, two systems" charted by Deng Xiaoping, that the understanding of "one country, two systems" of the central authority will prevail in Hong Kong, that the powers and responsibilities claimed by the central authority will be recognised by the majority of the Hong Kong people and that the central authority's interpretation of the Basic Law will be the most authoritative interpretation. On the other hand, the victory of the central authority does not mean that the opposition in Hong Kong will be destroyed or eliminated. The opposition will definitely continue to exist, only that their political clout and ideological

impact will weaken substantially.

When Wang Guangya (王光亞), Director of Hong Kong and Macao Affairs Office of the State Council, declared that the central authority would not allow anyone who was against the central authority to take up the post of the Chief Executive of the HKSAR; he also expressed the central authority's position toward the opposition. He said, "There are two categories of 'pan-democrats', as I have mentioned in various occasions. There are a small minority of people who have ulterior motives. They use the 'democracy' as cover and regard Hong Kong as an independent political entity. They wantonly distort the interpretation of the Basic Law, obstruct the administration of the HKSAR Government, obstinately fight against the governance of the central authority (and even collude with the outside forces), advocate and support the separatists including those who are in favour of 'Hong Kong independence', attempt to overthrow the ruling position of the CPC and the socialist system, both of which are established by the Constitution of China. Their views and actions are far beyond so-called 'free speech' and 'fighting for democracy'. Despite their small numbers, they have caused great damage. They are not only the 'opposition activists', but also the 'die-hards' and the 'stubbornites'. The central authority would struggle against this opposition in a firm and bold manner. Specifically, in regard to the design of the arrangements for the popular election of the chief executive, these people would be excluded from the electoral processes. Not only would members of the opposition be restricted from becoming candidates, they would also be blocked from victory. Even if they are fortunate enough to win the election for the chief executive, the central authority is determined not to appoint them. Otherwise, they will bring misfortune to Hong Kong and the whole country. In this regard, the central authority will stand by its principled position. As for the majority of the 'pan-democrats', most of them care about the development of the country and the future of Hong Kong. They are in favour of the restoration of the country's sovereignty over Hong Kong, and uphold the 'one country, two systems' policy and the national political system stipulated by the Constitution. Some of them may differ from us in their political views, democratic ideas and the ways of achieving democracy, but they approve of the 'one country, two systems' policy, the Constitution, the Basic Law and the national system and institutions. I wish to have more opportunities to communicate with these 'pan-democratic' friends and exchange our views in-depth on any issue based upon our common political basis. "[15]

The central authority has taken drastic measures to curb the opposition, bringing the major political issues under their control, and taking away from them the "powerful weapons" that can be used by them to mobilise protests. In this way the central authority can completely and finally counter-check the opposition. The central authority has clearly declared and stood by its position on the universal suffrage of the chief executive. By doing so, this major political issue has come under the jurisdiction and control of the central authority, and therefore the Hong Kong people will no longer have any "illusion" on that issue and refrain from being abetted or mobilised by the opposition. Accordingly, Hong Kong society can refocus on the social and economic issues.

What Wang Guangya said in the following has similar implications: "The biggest problem that Hong Kong faces currently is that a few people use the topic of universal suffrage to create social disputes and attract people's attention to the political issues, which have hindered both economic development and the improvement of people's livelihood. I have told my friends from the business community and the economic experts and professors that they should focus on how to further advance the Hong Kong economy and how to create better development opportunities for the next generation. I felt sorry that they had to spend lots of time and energy on those meaningless political disputes, thus resulting in the misallocation of resources. I have also told some grassroots friends in the pan-democratic camp that I understood their concern and eagerness to improve the livelihood at the grassroots. I was deeply touched by their humanistic concern, but if they thought that all

---

15 The speech of Wang Guangya, Director of Hong Kong and Macao Affairs Office of the State Council, on meeting with the members of the Legislative Council of Hong Kong in Shenzhen, on May 31th, 2015.

the problems were caused by the political system and regarded universal suffrage as the elixir to all problems, I was afraid they had come up with a wrong prescription. Most of the friends whom I have contacted with wished to end the dispute over universal suffrage. As they said, it was better for it to come to an end, because the endless political disputes had not only undermined social harmony in Hong Kong, but also had severe impact on economic development and the improvement of people's livelihood. The decision of the NPC Standing Committee demonstrates that the central authority is determined to solve the major problem that has haunted Hong Kong for a long time. The fundamental purpose is to bring Hong Kong back on track, to lead people from all walks of life to focus on the development issues, so as to provide better development opportunities for the younger generation, to maintain Hong Kong's advantage as a leader in development, and to create a better future for Hong Kong."[16] The speech above indicates that the central authority is firmly determined to bring the political issues that have haunted Hong Kong for a long time and been utilised by the opposition back under their control, and to encourage people of all sectors in Hong Kong to discuss the non-political issues instead.

(15) Preventing outside and internal forces from turning Hong Kong into "a base for subversion". For some time, the central authority has remained on high alert to the intentions and policies of the Western countries, particularly the U.S., towards Hong Kong. In recent years, the China-American relations have become increasingly tense as the U.S. is gradually adopting the "containment" strategy to curb China's rise. Whether the outside forces will step up efforts to turn Hong Kong into a "base for subversion" continues to be taken seriously by the central authority. Outside forces were faintly visible in the "Occupy Central" movement in 2014, which deepened the central authority's misgivings about the intentions of the U.S. and other outside forces towards Hong Kong. Recently, the central authority has upgraded its vigilant attitude towards the outside forces, from "opposing" any external interference to

---

16 Ibid.

"containing" it.

The *White Paper* clearly warns that "it is necessary to stay alert to the attempt of outside forces to use Hong Kong to interfere in China's domestic affairs, and prevent and repel the attempt made by a very small number of people who act in collusion with outside forces to interfere with the implementation of "one country, two systems" in Hong Kong."[17] Also, it is important that it prevents foreign forces from interfering in Hong Kong's affairs. "Hong Kong's affairs are internal affairs of China, and the Chinese central government has made timely representations with certain countries through diplomatic channels regarding their words and actions of interference."[18]

Due to the failure of the local legislation of Article 23 of the Basic Law in 2003, Hong Kong remains a lurking threat to national security since its return to China twenty years ago. With the deterioration of the international situation facing China and the rapid development of technology, Article 23 of the Basic Law is far from enough to deal with the severity and diversity of the national security threats that China encounters currently. To make matters worse, the HKSAR Government does not have the political will, energy and capability to accomplish the local legislation of Article 23 in the foreseeable future. As a result, the central authority is becoming increasingly anxious about the threats to national security posed by Hong Kong. In response to the future potential threats from the internal forces and outside forces, the central authority will not stand by idly and will definitely take precautions.

In fact, the mainland has conducted relevant studies on how to ensure national security in the case that Hong Kong is unable to legislate on Article 23. According to the mainland legal expert Rao Geping (饒戈平), "Mainland scholars have conducted studies on the [local] legislation of Article 23 of the Basic Law from multiple perspectives, considering various methods to implement Article 23 of the Basic

---

17 *White Paper* (《白皮書》), P40.

18 *White Paper* (《白皮書》), P10.

Law." "One method, definitely the best one, is that Hong Kong proactively implements Article 23 of the Basic Law on its own. If the political environment for legislation is not good, we may have to choose other methods. For instance, the SAR Government can make a new law similar to Article 23 based on the compilation of existing laws related to national security and adding some new regulations. Another option is to draw up the relevant law by using the security laws of other countries for reference, under the guidance of China's Constitution. If Hong Kong cannot solve the [local] legislation of Article 23 on its own, for the sake of protecting national interests, we may try out the National Security Law of the mainland in Hong Kong temporarily. Or, the central authority can work out and implement a temporary security law according to Hong Kong's current situation, until Article 23 is implemented in Hong Kong. All these are the methods discussed by scholars in the academic circles, not the central authority. [...]"[19]

Additionally, as the central authority is in charge of Hong Kong's national defence and foreign affairs, if an event endangering national security does occur in Hong Kong, the central authority can give executive orders to the chief executive, commanding him/her to take proper measures to remove the threats to national security. The Chinese Association of Hong Kong and Macao Studies specially established a study group on national security at the beginning of 2015, which signifies that the central authority takes seriously the relations between Hong Kong and national security.

Different Views on Implementation of "One China, Two Systems" Policy in Hong Kong

Since the return of Hong Kong to China, people from all sectors

---

19 Rao Geping (2014). "Article 23 Legislation is 'Part of the Comprehensive Implementation of the Basic Law'".("'23 條立法是全面實踐基本法的一部份'"). *Bauhinia Magazine* (《紫荊雜誌》). Vol. 282, P6-8, 8.

of Hong Kong have observed conflicts, difficulties and new challenges occurring in the implementation of the "one country, two systems" policy. By conducting analysis on the causes, content, severity and trend of those problems, they have put forward relevant solutions. They are as follows:

The Government of HKSAR and "the People Who Love China and Hong Kong"

The HKSAR Government and "the people who love China and Hong Kong" believe that many challenges are derived from structural and historical factors, mainly from the Hong Kong people's resistance to "the return to the motherland" and the CPC, their contemptuous attitude towards mainland compatriots, and their nostalgia for British colonial governance. Hence, various kinds of opposition forces exist in almost all social and political realms (such as the executive branch, the legislature, the judicial branch, political groups, social groups, the media, the Internet, the public, schools and council [legislative and district council] elections). They are well entrenched and can hardly be eliminated. As a result, such phenomena as the frictions between the executive and the legislature, low prestige and popularity of the chief executive, ineffective rule by the HKSAR Government, political struggles, social divisions, estrangement between the central authority and Hong Kong, and conflicts between the "two systems" are not only inevitable, but have also become the long-term political "new normal". Many officials and core elements among "the people who love China and Hong Kong" are pessimistic about the current situation, and some even consider it the "destiny" of Hong Kong.

Pessimistically, they conjecture that there is no force that has the ability to "turn things around". From their point of view, even if the central authority has great powers, it does not mean they will be able to change Hong Kong's situation, setting aside the question of whether the central authority is determined to solve the problems. They also doubt the central authority's ability to "administer Hong Kong" properly, and worry that the interference of the central authority in Hong Kong's affairs not only cannot reverse the situation, but may actually aggravate conflicts and stir antagonism in the society, leading to more severe and intractable consequences. In addition, they are afraid that if the central authority takes actions, these actions will later be turned into the central authority's long-term and active interference in Hong Kong's affairs. The "high degree of autonomy" will then remain in name only, and their power, status and already low prestige among the Hong Kong people will be further weakened. Although they do not say so openly, these people distrust the central authority.

Since the return of Hong Kong to China, many cases have shown that when the HKSAR officials and "the people who love China and Hong Kong" had to make choices between, on the one hand, upholding the authority of the Basic Law together with the national security and the central authority's interests, or on the other hand, pleasing the masses, they usually chose to evade the problems or postpone their treatment. The failure of the local legislation of the Basic Law Article 23 is a typical example. Even when the opposition lawmakers proposed amendments to the government bills and violated Article 74 of the Basic Law by "filibustering" in the legislative council, the HKSAR Government dared not stop them by legal means or seek help through the NPC Standing Committee's interpretation of the Basic Law, for fear of being criticised by the public for inviting the central authority to interfere in Hong Kong's affairs, to undermine the rule of law and to erode the judicial independence of Hong Kong.

In 2011, the decision of the Hong Kong Court of Final Appeal in the Chong Fung-yuen case obviously violated the NPC Standing Committee's 2009 interpretation of the right of abode in Hong Kong. The Legislative Affairs Commission of the NPC Standing Committee also made clear that it did not agree with the decision of the Court of Final Appeal. That decision later led to the influx of a large number of pregnant women who gave birth to children in Hong Kong without the right of abode, while also having a negative impact on the one-child policy on the mainland. Still, the HKSAR Government was not willing to discuss these problems with the central authority because the HKSAR Government was worried about being blamed for inviting the central authority to interfere in Hong Kong's affairs and for prompting the NPC Standing Committee to interpret the Basic Law.

Besides, the HKSAR Government and many politicians who qualify as "people who love China and Hong Kong" do not have sufficient confidence in their capability and cohesion. They understand that they can hardly enjoy high political "legitimacy" and esteem among the Hong Kong people, as they have committed the "original sin" of being "pro-Beijing" and as the chief executive is not elected by universal suffrage. Overall, their passivity and pessimism about the "one country, two systems" policy and Hong Kong's future has eroded their political self-confidence, their will to govern, and their ability to make a stand. A vicious circle thus appears. As the confidence of the Hong Kong people in the people who govern Hong Kong falls, the morale and solidarity of the HKSAR Government and the patriotic forces also drops, resulting in even poorer governing performance, which can hardly meet the rising expectations of the Hong Kong people. If the chief executive tries to be tough and draws a clear line between foes and friends in order to control the political situation, it may give rise to a more violent reaction and bitter resentment from the Hong Kong people, and the HKSAR Government will find itself under attack from all sides, becoming isolated and helpless. Such a dilemma is best exemplified in the governance of Leung Chun-ying.

On the whole, the HKSAR Government and many of "the people who love China and Hong Kong" believe that they are unable to change the current political situation as they are always in a weak position in Hong Kong's politics. Immersed in fatalism, they seldom solemnly and seriously put forward effective measures in response to the problems and challenges that arise in the implementation of the "one country, two systems" policy. But if those problems and challenges are not dealt with in a timely and effective manner by them, the much bigger problem of how to continue the "one country, two systems" after 2047 will receive even less attention from them. Meanwhile, they flatly turn down the suggestions from the opposition particularly regarding the reforms of political restructuring that would trigger reallocation of power and status, because such reforms would severely damage their vested interests. With regard to better implementing the "one country, two systems" policy, many HKSAR officials and "the people who love China and Hong Kong" are indeed pessimistic, indecisive and passive conservatives who are frustrated but feel helpless about Hong Kong's future.

The Opposition Forces

The various opposition forces in Hong Kong have different understandings of the problems and challenges encountered during the implementation of the "one country, two systems" policy and their solutions from the central authority, the HKSAR Government and "the people who love China Hong Kong". Actually, the opposition possesses greater theoretical sophistication and enjoys a greater say when analyzing and discussing the problems related to Hong Kong. They have constructed their own unique and relatively systematic viewpoints, and have received assistance from many scholars in research and in ideological propagation. In fact, many Hong Kong people are convinced by their argumentation. The core argument of the opposition is that Hong Kong would unavoidably face severe difficulties in governance after its return to the motherland. Some people even exaggerate the difficulty of governance to the extent of saying that Hong Kong is ungovernable. Moreover, some people believe that Hong Kong cannot be effectively governed under the "one country, two systems" policy as originally arranged by the central authority. To the opposition, the primary reasons for the difficulties in governance come from Hong Kong's "undemocratic" political system and the central authority's blatant interference in Hong Kong's affairs. The chief executive, not elected by universal suffrage, suffers from lack of political "legitimacy" in spite of his/her being duly elected under the Basic Law. They also deny the legitimacy of the central government. Accordingly, the chief executives "appointed" by the central government do not have legitimacy. On the contrary, the chief executives even lose some "legitimacy" because of the fact that their powers come from the central authority.

Second, the methods for selecting the chief executive endow the elite strata, particularly the business corporations with disproportionate political clout. As a result, the HKSAR Government's base of social support is too narrow, and the policies of the HKSAR Government are too biased in favour of the vested interests, leading inevitably to severe social and political inequality.

Third, since its return to the motherland, Hong Kong's capital structure has become more diversified and complicated, and the increasing internal conflicts within the bourgeoisie have further added difficulty to the governance in the HKSAR. During the "colonial" period, under the political domination of the colonial government and the big British corporations, the Chinese business elites played only a subsidiary role in the economy, and their political role was even more restricted. After the return of Hong Kong to China, under the policy of "one country, two systems", the task of preserving the original capitalist system is entrusted to the Chinese bourgeoisie in Hong Kong, who, therefore, are endowed with a set of political privileges and powers that they had never enjoyed during the "colonial" period. After 1997, with the decline of the British capital, the status of Hong Kong Chinese capital has been on the rise, and the rise in status of mainland capital (state-owned capital and private capital) is particularly dramatic. Consequently, competition and conflicts among the capitalists are growing fiercer and more intense. Generally speaking, the business community of Hong Kong does not have powerful and representative leaders. New rules of the game to regulate competition among business people have yet to be set up. And the HKSAR Government does not have the necessary authority and means to coordinate and reconcile the complicated interests of the business community. After the return of Hong Kong to China, the HKSAR Government's increased intervention in the economy has not only brought about more disputes with some business elites, but also weakened the support of the whole business community for the HKSAR Government.

Fourth, given the political influence of big business on the HKSAR Government, it is, with some truth, widely perceived by the public as in "collusion" with the business community. Such a government naturally does not have the will or the means to solve the serious and salient economic and livelihood problems afflicting Hong Kong, which thus leads to people's dissatisfaction with and resentment for the government. Since the return of Hong Kong, the severest problems that have attracted the most attention include: a narrow industrial structure with declining competitiveness, weak growth of the high value-added economic activities, deterioration of the business environment due to frequent political conflicts, stark income disparity, rising class conflicts, poverty, regression of the existential conditions and development opportunities for the middle classes, inadequate upward mobility for the youth, as well as deterioration of relationships between and trust in social institutions. Under "business-government collusion", the HKSAR Government dared not require the rich and other invested interests to assume additional social responsibilities and make increased sacrifices to reduce social conflicts.

Fifth, the leaders and officials of the HKSAR have limited political wisdom and ability and at the same time lack political solidarity. A "ruling party" or competitive "party politics" cannot appear because of opposition by the central authority. Under these conditions, the chief executive is politically isolated and helpless.

Sixth, after the return of Hong Kong to China, the HKSAR Governments have committed "favouritism in personnel appointment" and "drawing a clear line between friends and foes", to suppress and exclude the opposition forces. These actions have resulted in insoluble political conflicts and social cleavages, and the situation is becoming increasingly intractable. Gradually, the chief executive primarily takes orders from the central authority without regard for Hong Kong people's needs, demands and feelings, places the interests of the central authority and the mainland above those of the Hong Kong people, and damages Hong Kong's core values, institutional rationality and procedural justice, and consequently, not only intensifies the conflicts between the central authority and the Hong Kong people, but arouses the Hong Kong people's anxieties concerns about the "mainlandisation" (大陸化) of Hong Kong or "reddening" (赤化) of Hong Kong, and also inciting Hong Kong people's antagonism towards the chief executive, and lead to severer political polarisation in Hong Kong.

According to the opposition, after the return of Hong Kong to China, the difficulties of governance in Hong Kong lies at the root of all political, social and economic problems. To protect its powers and interests, as well as to strengthen the HKSAR Government's ability in governance, the central authority has increasingly interfered in Hong Kong's internal affairs. But the interference in Hong Kong's affairs not only worsens the governance in Hong Kong, undermines the principle of "the people of Hong Kong governing Hong Kong" and a "high degree of autonomy", but also damages the relations between the central authority and Hong Kong people, and further weakens the confidence of the Hong Kong people and the international community in the "one country, two systems" policy.

In the view of the opposition, the central authority has undermined "one country, two systems", "the people of Hong Kong governing Hong Kong" and "a high degree of autonomy" as proven by their actions. These actions include the NPC Standing Committee's interpretations of the Basic Law, interference in Hong Kong's governance by the Liaison Office of the Central People's Government in the HKSAR, the central authority's "calling the shots" (指手劃腳) in Hong Kong's affairs, the central authority interfering with elections of chief executive, Legislative Council and District Council, the central authority's support for the unpopular chief executives, siding with "the people who love China and Hong Kong", preference for Hong Kong's capitalists, oppressing the opposition, subduing the anti-CPC and anti-government media, thwarting the democratic development in Hong Kong, making the "evil law" of Article 23 of the Basic Law, and promoting national education in Hong Kong.

The opposition believes that the demonstration of tens of thousands of people triggered by the local legislation of the Basic Law Article 23 in 2003 should have allowed the Hong Kong people to gain more power in political participation, or even provided the opposition with the opportunity to acquire the power to govern Hong Kong. But, in fact, it instead is leading to the central authority's forceful interference in the Hong Kong's affairs, which is particularly reflected in the favourable economic policies to "support Hong Kong", which actually are efforts to "support Tung Chee-hwa", as well as the active measures to support the "people who love China and Hong Kong". The opposition claims that the central authority's support for the HKSAR Government does little to improve its political prestige or authority, but instead gives rise to Hong Kong people's serious misgivings and concerns about the central authority's "interference" in Hong Kong's internal affairs. They are fearful that the central authority will tighten its policy toward Hong Kong, strengthen the control over Hong Kong and lead Hong Kong toward "mainlandisation" or "socialism".

When Leung Chun-ying comes into power, because many people in Hong Kong believe that Leung has a strong "pro-Beijing" background (some even believe that he is a communist) and that he always take orders from the central authority, the political misgivings and fear in society rose abruptly, leading to the intensification of antagonism between the Hong Kong people on one side, and the HKSAR Government and the central authority on the other. In the view of the opposition, the publication of the *White Paper* signifies that the central authority is resolved to control Hong Kong and demand obedience to the executive branch from the Legislative Council and the judiciary. The central authority's firm and tough stance in the dispute over the arrangements for the popular election of the chief executive has destroyed the opposition's "dream of universal suffrage", and is considered by the opposition forces as the demonstration of central authority's "treachery", violating their "promise" of democracy for the Hong Kong people.

For the opposition, in order to break the current dilemma, stop the central authority from further "interfering" in Hong Kong's affairs and protect Hong Kong's "core values" and institutions, they have to, with a sense of crisis and political urgency, promptly mobilise people to "declare war" on the central authority. The opposition and the central authority share the "common point" that it is time for the "decisive battle" between them. As promised in 2007 that the chief executive of Hong Kong could be elected by universal suffrage in 2017, all opposition forces agreed to join forces to fight for the common goal of "real universal suffrage". They threatened to compel the central authority to "surrender" by radical or even violent actions including the "Occupy Central" movement. The calculation of the opposition is this: in order to defend Hong Kong, the Hong Kong people must place Hong Kong's power and fate in their own hands, but particularly in the hands of the opposition. Under this line of thinking, striving for "real universal suffrage" means striving for the power of control over the HKSAR. Only when the power of governance over the HKSAR is taken away from the central authority will Hong Kong have hopes.

Because of the central authority's firm and tough position and the disapproval of the Hong Kong people, the illegal "Occupy Central" movement, which lasted for 79 days, ended in failure. Nevertheless, though most Hong Kong people hoped that the universal suffrage for the chief executive made possible by the NPC Standing Committee's "8.31 Decision" could be passed, the opposition vetoed the electoral reform bill in the Legislative Council, which showed that many of them were determined to continue to confront the central authority. The opposition will continue to struggle over political issues, including restarting the "five-steps" political reform process, revising of the Basic Law, "constitution-making by the people" [making a new Basic Law by the Hong Kong people alone], and criticising the existing electoral and political systems. They will continue to wage struggles inside the Legislative Council and in the society, raising all kinds of radical demands regarding people's livelihood, condemning social inequality, "politicising" economic and livelihood issues, adopting some of the "nativist" ideas and pitting the Hong Kong people against the mainland compatriots, obstructing the economic integration between Hong Kong and the mainland, objecting to Leung Chun-ying's "reappointment" as the chief executive of the HKSAR, and defaming the "people who love China and Hong Kong", all for the sake of challenging the central authority and the chief executive. They will also take their own actions during the election of the next chief executive, aiming to weaken the "legitimacy" of the election. The more Hong Kong people are dissatisfied with the central authority and the chief executive, the more they can win the favour of the masses and dilute people's discontent over the opposition's rejection of the arrangements for the popular election of the chief executive in the legislature, and the more benefits they hope they can gain in the Legislative Council and District Council elections.

The political struggles of the opposition have the following features: radicalisation, diversification, persistence, and fragmentation, with mounting violence. Although there is little chance for the occurrence of large-scale and sustained mass movements, the proliferation of protest actions, which are mainly for unleashing fury and anger, will become the "new normal" in Hong Kong politics. These protests have the following characteristics: large quantity, diverse forms, small-scale, a high degree of spontaneity, poor or loose organisation, inconspicuous or no leadership, online mobilisation, transiency and wildcat-style. The lethality of the individual protests is limited, but as the protests grow in frequency and intensity, they will accumulatively do great harm to Hong Kong's stability and governance. Unless some major political events take place and ignite great outbursts of public outrage, it will be difficult for the opposition to launch or replicate social movements like the "Occupy Central" movement in 2014.

Meanwhile, the opposition has to face the threats of division and reorganisation, the rise of radical forces, the emergence of extreme political demands such as "self-determination" and "the independence of Hong Kong", the loss of public support, the estrangement and hostility of the youth, and the decline of their combat capacity. Their rejection of the moderate electoral reform endorsed by the central authority actually does no good to the development of the opposition and Hong Kong's democratic movements in the future. By continuing the strategy of confrontation, the opposition has lost a rare opportunity to improve their relationship with the central authority, and as a result, the radical forces among them will face greater pressure from the central authority. Also, the opposition has lost the opportunity to influence the outcome of the election of the next chief executive and of playing a more active role in the HKSAR governance. The Hong Kong people are dissatisfied with the opposition for their rejection of the electoral reform, and they feel politically powerless and alienated. As a result, the power and appeal of the opposition have been weakened. The further rise of various radical "nativist" groups and other extremists may lead to more intense and violent acts of civil disobedience, further alienating political moderates. On the other hand, the tough and uncompromising attitude of the central authority on the most important political issues, and the failure of the opposition forces' radical actions to achieve their goals will inevitably compel some opposition activists to reflect and pull back, particularly in regard to their relations with the central authority as well as their political tactics. The opposition activists who are not willing to continue intense confrontation will choose to stand aside, or try to find out the so called "middle way" or "third way". For the radical oppositionists, particularly the youth, their battlefields of political mobilisation will be moved to "the civil society", schools and civil groups in particular, and they will turn their attention to the future of Hong Kong after 2047.

The major divergence among different parties, however, is still their attitude toward the CPC: dialogue and cooperation, or confrontation. In this sense, there is hardly a "middle way" or "third way" at all. As I can see, those who neither choose the more radical path of the mainstream opposition, nor the path of "the people who love China and Hong Kong" will eventually become the "oppositionists" or the "reformists" within the patriotic camp of "the people who love China and Hong Kong"; in other words, become the reformist arm of the pro-establishment camp, or transform themselves into the "loyal opposition".[20] The "loyal opposition" recognises and accepts the constitu-

---

20   See Lau Siu-kai (2015). "Can Hong Kong's Opposition Transform Themselves into 'the

tional and political systems of China and Hong Kong and is willing to work within these systems to gain power or seek reforms. The so called "middle way" or "third way" is unviable in a context of stark political polarisation, and most likely is only a transitional political path, the end point being the establishment camp.[21] In any case, the division and reorganisation of the "mainstream" opposition camp and the entry of more extreme and radical forces into the political fray will inevitably lead to political frictions among various opposition forces and the estrangement of the Hong Kong people. The overall strength of the opposition will decline. The differential treatment of various opposition forces and figures by the central authority, with the purpose of enticing the moderates among them to change their stance toward the CPC, will accelerate the divisions within the opposition. In the meantime, what binds them loosely together is still their common animosity toward chief executive Leung Chun-ying.

It is of great concern that the "marginal" or "non-mainstream" opposition forces tend to advocate various forms of "nativism" and "separatism", and attempt to use these new political views to reconstruct the "identity" of the Hong Kong people, strengthen their capacity of political mobilisation, improve the theoretical level of their "doctrines", reinforce their ideological appeal, and win recognition from the West and the forces in Taiwan that advocate Taiwan independence. Some "non-mainstream" opposition forces even advocate "defending Hong Kong" in a radical and violent manner. They are pursuing different political goals, including striving for "maximal" autonomy under the "one country, two systems" framework, strengthening Hong Kong people's "awareness of Hong Kong as a separate entity", improving Hong Kong people's "sense of nativism", minimising cooperation and exchanges between Hong Kong and the mainland, opposing economic integration between Hong Kong and the mainland, resisting the popular-

---

Loyal Opposition?" ("香港反對派能否轉型為'忠誠反對派'？".Bauhinia Magazine (《紫荊雜誌》. Vol. 300, PP44-47.

21   See Lau Siu-kai (2016). "Why is the 'Middle Road' Difficult to Tread in Hong Kong" ("香港政壇'中間路線'難走"). Hong Kong & Macao Affairs (《紫荊論壇》), Vol. 27, PP16-23.

isation of Mandarin Chinese (*putonghua*) and the promotion of national education in any form, and even calling for "self-determination" by Hong Kong people or even the "independence of Hong Kong".

With antagonism towards the People's Republic of China and the Chinese nation, the "non-mainstream" opposition forces, with the angry and radical youth at their core, accept neither the CPC's ruling status in China, nor the central authority's jurisdiction over Hong Kong under the "one country, two systems" policy. Some of them even question China's sovereignty over Hong Kong. They are proud of Hong Kong's system, institutions and values, engage in halcyon revisionism of the colonial rule in the past, and yearn for Western institutional systems. They are afraid that the central authority's "Sinocentrism" (天朝主義) ["overweening dominance by the imperial court] will recklessly destroy Hong Kong's core institutions and values, therefore Hong Kong people must rise up to defend Hong Kong against the challenges.[22] These "non-mainstream" opposition forces are not willing to entrust Hong Kong's future to Chinese (even if it's western-style) democracy, and in fact they do not believe that China will veer toward "peaceful evolution". Very few of them even predict that the CPC will eventually collapse, or as the domestic conflicts worsen, provide Hong Kong with the chance to develop democratically on its own without facing the opposition of the central authority.

The overall "non-mainstream" opposition has a weak national awareness but a strong "sense of nativism", "awareness of Hong Kong as a separate entity" and "sense of separatism". They deny the "mainstream opposition's" advocacy of "returning to China in exchange for democracy in Hong Kong" ("民主回歸"), deny that Hong Kong has the responsibility or the ability to promote China's democratic development via Hong Kong's democratic demonstration, and disagree that Hong Kong should interfere in China's affairs. In their view, the Hong Kong people should not bother themselves with the development of

---

22　Chan Koon-chung (2012). China's Sinocentrism and Hong Kong (《中國天朝主義與香港》). Hong Kong: Oxford University Press.

the country, but instead should focus on its own interests and security, resist "interference" from the central authority, and lean towards "independence". Dissatisfied with the current situation in Hong Kong, some of the young people flock to the ideas of "nativism" and "separatism".

My opinion is that, as the opposition as a whole is not willing to recognise the central authority's "one country, two systems" policy and the Basic Law, and insists on fighting against the central authority, their proposals for breaking the current impasse not only are unrealistic, but also will bring disaster to Hong Kong. These proposals are merely ways to vent people's anger and anxieties, and those people who promote them have neither concrete plans nor confidence in their ultimate success. These suggestions are merely the "products of thoughts" of some intellectuals in Hong Kong, mostly idealistic wishful thinking and illusions that can hardly be understood by Hong Kong people or be said to resonate with them. Although these suggestions are favoured by some media, largely because of their newsworthiness or sensationalism, they cannot form a powerful political current, but only to give rise to backlashes from the central authority, antagonise mainland compatriots, and even cause resentment and revulsion among Hong Kong people.

Since the Second World War, the international community basically holds a conservative and vigilant attitude toward the boundary changes of countries and the emergence of new countries, mainly for fear of arousing severe conflicts or even wars among countries. Unless extreme humanitarian catastrophes or brutal regimes occur in some regions, the Western countries usually do not "interfere" in the internal affairs of other countries, overthrow the regimes of other countries, or divide or split other countries. Even if an extreme situation does occur, whether the West will take actions or not depends very much on the price that the West has to pay. Generally speaking, the Western countries often adopt an ambiguous and indecisive stance toward independence movements that occur in different regions around the world, particularly if those actions would bring about conflicts or even wars among the major powers and hence pose threats to regional or world peace. Accordingly, even if Hong Kong's "nativism" and "separatism" win sympathy from the West, they won't likely gain any strong or substantial support, particularly given that the West would not be willing to become the enemy of a rising and powerful China.

No favourable conditions exist for Hong Kong to gain independence, to separate from the country, or to fight against the country, from both subjective and objective perspectives. Given Chinese people's deeply rooted cultural tradition of "grand unification" ("大一統"), along with most Hong Kong people's identification with the Chinese nation, plus the rise of China over the past two decades and the general sense of mounting national pride, nationalism and patriotism, it is a foregone conclusion that the central authority and mainland compatriots will not stand by or remain indifferent to "separatism" in Hong Kong or actions that challenge the central authority's authority while excluding mainland compatriots. Moreover, any remark or action that attempts to enlist the aid of foreign forces with an aim of establishing an independent country or an independent political entity will incur severe counteraction on the part of the central authority and censure of the Chinese people, thus putting Hong Kong at extreme risk. If the forces advocating "separatism" do become great forces and "separatism" were put into practice, the implementation of the "one country, two systems" policy would become inordinately difficult because it would make Hong Kong a severe political threat to the central authority and the country. Hong Kong people would also be disdained by mainland compatriots. Under such grave circumstances, it is a certain outcome that governance of Hong Kong would be completely taken up by the central authority, under the policy of "one country, one system".

The Hong Kong People

To be fair, the "one country, two systems" policy comes so easily that many Hong Kong people simply take it for granted. They gain the favourable treatment envied by the mainland compatriots from the central authority without strenuous efforts or even significant sacrifices.

As a result, many Hong Kong people don't know how to appreciate and cherish it, and it brings about some consequences that are detrimental to the long-term and successful implementation of the "one country, two systems" policy.

First of all, the Hong Kong people would easily develop an "unrealistic" sense of self-importance and arrogance, thinking that Hong Kong plays an extremely crucial and indispensable role in the China's development, elevating China's international reputation and facilitating cross-straits reunification, thus taking the preferential treatment from the central authority for granted. They see nothing wrong in the gap between Hong Kong people and the mainland compatriots in terms of rights, obligations and treatment, thinking that they are entitled to them or deserve them. No matter what difficulty Hong Kong is faced with, many Hong Kong people think that it is the central authority's responsibility to come to Hong Kong's help or rescue, because it is about the country's interests and the nation's honour. On that ground, the Hong Kong people show little gratitude to the central authority and the mainland compatriots, not to mention serving the motherland voluntarily and faithfully.

Second, there is a tendency for the Hong Kong people to develop unrealistic expectation of the central authority. They are prone to think that in order to maintain the prosperity and stability of Hong Kong, the central authority will make all efforts to "please" or "charm" the Hong Kong people and avoid conflicts with them in Hong Kong, even at the cost of violating the basic principles of the "one country, two systems" policy and the Basic Law.

Third, many Hong Kong people tend to believe that the "one country, two systems" policy is custom-made for them, and ignore its even greater importance in the country's interests and needs as a principal national policy. As a result, the Hong Kong people often neglect their responsibilities and duties in the implementation of the "one country, two systems" policy, particularly their responsibilities for safeguarding national sovereignty and security. They do not, and are not willing to understand such a simple fact: if Hong Kong cannot safeguard the interests of the country and the central authority under the "one country, two systems", this policy will no longer be a wise national policy that is beneficial to our country and our nation. Consequently, it will be hard for the central authority to continue this national policy. The narrow-mindedness of the Hong Kong people is clearly exposed in two political events—the failure of the local legislation of Article 23 of the Basic Law and the rejection of the electoral arrangements for the popular election of the chief executive prescribed by the "8.31" decision of the Standing Committee of the NPC.

Fourth, the Hong Kong people assume that Hong Kong will "always" play an indispensable role in national development and cross-straits reunification. They do not have a sense of urgency, and are not psychologically prepared to accept the decline of Hong Kong's status as well as its role in China's development. Most Hong Kong people do not bother to take active measures to maintain, or to improve the importance of Hong Kong to the country. Instead, they feel resentful of the rise of the country, the improvement of the mainland compatriots' standard of living, and the increase in the major mainland cities' economic competitiveness. They have even comported themselves with hostility toward the mainland compatriots, considering them rivals. After the return of Hong Kong to China, due to the frequent political conflicts, the difficulties in industrial restructuring, and the decline of Hong Kong's competitiveness in the international economy, the self-confidence and pride of the Hong Kong people declined as well. Also, their voiced resentment and discrimination against the mainland compatriots, based on a sense of false superiority, is not only becoming increasingly hollow, damages the relations toward compatriots in Hong Kong and the mainland. Hong Kong people's resentment of the mainland compatriots has subsequently evolved into their doubts about the central authority, and they further question the central authority's intention with regard to the implementation of the "one country, two systems" policy.

The Hong Kong people's conservative mentality and narrow-mindedness, their decreasing self-confidence and self-esteem, their anxieties about Hong Kong's future and their rejection of the central authority and the mainland compatriots stand as serious obstacles to any efforts to try to seize the opportunities provided by national development. Any policy or measure that purports to promote the economic "integration" between Hong Kong and the mainland will be castigated as the "betrayal of Hong Kong". Such attitudes and behaviours have already greatly hurt the feelings of the mainland compatriots and the relations between Hong Kong and the mainland.

The influx of capital and expertise into Hong Kong from the mainland pose great challenges to the ordinary Hong Kong people who do not have a competitive edge, particularly the youth. Unlike the global cities such as New York, London and Tokyo, those who are not competitive are not willing to leave Hong Kong for other places. It is true that these people have a limited ability to survive and prosper in other places. A small group of people who are stuck between the elites and the lower-classes are living a difficult life. It's not easy to retrain and re-equip those people, and in any case, it is impossible to achieve these goals in the short term. As for those who are getting old, it's more unlikely for them to learn new knowledge and skills. This portion of the population are encountering difficulties in Hong Kong and face the prospect of an inauspicious future for themselves and their next generation. Public dissatisfaction with the current situation and pessimistic feelings continue to rise, which forms the basis for all kinds of "nativism" and protectionism. The "Hong Kong-centred" mentality, the lingering sense of arrogance and the resentment towards the central authority and the mainland compatriots of the Hong Kong people have, to some degree, provoked adverse reactions on the part of the central authority and the mainland compatriots.

Strictly speaking, under the "one country, two systems" policy, the central authority does not require the Hong Kong people to "love the country" and to love it inordinately, but only urge them not to do harm to the nation, to the country or to Hong Kong. Deng Xiaoping once said: "What's a patriot? The standard for a patriot is to respect our nation, to sincerely uphold the country's resumption of sovereignty over Hong Kong, and abstain from damaging the prosperity and stability of Hong

Kong. Anyone with these qualities can be called a patriot, regardless of their belief in capitalism, feudalism, or even slavery. We do not require them to approve of China's socialist system, as long as they love the country and Hong Kong."[23]

He added: "After 1997, some people in Hong Kong will scold the CPC and China, and we will allow them to do so. But if they turn it into action, and want to turn Hong Kong into a base against the mainland under the guise of 'democracy', we will have to intervene."[24] Under the "one country, two systems" policy, the Hong Kong people will enjoy much more individual rights than responsibilities and obligations to the country. The country does not require the Hong Kong people to "love the country" and to "love the CPC" as the mainland compatriots are doing. Nevertheless, from the perspectives of "human feelings" and "political ethics", the central authority and the mainland compatriots do hope that Hong Kong people will be grateful to the country for the favourable treatment by the central authority, thus improving their relations with the central authority and the mainland compatriots, and proactively safeguarding the interests of the central authority and the mainland compatriots, rather than exclusively minding their own positions and interests. In this regard, the central authority and the mainland compatriots must be rather disappointed with the Hong Kong people.

The events that occurred in Hong Kong after 1997, including the failure of the local legislation of the Basic Law Article 23, the HKSAR Government's sudden restriction on the purchase of baby milk powder by outsiders, restriction on outsiders' (mainly the mainland compatriots) purchase of properties in Hong Kong, the "Occupy Central" movement, the rise of "nativism", the call for "Hong Kong independence" and the demonstrations against the mainland compatriots have greatly damaged Hong Kong's relations with the central authority and the mainland compatriots. At the beginning, immediately after Hong Kong's return to China, the national leaders and central government

---

23 Deng Xiaoping (1993). *How Deng Xiaoping Views Hong Kong Issue* (《鄧小平論香港問題》). Hong Kong: Joint Publishing. P8.

24 Ibid, Hong Kong: Joint Publishing. P8.

officials did not harbour any hope that the Hong Kong people will "love the country". However, as the resistance of the Hong Kong people to the central authority and disputes with mainland compatriots continue to increase, and actions that challenge the central authority's authority and powers are on the rise, the central authority has gradually expressed their expectation that Hong Kong people, particularly the young people "love the country". With the sharp increase of the national consciousness and patriotism on the mainland, mainland compatriots are disposed to become angry and dissatisfied with the Hong Kong people for their "nativism" and provocation of the central authority and the mainland compatriots.

There is no doubt that the Hong Kong people are dissatisfied with the current situation in Hong Kong, but as for how to change the situation, "different people give different views". They neither agree with the central authority's position, nor approve of the solutions raised by the opposition. As a result, they can only live a life with anxiety and restlessness.

"One Country, Two Systems" and the Mainland Compatriots

After the return of Hong Kong to China, due to the disorderly politics, ineffective governance, sluggish economic growth and declining competitiveness of Hong Kong, some mainland compatriots (including officials at all levels and ordinary people) have changed their impression of Hong Kong, and begun to question the effectiveness and value of the "one country, two systems". After all, in the view of the mainland compatriots, the reason why Hong Kong people can enjoy preferential and favourable treatment is that, under the "one country, two systems" policy, Hong Kong can play its unique role in China's "reform and opening up" strategy, becoming a driving force for the economic modernisation of the country and the improvement of people's living standard. Meanwhile, politically Hong Kong should not become a threat to the country and the central authority. To be more specific, Hong Kong should not be turned into a place that threatens national security. The mainland compatriots consider the "one country, two systems" as a "contract" between the country and the Hong Kong people. Under this contract, Hong Kong should assume economic and political responsibilities for the country, in exchange for the preferential treatment from the central authority.

In the past nineteen years since the return of Hong Kong to China, China has been rising and developing. In the mainland, with the long-term and rapid growth of the economy, the people's living standards have largely improved, the coastal big cities have become prosperous and wealthy, and China's contact with other countries become more frequent and closer. As a result, the mainland compatriots' national pride keeps soaring, and they are optimistic about their own future and the future of the country. A feeling known as "self-confidence in the institutions" (制度自信) and "self-confidence in the chosen path" (路線自信) and "self-confidence in the theory" (理論自信) continues to rise. In contrast, Hong Kong has been stricken twice by international financial crises and twice by the major epidemics (avian influenza and SARS) since its return to the motherland. Furthermore, because Hong Kong's narrowly based industrial structure has encountered difficulties in its adjustment, and the HKSAR Government is facing great challenges in governance, along with endless political conflicts and social frictions, a large proportion of Hong Kong people are confused and worried about the future of Hong Kong.

Although Hong Kong has gone through all kinds of hardship and achieved an economic growth that is faster than the Western countries through their own efforts and because of various favourable economic policies extended by the central authority, as a mature economy closely related to the West and affected by the sluggish economy in the West, Hong Kong's growth largely pales in comparison with that of the mainland. The favourable policies of the central authority have indeed brought great benefits to the Hong Kong's economy on the whole, but the fact is that not everyone has access to the substantial benefits brought by those policies. Some business corporations, financial institutions, professionals and particular industries (real estate, tourism, retail trade, and hotels) have derived great benefits, but a majority of people, in particular the youth in Hong Kong, have gained little or nothing. Some even think that their interests, daily life and living environment have suffered. Furthermore, they are discontented about the mainland compatriots' coming to Hong Kong and enjoying the cheap public services there.

Today, there are remarkable differences in the economic growth and prospects between Hong Kong and the mainland, which, on the one hand, has reduced Hong Kong people's long-held superiority and arrogance towards the mainland compatriots, and on the other hand, has also changed the mainland compatriots' impression and attitude towards the Hong Kong people. Some mainland compatriots have begun to look down on the Hong Kong people, regarding Hong Kong people as complainers who can make no progress on their own. In addition, the Hong Kong people are not only reticent about carrying out their responsibilities for the development of the country under "one country, two systems", but they are also not willing to undertake the important mission of safeguarding national security. Moreover, they are quick to ask the central authority for help, knowing that the country will not let Hong Kong down for fear of bringing shame on the Chinese nation. Even so, the Hong Kong people have not expressed their gratitude to the country and the mainland compatriots sincerely and openly. Instead, they care more about their own interests, eulogise the colonial administration of the past, challenge the central authority's authority again and again, discriminate against and reject the mainland compatriots, and even put forward the ideas of "nativism" and "Hong Kong independence" in order to separate Hong Kong from the mainland.

What the Hong Kong people have done, in the view of some mainland compatriots, shows that the Hong Kong people are deeply under the spell of colonialism, possess a feeble national awareness and exude great arrogance. Moreover, as they are not willing to see the mainland compatriots as equals, they tend to exaggerate Hong Kong's "superiority" in order to belittle the mainland and elevate their own status, in order to console their frustrated dignity and reconstruct a proud identity for themselves. In turn, many mainland compatriots regard Hong Kong as less valuable to the country than it was in the past, and some even consider Hong Kong a burden to the country; as a result, some are beginning to question the correctness and effectiveness of the "one country, two systems" policy, leading to a feeling of dissatisfaction with the central authority's "preference" for Hong Kong. The mutually disapproving viewpoints of Hong Kong people and mainland compatriots will undoubtedly have an adverse impact on the implementation of the "one country, two systems" policy and the relations between the "two systems".

It is difficult to speculate upon the proportion of mainland compatriots who sincerely understood and accepted the national policy of "one country, two systems" as put forward by Deng Xiaoping, a policy tilted to favour Hong Kong, as mainland politics at that time were less transparent. One certainty is that, whether or not the mainland compatriots agreed with the policy of the Chinese Government, they had neither the power nor the channels to oppose the "one country, two systems" policy and they had little choice but to allow it.

The former director of the Hong Kong and Macao Affairs Office of the State Council, Lu Ping, recalled that the central authority did encounter some difficulties when they initially tried to persuade the mainland compatriots to accept the "one country, two systems" policy. He said, "At the very beginning, some people in the mainland did not understand why the central authority offered such favourable treatment to Hong Kong. To be specific, Hong Kong did not need to turn over tax revenues to the central authority. As Shanghai and Guangdong at that time bore heavy financial burdens, why did Hong Kong get such favourable treatment? Fortunately, the mainland compatriots later understood the decision made by the central authority, and generously took the interests of the whole country into consideration."[25] Under the strong

---

25 Lu Ping (2009). *Lu Ping's Verbal Recollections of the Return of Hong Kong to China* (《魯平口述香港回歸》). Hong Kong: Joint Publishing. P65.

leadership of Deng Xiaoping (and channels for the participation or influence on government of ordinary people, whatever their objections, were limited) the central authority was able to overcome all resistance in order to establish the "one country, two systems" policy.

Great changes have taken place in today's political ecology in the mainland, mainly represented by the following features. Firstly, the leadership system of the central authority has become more collective, transparent, institutionalised, rationalised and democratic. Secondly, the number of institutions and groups in the party, in the government, and in the society that have political influence on the central authority's decisions has been on the rise.[26] Thirdly, there are more and more channels for mainland compatriots to express their opinions, suggestions, dissatisfaction and opposition to governmental policy. Fourthly, the central authority has to increasingly taken into account the reactions of mainland compatriots in their policies toward Hong Kong. It becomes much harder for the central authority to make policies in favour of Hong Kong that are excessively against the will of the mainland compatriots. In fact, as domestic politics in China become much more open, the mainland compatriots' ability to influence the central authority's decisions via various channels increases. The popularisation of the Internet

---

26 Hu Angang (2014). *Research on the Collective Leadership of China* (《中國集體領導體制探究》). Hong Kong: Chung Hwa Book Company; Wang Shaoguang, Fan Peng (2013) *The Chinese Model of Consensus Decision-Making: A Case Study of Healthcare Reform* (《中國式共識型決策: "關門" 與 "磨合"》).Beijing: China Renmin University Press. David Lampton, an American who is knowledgeable about China, described China's collective leadership: "[O]ne great change over the past four decades is that the idea of global interdependence is increasingly recognised and accepted, not only by elites but by ordinary citizens as well. The biggest change is the development of a domestic social and political system characterised by a weaker, less cohesive leadership group, a more pluralised society and bureaucracy, and substantial actors in government, society, and the economy with more resources to promote their interests." (Lampton, 3) "Beijing is increasingly comfortable with the institutions of the global economic order—wanting to boost its influence within them—and uncomfortable with the U.S.-led security order, founded as it is on bilateral and multilateral alliances of which China is not a part. The United Nations-led security institutions are much more congenial to Beijing." (Ibid, P135) David Lampton (2014). *Following the Leader: Ruling China, from Deng Xiaoping to Xi Jinping.* Berkeley and LA: University of California Press. P3 and P135 respectively. Archie Brown, a British scholar who is an expert in the study of political leadership also claimed that collective leadership makes fewer mistakes, and is more effective than any other style of leadership. Archie Brown (2014). *The Myth of the Strong Leader: Political Leadership in the Modern Age.* New York: Basic Books.

has also greatly enhanced the mainland compatriots' participation and influence in politics, especially in the Chinese Government's foreign policies.[27] Fifthly, with the development of the mainland's economy, as well as the transformation and upgrading of the industrial structure, the relations between the mainland and Hong Kong have palpably changed. In the past, the relations between the mainland and Hong Kong mainly consisted of cooperation, complementation, and a win-win outcome; however, current relations increasingly reflect a sense of competition, contradiction and friction. Increasingly, local governments, enterprises and professional organisations discover conflicts and opposing interests between Hong Kong and the mainland in a wide number of fields.

For instance, Shanghai and Shenzhen are developing financial and high-end professional services, which will undoubtedly weaken Hong Kong's advantages in these fields. In addition, as Hong Kong enterprises enter the mainland market, and mainland enterprises march into Hong Kong, intense competition between them is hard to avoid. The professional groups of Hong Kong, unwilling to open Hong Kong's market to mainland professionals, want to safeguard their interests in Hong Kong via "protectionism"; however, they ironically hope that the professional services market on the mainland can be opened to them, which definitely will infuriate the professionals of the mainland. More-over, colleges and universities on the mainland become anxious, as many excellent students are attracted by the universities of Hong Kong. With all these contradictions, the pressure on the central authority is mounting. The central authority has shown partiality for Hong Kong so far when dealing with the relations and conflicts between Hong Kong and the mainland, but as the influence of the mainland compatriots on the central authority's policies towards Hong Kong increases, it is not hard to imagine that the central authority will have to formulate a more equitable balance between the mainland and Hong Kong in the future. Likely, the Hong Kong people who have taken for granted the central

---

27 See Yang Guobin (2009). *The Power of the Internet in China: Citizen Activism Online.* New York: Columbia University Press.

authority's favourable and preferential treatment will react with chagrin, complaining that the central authority is "changing" their policies towards Hong Kong, and questioning their sincerity in the implementation of the "one country, two systems" policy.

Hong Kong's Critical Moment

The current situation is quite clear. The "one country, two systems" has achieved decent results since its implementation in Hong Kong twenty years ago, but the central authority, the mainland compatriots, different political forces in Hong Kong, and even the Hong Kong people are not so satisfied with some phenomena in Hong Kong under the "one country, two systems" policy. The dissatisfaction of all parties will continue to grow if we do nothing to change them, and, as a result, the "one country, two systems" policy will run into more difficulties in its future development, and the future of the "one country, two systems" policy after 2047 will be full of uncertainties. Some people of the mainland and Hong Kong have already begun to think about Hong Kong's future after its status quo "remains unchanged for fifty years". They are particularly concerned about whether the "one country, two systems" policy can continue and how it will be continued. It is conceivable that, if the current situation in Hong Kong remains unchanged or even gets worse, and the dissatisfaction of all parties keeps growing, the so-called issue of "Hong Kong's future" will re-emerge, and new political struggles will come into being again, which will inflict immeasurable damage on Hong Kong's prosperity and development.

The core of the current problems, again, is that the central authority and Hong Kong's opposition forces hold completely different interpretations of the problems associated with the implementation of the "one country, two systems". Due to the disputes about the universal suffrage of the chief executive and the failure of the electoral reform, the central authority and the opposition are now standing against each other. In view of the lack of mutual respect and trust, it is unrealistic to expect that the two sides can jointly overcome the current difficulties in the implementation of the "one country, two systems" policy by means of negotiation or consultation in the foreseeable future. Therefore, the central authority and the opposition forces of Hong Kong are now embroiled in a "decisive battle", or a "showdown". In other words, Hong Kong is now at a critical crossroads. The "decisive battle" may be protracted, with high prices paid by all parties, and it will end only when one side has decisively defeated the other. The defeated side will eventually be forced to adjust its position and strategies and face reality, and with the pressure of Hong Kong people, begin to strive for its interests and goals in a more effective way in a new and less favourable political landscape. Taking into consideration the power disparity between Beijing and the opposition, I believe that the central authority will win the battle in the end. The final outcome, of course, will not be a complete victory for one side and a defeat for the other. The victory of the central authority means that their version of "one country, two systems" will prevail in Hong Kong hereafter. After the battle, the "winning" side will, to some extent, adjust its original position toward a "soft-line" approach, so that the "defeated" side can be placated and be willing to resurrect some form of cooperation with the "winning" side.

The central authority has definitely decided to defend the nation's security and interests, as well as their powers under "one country, two systems", on its own, and are willing to use the powers at their disposal to achieve their goals. This has been amply demonstrated in the dispute over the universal suffrage of the chief executive. The central authority had stated its position on the universal suffrage of the chief executive very early, even before the formal consultation process began in Hong Kong, and remained unswerving even under the violent attack of the opposition forces and some pressure from the West. By virtue of this struggle, the central authority has shown the Hong Kong people that it would no longer be afraid of expressing its thoughts or committing to a position, and that it would no longer remain indifferent, passive or timid. The opposition forces and the Hong Kong people have not been psychologically prepared for the sudden change of the central authority's political attitude and felt at a loss all of a sudden. At the beginning, the opposition didn't believe the central authority would change its political attitude, so they tried to intensify and scale up the struggles, and of course they ended in failure. In my opinion, since the central authority has made up its mind to bring the implementation of the "one country, two systems" policy back on track after careful and serious consideration, it will unswervingly execute the new policies towards Hong Kong that aim at "bringing order out of chaos" ("撥亂反正"). The central authority will declare their position on major political issues in the future, so as to thoroughly settle the political disputes in Hong Kong, and by doing so redesign the political game rules in Hong Kong. The publication of the *White Paper*, the issuing of the NPC Standing Committee's "8.31 Decision" in 2014 and the interpretation of article 104 of the Basic Law in 2016 by NPC Standing Committee demonstrate the central authority's resolution and courage. Only when the central authority's strategies and basic positions become clear, will it be possible to empower "the people who love China and Hong Kong" and enhance the HKSAR Government's ability and effectiveness in governance.

The opposition is anxious and angry about the decisions and the new approach of the central authority, particularly the rejection of their demands for the arrangements for the "real popular election" of the chief executive. In the view of the opposition, the change of the central authority's policies toward Hong Kong means that the central authority is ready to fully take over Hong Kong, and the chief executive Leung Chun-ying and the HKSAR Government are now "accomplices". They think that Hong Kong is now facing a life-and-death moment, so Hong Kong people must rise up to fight against the central authority, in order to safeguard "our city" . The opposition believes that if they intensify the political struggles or change the techniques of struggle, the central authority will eventually succumb under the political pressure of the Hong Kong people. In the coming years, Hong Kong will continue to be mired in a quagmire where the struggles between the opposition forces and the central authority constantly occur.

Admittedly, the central authority's new policies towards Hong Kong are not likely to be brought to a successful completion, because difficulties and obstacles are bound to occur during the implementation of the new policies. One reason is that the opposition and some Hong Kong people will resist the new policies. Besides, it is not easy to strengthen "the people who love China and Hong Kong" and enhance their ability to encounter the opposition. However, since China is a great power, and the central authority has the legal powers to shape the political system and the game rules of Hong Kong, the central authority should to a great extent be able to bring Hong Kong's politics back on track. The central authority has succeeded in bringing the extremely controversial issue of electoral reform "under their control" by promulgating the "8.31 Decision" on the universal suffrage of the chief executive. If the central authority in the future declare its "final" position on some other controversial political issues, it means all the major political issues will be brought under the control of the central authority as well. This "de-politicisation" plan spearheaded and promoted by the central authority will inevitably instigate severe political conflicts in Hong Kong in the short term. As the opposition sees their political interests, political future and ability to mobilise be weakened by the central authority, it will put up a desperate struggle against the central authority, but unfortunately, they will not be able to make a change. The fierce political struggles will last for some time even after the major politic issues are "brought under the control of the central authority". Be that as it may, Hong Kong from then on will suffer less from the fierce political disputes. Realising the futility of confrontation, some opposition activists will reluctantly accept the reality and gradually seek a proper relationship with the central authority on the basis of recognising and respecting central authority's power. As most Hong Kong people have no intention to fight against the central authority and regard prosperity and stability as the ultimate goals, they will gradually boycott and resist the struggles launched by the opposition. The departure of Leung Chun-ying and his replacement by Carrie Lam (林鄭月娥) will bring about a less confrontational situation. The radical forces will likely fade away along with the opposition's ability to make a stand in politics, and eventually, it is expected that Hong Kong's society will again be guided by pragmatism and rationality. If so, the major tasks for the HKSAR Government in the future will be economy development, social harmony and improvement of people's livelihood. The social basis of the governing coalition will be broadened, incorporating the moderates (as well as some opposition figures) who are in favour of social, political and economic reforms. A new governance strategy that balances the interests of people from different classes and from all walks of life will be implemented. In addition, the governing coalition will gather enough political strength to promote all sorts of reforms, so that the social, economic and livelihood issues that are the concerns of some "nativists" too can be solved in a rational manner and in a peaceful political environment. The most ideal situation is that the opposition activists and "the people who love China and Hong Kong" lessen their disputes and enhance mutual trust by cooperation on the practical issues, so as to remove the obstacles for effective HKSAR governance, and bring new space and hopes for Hong Kong's democratic development in the future. All these eventually will fundamentally change the gloomy political atmosphere in Hong Kong since its return to the motherland. In that case, the central authority can gradually relieve its misgivings and fears about Hong Kong's politics, and gradually reduce its interference in Hong Kong's affairs. As a result, the relations between the central authority and Hong Kong people will improve permanently as well.

CONCLUSION

The "one country, two systems" policy is unique in human history as a mechanism for duly handling the conflicts within a country and promoting national reunification. It allows the Hong Kong people who mistrust the central authority to continue living and developing in a different political, social and economic system following the return of Hong Kong to China. Meanwhile, it expects Hong Kong to contribute economically to the country while refraining from posing a political threat. If Hong Kong is economically and politically cut off from or even completely isolated from the mainland after its return to China, in other words, if the "two systems" are indeed completely separate, it might seem to follow that practising "one country, two systems" would be accomplished more easily. However, the reality is less simple. In this account, I have presented some of my observations about the experiences and lessons learned from the actual implementation of the "one country, two systems" policy following the return of Hong Kong to China.

Since the return of Hong Kong, the practice of the "one country, two systems" policy has encountered a multitude of difficulties and challenges. Although the strategic and economic value of Hong Kong to the rising China is lower than it was in the past, and although some Hong Kong people and mainland compatriots retain doubts about the practical significance and utility of the "one country, two systems" policy to Hong Kong and the country, so far, few have flatly repudiated the policy. On the other hand, many people have put forward various suggestions, stemming from their different interests and positions, to "improve" the "one country, two systems" policy—without anyone in fact setting out some better arrangement to resolve the problems related to "Hong Kong's future".

The biggest difficulty in the implementation of the "one country, two systems" policy in Hong Kong derives from the fact that "one country, two systems" is a sort of "unity of contradictions" ("矛盾統一體"), and it's hard to integrate "contradictions" with "unity". With the unrelenting changes in the international and domestic political landscapes, it is even harder to resolve incongruences between the "contradictions" and "unity". The biggest contradiction hidden behind the "one country, two systems" policy is the lack of mutual trust between the central authority and the Hong Kong people following the return of Hong Kong to China. The central authority always worries about political threats from Hong Kong, and the Hong Kong people remain suspicious of the CPC. This long-standing contradiction has led many Hong Kong people to be convinced by the "alternative" interpretations of the "one country, two systems" policy, thus creating serious political obstacles to the implementation of the policy following Hong Kong's return.

Due to the fact that, during the transitional period, the British installed a range of political measures that violated the "one country, two systems" policy, combined with the fact that, following the handover, other obstacles have continued to impede the implementation of "one country, two systems", the proper atmosphere and conditions for launching and implementing the policy have been lacking right from the beginning. As the policy was put into practice under such inauspicious circumstances, it has not only failed to develop in accordance with the original design and blueprint of the Chinese Government but has also led to many undesirable consequences. They include the emergence of a cluster of internal conflicts, frequent conflicts between the HKSAR and the central authority, the erosion of Hong Kong's prosperity and stability, and Hong Kong becoming a potential political threat to the country. As a result, it remains rather difficult to achieve the strategic goals of the "one country, two systems" policy.

After the return of Hong Kong to China, the central authority's understanding of the "one country, two systems" policy remained quite different from that of most Hong Kong people, particularly with respect to the powers and responsibilities of the central authority under the "one country, two systems", giving rise to many difficulties in the implementation of the policy. The policy of "one country, two systems", in itself, is based on the assumption that this contradiction would endure for a long time to come, and therefore, the central authority has allowed the Hong Kong people, who are suspicious and fearful of the CPC, to continue living under the original institutions and values in Hong Kong, solemnly promising that it would not damage or interfere in Hong Kong's "one system".

The essence of the "one country, two systems" policy is simple and clear, and on the surface seems easy to execute. However, the central authority's "one country, two systems" policy contains three attached conditions. The first condition is that Hong Kong should perform an "indispensable" function in promoting China's economic development through forging close economic ties with the mainland. The second condition is that Hong Kong cannot interfere in the mainland's politics, or try to change the ruling status of the CPC. In a word, Hong Kong cannot become a base for anti-communism or subversion. The second condition has always referred to the fact that the British served as a barrier preventing anti-communist forces in Hong Kong from interfering in mainland politics. So, what is meant by maintaining Hong Kong status quo for fifty years also implies that Hong Kong should not become a "base of subversion" following its return to China. The third condition (that of the central authority's final word) is based on the principle that the "one country, two systems" policy is not custom-made for the interests of Hong Kong people alone. On the contrary, the policy is part and parcel of the most important national strategy of building socialism with Chinese characteristics. Accordingly, the policy is intimately related to the country's sovereignty, security and development. In this regard, the central authority has reserved important powers under the "one country, two systems" policy, not only in terms of national defence and foreign affairs, but also extending to Hong Kong's political system and the comprehensive and accurate implementation of the "one country, two systems" policy in Hong Kong.

Since neither the British nor Hong Kong's opposition forces agree with the central authority's interpretation of the "one country, two systems" policy, they do not recognise or respect the power and responsibilities of the central authority under the policy. The competition between China and Britain before the return of Hong Kong to China and the struggles between the central authority and Hong Kong's opposition forces after the return are inevitable. There are several reasons why the opposition forces do not agree with the central authority's "one country, two systems" policy. The first reason is because of their anti-communist bias. Secondly, those in favour of "return of Hong Kong with democratisation" aim to establish a Western-style democratic regime in Hong Kong and to push China to follow the pro-Western path of "peaceful evolution" through setting Hong Kong's democracy as an example for the mainland. Thirdly, the opposition activists who advocate reforms in Hong Kong reject the political system and public policies of the HKSAR for being biased in favour of the bourgeoisie, and they hope to construct a fairer and more inclusive society in Hong Kong. The fourth reason is that the opposition activists who harbour "nativism" or "awareness of Hong Kong as a separate entity" disapprove of close connections between Hong Kong and the mainland. They believe that the "inferior" culture and system of the mainland will eventually destroy and replace Hong Kong's institutions and values.

Some Hong Kong people believe that, as a part of the Chinese people, they have the right to participate in the central authority's affairs, and "influence" the country's development in line with their political views. Such a self-righteous attitude has given rise to resentment among mainland compatriots, because their "instructions" and "demands" concerning national development, which remain only slogans and stem from their anti-communist sentiments and lionisation of Western values, can hardly convince their mainland compatriots.

Following the return of Hong Kong to China, there remains a method that could be adopted by the opposition in Hong Kong in the pursuit of their inclinations. One is to sincerely discuss problems with the central authority within the framework of the Basic Law in order to make progress in Hong Kong in all areas through the NPC Standing Committee's interpretations of the Basic Law or even the revision of the Basic Law. This method is based on the premise that the opposition accepts the central authority's "one country, two systems" policy, recognises its powers and responsibilities, understands its misgivings and anxieties, and respects national security and interests. Given that the central authority is aware of the dissatisfaction of many Hong Kong people with the current situation, it is very likely that it would follow some suggestions offered by the opposition. If such a "mutual benefit and win-win" result does come about, it will certainly deliver benefits to Hong Kong and provide vital force to the "one country, two systems" policy. However, the opposition does not trust the CPC. Instead, it is driven by a sense of self-importance and is prone to overestimate the support of the Hong Kong people. The opposition forces choose to deal with the central authority in a rebellious way by mobilising the masses in struggle against the central authority, with the illusory hope of forcing the central authority to submit to them. As a result, Hong Kong is trapped in quagmire of endless political struggles that are pernicious to Hong Kong's long-term development and that pose insuperable obstacles to the implementation of the "one country, two systems" policy in Hong Kong. Looking ahead, as the Hong Kong people gradually tire of political struggles and realise that Hong Kong and the country are closely related and mutually dependent in their fate and interests, the resources for the opposition to launch further large-scale struggles will diminish.

Under the principles of "Hong Kong people governing Hong Kong" and "a high degree of autonomy", the Chinese Government is faced with a serious dilemma. The central authority is worried about the deviations from the "one country, two systems" policy as originally understood by the makers of that policy but is at a loss as to how to rectify the situation. If the central authority chooses not to interfere, the practice of "one country, two systems" policy will go astray, to the detriment of the interests of both Hong Kong and the country as a whole. If the central authority interferes in Hong Kong's affairs, it will cause people to claim that it is undermining Hong Kong's high degree of autonomy. The central authority at the beginning has adopted the strategy of "non-interference", hoping that the situation in Hong Kong would soon be back on track. Nevertheless, Hong Kong actually shows a worsening trend. So the central authority feels an imperative "to impose order on chaos" with regard to the implementation of the "one country, two systems" policy. From an objective perspective, the Chinese Government cannot avoid positive "interference" in Hong Kong's affairs in order to create the favourable and indispensable conditions for the successful implementation of the "one country, two systems" policy. Its only option is to leave the scene after these conditions are in place, so as to fully realise the principles of "Hong Kong people governing Hong Kong" and "a high degree of autonomy". In other words, the short-term "interference" as "a necessary evil" is to serve the long-term goal of "non-interference". The arduous task of bringing order is still in its early stages, but so far it has triggered many political conflicts in Hong Kong despite also making some achievements.

When the Basic Law was first promulgated in Hong Kong in 1990, not all Hong Kong people accepted the goal of maintaining the political, economic and social "status quo" of Hong Kong through the policy of "one country, two systems", so there have always been voices calling for reform. After the return of Hong Kong to China, a range of deeply rooted problems worsened, with disputes becoming increasingly frequent, vociferous and acrimonious with regard to such issues as the future development of Hong Kong, the functions and role of the government in social and economic development, the relations between Hong Kong and the mainland, the political system of Hong Kong, the nature of public policies as they relate to different fields, and lately, the future of Hong Kong after 2047. If political mutual trust is established among all parties, then finding a way of pursuing reforms under the Basic Law, or even further expanding the scale of reforms by revising the Basic Law, will become possible. After all, both the central authority and Hong Kong people are aware that the world, the nation, as well as Hong Kong are undergoing dramatic changes. It is unrealistic to maintain Hong Kong's "status quo" as it was in the mid-1980s, and therefore, the "one country, two systems" policy should be implemented wisely and flexibly with a strategic emphasis on development and dynamism. Undoubtedly, we should maintain the basic principles of the "one country, two systems" policy, including "Hong Kong people governing Hong Kong", "a high degree of autonomy" and "Hong Kong not becoming "a base for subversion". Otherwise, the "one country, two systems" policy would go astray, harming the interests of the nation as a whole and Hong Kong in particular.

Actually, changes can be made in the division of the powers between the central authority and the HKSAR Government, and in the realms of politics, society and economy in Hong Kong. For example, the central authority can grant more power to Hong Kong when appropriate; the role of the HKSAR Government in the economic and social development can be broadened; the fiscal policy of the HKSAR Government can be adjusted to reduce inequality in Hong Kong's society; the people's living standards can be improved; Hong Kong can choose a more balanced and more sustainable road of development; a more inclusive political system and style of governance can be created; and the policies of the HKSAR Government can lay more emphasis on reducing disputes and conflicts among different generations so as to mitigate the grievances of young people. In actuality, if the central authority and Hong Kong people lack mutual trust and respect, and political divisions and conflicts continue to fester in Hong Kong, implementing the "one country, two systems" policy from perspectives of development and dynamism will become unfeasible. Not only will Hong Kong's "status quo" remain unchanged, but it will become the catalyst for more intense conflicts.

After nineteen years of experience, more and more national leaders, central government officials, mainland compatriots and Hong Kong people have become aware that they should approach the "one country, two systems" policy from the perspective of development and dynamism. Instead of understanding the principle of "Hong Kong remaining unchanged for fifty years" from a narrow and static perspective, they should observe the changes in Hong Kong in a conscientious, rational, patient and unruffled manner, taking proper measures to deal with these changes, proactively responding to the changing demands of the Hong Kong people, particularly the young people. The *White Paper* rightly points out: "The endeavour to further the practice of the 'one country, two systems' requires both a comprehensive and accurate understanding and implementation of the policy to ensure that the practice moves forward on the right track and a proactive and effective response to the difficulties and challenges confronting Hong Kong in its development."[1]

After all, it is unrealistic to require Hong Kong people of all generations to wholeheartedly accept the principle of Hong Kong "remaining unchanged for fifty years" with all the embedded injustice and contradictions. As for how to properly modify the "one country, two systems" policy so that it will continue to gain recognition from all parties in Hong Kong and do no harm to the security and interests of the central authority and the country, all parties of Hong Kong and the mainland will be required to adopt a humble, open and tolerant mindset. To be sure, if proper adjustments can be made to the "one country, two systems" policy in line with the changing circumstances, all parties can yet maintain a high level of respect and confidence in the practice of the "one country, two systems" policy.

With the development of the "one country, two systems" policy, more and more categories of people and interests will get involved and will hold varying opinions concerning the policy. Some might become more critical. At the beginning of the 1980s, not many people had the opportunity to participate in formulating the "one country, two systems" policy, so the interests and opinions taken into consideration were not broad enough. Today, the country has achieved great success in its development and progress in all fields, leading to much closer connections between the "one country, two systems" policy and the interests of the compatriots, business, professions, civil groups, local governments,

---

1   *White Paper* (《白皮書》). P40.

as well as many other interest groups on the mainland. They can now put forward their views, suggestions and demands pertaining to the "one country, two systems" policy via various channels. Though the bourgeoisie in Hong Kong still has considerable influence over the central authority's policies towards Hong Kong, other social and political forces in Hong Kong are also gaining power, and much dissatisfaction has been aroused because of the central authority's excessive partiality to vested interests. In the future, the number of people who regard themselves as stakeholders in the "one country, two systems" policy will continue to grow both on the mainland and in Hong Kong, adding difficulty and complexity to the practice of the "one country, two systems" policy. It is very likely that Hong Kong will have to go through great changes after "remaining unchanged for fifty years". Whether the "one country, two systems" policy will continue and how it will be continued remain important issues that the central authority and Hong Kong people have to face together. In fact, some people, particularly the youth in Hong Kong are already trying to place the future of Hong Kong after 2014 on the public agenda. By doing so they want to make known their discontent and put forward their different visions of Hong Kong's post-2047 future.

The "Hong Kong problem" and the central authority's policies towards Hong Kong are receiving increasing attention from the mainland compatriots. The central authority needs to give greater consideration to the mainland compatriots' feelings towards Hong Kong, as well as their opinions about the central authority's policies towards Hong Kong when they handle Hong Kong affairs and as various interests on the mainland become more complicated and more channels are made available for people to express their views and vent discontent. Since the central authority has become more open and transparent in its policies towards Hong Kong, and the number and diversity of institutions and people participating in the decision-making process continue to rise (among which a greater number have mixed feelings concerning Hong Kong) the central authority must better balance the interests of Hong Kong with that of various parties on the mainland and must invest greater concern in the mainland compatriots' mixed feelings and attitudes towards Hong Kong. In the past, when Deng Xiaoping and the central authority he led enjoyed great political authority, and the central authority could formulate policies concerning Hong Kong quite autonomously, without much interference by the mainland compatriots. However, as the political ecology on the mainland changes and people with different degrees of influence have more opportunities to participate in politics, China's leaders can no longer be "dictatorial" and "authoritative" in making decisions. Should a majority of mainland compatriots develop an unfavourable impression of Hong Kong people, thinking of them as unfairly yet ungratefully enjoying disproportionate rights and privileges without assuming responsibilities towards the country, the central authority may come to regret being too lenient and generous in their handling of Hong Kong affairs and showing partiality for the Hong Kong people.

The principle behind the "one country, two systems" policy is simple, but its practical operation is in fact extremely difficult, particularly when it is implemented in a time of great changes and in an atmosphere full of doubts and distrust between the Hong Kong people and the central authority. To properly handle the relations between the "two systems", the two sides must continually keep in mind the basic principles and strategic goals of the "one country, two systems" policy, must maintain a strong sense of reality, must understand the historical background of the policy, must consider the overall situation, must learn about each other's concerns, plight, difficulties and worries, must be willing to respect each other's interests, position and feelings, must provide each other assistance, must try to avoid conflicts and friction, must be self-disciplined, and in particular, must resist pressure from outside forces hostile to the central authority.

Although the opposition often questions the central authority's sincerity in maintaining Hong Kong's original system, institutions and "core values" and frequently accuses Chief Executive Leung Chunying of taking the orders from the central authority to destroy Hong Kong's system, institutions and "core values", such misgivings are groundless because it is in the best interests of China to maintain Hong Kong's original system and values. As relations between Hong Kong and the mainland grow closer, Hong Kong will inevitably be affected by the mainland, particularly when Hong Kong people become less confident in their original system, institutions and values and consider that the institutions and ways of doing things in the mainland are worth emulating and learning from. The fact that some Hong Kong people believe that the central authority plans to destroy Hong Kong's "system" and replace it with the mainland's "system" in fact shows their mistrust of the central authority and their declining sense of superiority and confidence in Hong Kong's own system and values. While Hong Kong people do not approve of the behaviour of individual chief executives, they should not then come to the conclusion that the central authority intends to "mainlandise" Hong Kong. Based on my contact with some mainland leaders, officials and scholars, the mainland elites' "confidence in the political system of the mainland" is continuing to rise, and their appreciation of Hong Kong's system today is lower than in the past. Nevertheless, they still find much merit in Hong Kong's system, notably, the rule of law and the rationality of institutions.

The distinctions between socialism and capitalism were quite stark at the beginning of the 1980s (though different forms of socialism and capitalism could at that time be found around the world) especially in terms of the relative roles of the government and the market pertaining to resource allocation, the importance of the private sector in the economic system, and the configurations of the welfare system and the forms of welfare provision. From the remarks of the central authority, I have noticed they hold an indefinite conception that, with time, the gap between socialism with Chinese characteristics and Hong Kong's capitalism will gradually narrow because of the following factors: the increasingly important role played by the market mechanism and private sectors in the economic system of the mainland, the increasingly open and progressive political and legal systems on the mainland, the closer and closer economic relationship between the mainland and Hong Kong, together with China's plan to create several more cities like Hong Kong on the mainland. With the shrinkage of differences between the mainland and Hong Kong, setting the mainland's system against that of Hong Kong and see them as socio-economic models opposed to each other will no longer be appropriate. Looking into the future, the boundary between the "two systems" in "one country" will become obscure, and as a result, a "one country, one system" will not be far behind. Under such circumstances, even if Hong Kong's system remains unchanged for fifty years or even longer, the general trend of "unification" of the system and governance of the whole country will not be reversed. The "short-term" estrangement, differences, unfairness and even damaged feelings between the mainland compatriots and Hong Kong people brought about by the "one country, two systems" policy will prove ephemeral.

The "one country, two systems" is an important national policy that in a particular historical moment does adequately address the unique problems related to "Hong Kong's future". It will definitely not last forever, because it contains the seeds of self-termination. Under the "one country, two systems" policy, we can look forward with optimism to the country's future and expect great successes from China's modernisation efforts. Although the gap between the mainland's "system" and Hong Kong's "system" will be further narrowed under their mutual influence in various fields, the "two systems" will not completely merge into "one system" in the end. The rise of the country has enhanced the mainland compatriots' confidence in the mainland's "system". In contrast, many Hong Kong people, to some degree, have become less confident in their "system". Many people in the past predicted that the convergence of the "two systems" would come about as a result of the economic integration between the mainland and Hong Kong, and as a result, the mainland's "system" would gradually draw closer to Hong Kong's "system". But in the future, Hong Kong's "system" may also have to learn from the "system" of the mainland in some areas. In addition, the idea of a "grand unification" of the Chinese people will not allow for the permanent co-existence of "two systems" in China. In any event, whether under the profound influence of the "grand unification" idea, or by the natural convergence of the "two systems", the "one country, two systems" policy will come to an end someday. If that day does come, it will be a happy occasion and not necessarily against the will of the Hong Kong people.